THE
COMPLETE
MODERN
BLACKSMITH

THE COMPLETE MODERN BLACKSMITH

ALEXANDER G. WEYGERS

Illustrations by the author

Ten Speed Press
Berkeley, California

"The Making of Tools" originally published by Van Nostrand Reinhold in 1973.
"The Modern Blacksmith" originally published by Van Nostrand Reinhold in 1974.
"The Recycling, Use, and Repair of Tools" originally published by Van Nostrand Reinhold in 1978.

Ten Speed Press
P.O. Box 7123
Berkeley, California 94707
www.tenspeed.com

Distributed in Australia by Simon and Schuster Australia, in Canada by Ten Speed Press Canada, in New Zealand by Southern Publishers Group, in South Africa by Real Books, and in the United Kingdom and Europe by Airlift Book Company.

Cover design by Catherine Jacobes
Illustrations by Alexander G. Weygers

First Ten Speed Press edition, 1997
Printed in the United States of America

ISBN 0-89815-896-6

8 9 10 11 — 08 07 06 05

CONTENTS

NOTE ON SAFETY

Before attempting any of the procedures described in this book, please be sure to familiarize yourself thoroughly with the "Hints on Using Power Tools and Other Admonitions" on page 292. When working with tools, always observe shop safety practices: the safe toolmaker and blacksmith is always careful, systematic, and in control of his tools and materials, avoiding injury to himself, other people, and his tools.

The publisher, copyright holder, and estate of Alexander G. Weygers are not liable for accidents or injury caused by unsafe use of the tools and techniques described in *The Complete Modern Blacksmith*.

THE
MAKING
OF TOOLS

CONTENTS

Introduction

This book teaches the artist and craftsman how to make his own tools: how to design, sharpen, and temper them.

Having made tools (for myself and for others) for most of my life, I have also enjoyed teaching this very rewarding craft, finding that anyone who is naturally handy can readily succeed in toolmaking. The student can begin with a minimum of equipment, at little expense. Using scrap steel (often available at no cost), he can start by making the simplest tools and gradually progress to more and more difficult ones. Once a student has learned to make his own tools, he will be forever independent of having to buy those not specifically designed for his purpose.

Many students, after three weeks of intensive training under my guidance, have produced sets of wood- and stonecarving tools as fine as my own. And, over the years, the students invariably observe that the value of the fine tools they have made during the short course exceeds the cost of their tuition. One might say they were taught for free, emerging as full-fledged toolmakers, each one carrying home a beautiful set of tools to boot.

The guidelines to toolmaking in this book are identical to the information and explanations I give in class. The drawings are "picture translations" of my personal demonstrations to students, showing the step-by-step progression from the raw material to the finished product — the handmade tool.

1. A Beginner's Workshop

A beginner's workshop ought to include the following basic equipment:

A sturdy workbench, onto which are bolted:

  Grinder (¼ hp) and *chuck extension*

  Dry grinding wheel assembly, based on a ½ hp motor grinder

  Table-model drill press (½ hp; ½-inch-diameter drill capacity)

  Side grinder (¼ hp)

  Rubber abrasive wheel (¼ hp; can of water adjacent, to cool work-pieces)

  Buffer (¼ hp)

  Vise (35-pound, or heavier)

  Anvil (25-pound, or heavier; preferably mounted separately, on a freestanding wood stump)

A large drawer under bench, in which to store a cumulative collection of accessory tools: wrenches, files, drills, hammers, center punch, cold chisels, hacksaw, assorted grinding wheels, sanding discs, drill vise, and chuck key, etc.

A small forge, in which to shape and temper your tools (a forge is preferred, though a *charcoal brazier* will do). To draw temper color, use either a gas flame (as in a kitchen stove), a Bunsen burner, or the blue flame of a plumber's blowtorch fueled with gasoline.

A quenching bucket

A coal bin

This minimum workshop can be equipped very satisfactorily with secondhand articles, provided they are in good working order. Therefore, look for discarded washing-machine motors, surplus-store grinding wheels, hammers, files, etc. Flea markets and garage sales frequently will provide surprisingly useful items with which to complete your shop.

INSTALLATION OF EQUIPMENT

The Workbench

In nine out of ten hobby shops, workbenches are too flimsy. Instead, they should be as sturdy as possible, to support equipment which is put under great strain. Everything fastened to a *heavy* workbench benefits from its supporting mass and rigidity. Obviously, for instance, much energy will be wasted if hammer blows are delivered onto workpieces held in an insecure vise on a wobbly or lightweight bench.

The equipment arrangement here is mainly to guide those who have limited space in their workshop. A long, roomy bench built against a strong wall is the ideal arrangement: it allows the various machine units to be spaced out in the most workable and orderly way, while

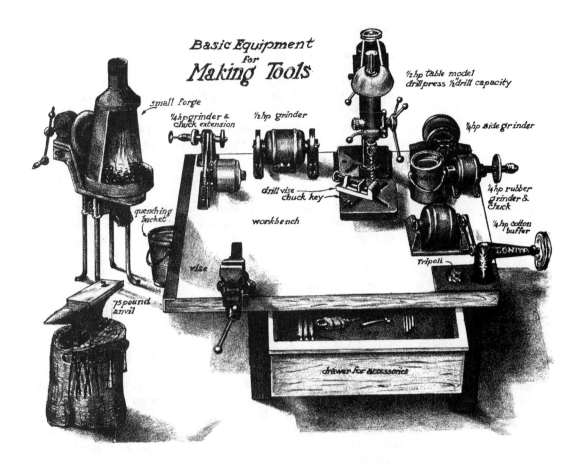

Basic Equipment
for
Making Tools

small forge

¼hp grinder &
chuck extension

½ hp grinder

½ hp table model
drill press ½" drill capacity

¼hp side grinder

drill vise
chuck key

¼hp rubber
grinder &
chuck

¼hp cotton
buffer

quenching
bucket

workbench

BONIT?

Tripoli

vise

75 pound
anvil

drawer for accessories

utilizing the wall as additional reinforcement. (Make sure though, that noise will not carry through the wall and disturb neighbors.)

The Bench Vise

The vise shown is not adjustable at its base, but those that are — ones that can be turned around and clamped at any angle — ought to be bolted at the exact corner of the bench.

Bench vises should not weigh less than twenty-five pounds — and the heavier, the better; with luck, you may even find a good secondhand seventy-five-pound monster. Large vises, with wider jaws, can hold bigger workpieces, a great advantage when woodcarvings are to be clamped in them.

The Drill Press

As the most important piece of equipment in your shop, the drill press should not be smaller than that shown here. Bolted onto the bench, its swivel table should be adjustable in every position, and high enough in its lowest position to clear all permanent bench machinery when a long workpiece is clamped on it.

A versatile machine, the drill press can do routing, filing, rasping, and grinding, in addition to the drilling it is designed to do. It can also be adapted to function as a wood lathe (see Chapter 10). Once you become familiar with the drill press, your own inventiveness will find many uses for it.

Accessories to the Drill Press

In addition to assorted metal- and wood-cutting drills stored in the bench drawer, one major accessory is the drill-press vise (as shown). Placed on the drill press's swivel base, it is used to hold a workpiece that has previously been marked by a center punch. Firmly held in the vise, it prevents the drill from wandering off these marks.

A speed-reduction attachment for your drill press is very useful, if you can afford one. For much heavy-duty work, large drills should be slowed down, to save wear and tear on both drills and press.

Other small tools, such as routers, rotary files and rasps, drum sanders and grinding points, special woodcutters, chuck inserts for special tasks, and various others will gradually be added to your basic drill-press equipment.

As you will be learning to make many of these accessory tools yourself as you proceed, try to resist the temptation to buy them before they are needed.

The ½ hp Grinder

This grinder is the next most important piece of equipment for your workbench.

If you use a fast 3600 rpm electric motor, it is safest to have the grinding wheels not larger than 6 inches in diameter. A slower motor of 1750 rpm may safely have wheels up to 8 inches in diameter and not less than ¾ inch thick. These safety factors are purposely more severe than those required by industry because we shall eliminate the wheelguards in order to have maximum access to the fully exposed wheels.

It should be noted that *extreme caution* is needed when using a high-speed electric motor. While you may be tempted to use one should it come your way, you should realize that it is much safer to work with slower-speed tools in hobby shops. Injury can occur if instruments are inserted off-center when cutting, grinding, or rubbing. Centrifugal force — greater at high speed than at low — may cause inaccurate instruments to fly apart, bend, and whip around, causing serious bodily harm. A little less speed equals a little more safety, and you can still progress fast enough, since the machine saves time and muscle.

All rough work (requiring a large quantity of metal removal by means of grinding) should be done on the ½ hp unit, using coarse, hard, and thick stones. If you decide to buy a new motor, choose a model with a double-shaft arbor like the one shown, which is designed to receive a grinding wheel on each end of the shaft.

First, place an extension adapter on the shaft as shown. Secure it firmly with two setscrews, which should be tightly seated in $1/16$-inch-deep holes and tightened with an Allen-wrench key. These adapters, in turn, are to hold a 6-inch wheel, 2 inches thick, fixed

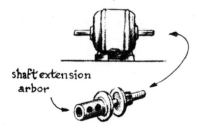

¼ hp electric motors sufficient for toolmaking tasks

shaft extension arbor

between washers and nuts. (I have threaded my motor shafts to eliminate the need for adapters, using washers and nuts to lock wheels. As both ends of the shaft have right-hand threads, the left-hand wheel must be locked with two nuts, since, when it rotates toward you during grinding, it tends to loosen itself.

Tool Rests

Bolted separately on the bench below the wheels, tool rests are adjustable simply by tapping the hammer against them to close the gap that forms as the wheels are worn down (especially after redressing the wheels). If it seems too difficult to make this tool rest, have a welding shop do it for you.

The ¼ hp Grinder and Chuck Extension

This grinding unit is used mainly to receive chuck inserts of small caliber, which do the more refined grinding, sanding, and polishing, especially of delicate tools.

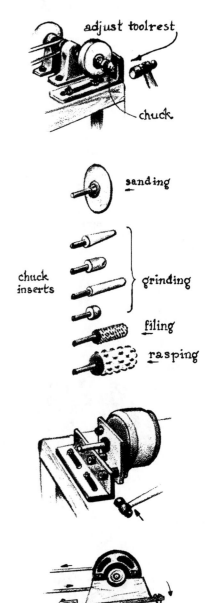

Since ¼ hp electric motors of 1750 rpm are standard in household utility machines such as laundry- and dishwashers, these motors are often discarded or available secondhand. As some have odd foundation bolts through silencing rubber cushions, a little inventiveness is called for in mounting them. With the one shown here, I simply tap a hammer against the frame on which the motor is bolted; this tightens the belt whenever necessary, even with the motor running.

The ¼ hp Buffer

Here, the ¼ hp motor (one salvaged, without bolt holes or base) is cradled in two wooden blocks and strapped down with a piece of band steel. The blocks, in turn, are screwed on a plywood baseboard, and the entire assembly is then hinged onto the bench. This method allows the motor's weight to "hang" in the belt, thus automatically keeping the belt tightened against slippage.

Sometimes, a slight unevenness in the belt (or a slightly eccentric pulley) can cause vibration. The motor baseboard can then be anchored to the workbench with a long wood screw, which steadies it but still allows some play.

polishing
(buffing)

buffing compound

the side grinder

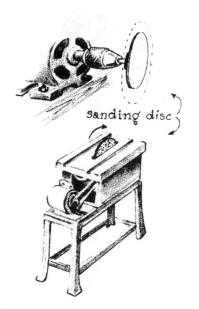

sanding disc

The Cotton Buffer

The 8- to 10-inch-diameter cotton buffing wheel is mounted on the shaft of a housing salvaged from an old-fashioned sewing machine. Stripped of its other parts with hacksaw or abrasive cutoff wheel, this housing is screwed onto a steel base plate, which, in turn, is bolted to one corner of the workbench. This positioning allows maximum access to the buffing wheel.

If this type of shop-made buffing unit seems too much of an undertaking at the outset, compromise by purchasing a ready-made buffer on a small shaft as a chuck insert; then clamp it in the drill press chuck or in the chuck on the extension of one of the ¼ hp grinding units. Such chuck inserts are readily available in hardware stores, mail-order houses, or through hobbyists' catalogs.

Buffing Compound

The best buffing compound is an abrasive wax mix called *tripoli,* found in any hardware store. Choose coarse grit for toolmaking jobs. A one- or two-pound block of tripoli, held against the rapidly rotating cotton buffer, momentarily melts the wax in the mix and thus adheres to the buffer. Once cooled and hardened, the coated buffer becomes a polishing tool, as well as a mechanical sharpening strop. It will buff all finely ground steel mirror-smooth, and, in short, prove altogether indispensable in your toolmaking.

The ¼ hp Side Grinder and Sanding Discs

This grinder is built from a discarded motor and an industrial cutoff-wheel remnant.

Steel-construction plants use cutoff wheels that are made from an abrasive mix, sometimes fiber-bonded to make them shatterproof. When new, they measure approximately 24 inches in diameter and ⅛ inch to ³/₁₆ inch thick. They are clamped between 6- to 8-inch-diameter washers. When they become worn down to about 10 inches in diameter, so little remains outside the washer to cut with that a new wheel must replace it and the remnant becomes a waste product.

If you can locate a steel-construction plant nearby that uses this cutting method, they will surely have some 8- to 10-inch discards to give you or sell for little. (See Chapter 6 for making a side grinder.)

If you have a table saw, you can mount side-grinding abrasive discs on it in place of the saw blade. First, remove all sawdust that might be ignited by sparks coming from the disc. Second, be sure to protect your machine from abrasive dust that might reach any bearings that have not been properly sealed. Such discs are, as a rule, made of ³/₁₆-inch-thick steel, to which is glued a tough sheet of coarse- or fine-grit abrasive.

Another side grinder setup employs a slightly flexible, plastic-bonded abrasive disc as a chuck-insert accessory. Several kinds are offered in hardware stores; their advantage is that the whole disc can be utilized, instead of only the half rotating above the saw table.

The Abrasive Cutoff Wheel

If you have an 8- to 10-inch circular table saw, mounting an abrasive cutoff wheel in place of the saw blade will enable you to cut hardened steel. This piece of equipment becomes a great help early in your toolmaking education, especially if scrap steel is to be your major source of material. All hardened steel can be cut cleanly with it, without having to anneal the steel beforehand.

abrasive cut off wheel

The Rubber Abrasive Wheel

The rubber abrasive wheel is indispensable but difficult to master. A coarse-grit, 6- to 7-inch-diameter wheel, ½- to ¾-inch-thick, is good enough; a 1750 rpm salvaged washing-machine motor is fast enough. Mount it between washers on a ½-inch-diameter standard extension shaft, thus leaving room, after mounting the rubber wheel, for a standard ½-inch auxiliary chuck, if needed. Both extension shaft and chuck are available at hardware stores that handle power tools and their accessories.

rubber abrasive wheel

The potential harm that can come to the rubber abrasive wheel as well as to yourself must never be ignored or underestimated (this is described at length in Chapter 6). It can be perfectly safely used once its correct handling is clearly understood, and its advantages outweigh both the risk of accident and any consequent cost of repair. Its careful use, once mastered, can produce wondrous results.

The Anvil

A small anvil is shown here. Anvils must be bolted onto a freestanding wood stump next to your forge. Ideally, anvil, forge, and water-quenching trough should be clustered in the darkest corner of your shop (for only in the semi-dark should you attempt to judge the color of heated steel).

The Forge

An old, or even antique, little outdoor riveting forge is excellent for a beginner's shop. Blacksmith coal, which can be bought wherever horseshoers still ply their trade, is the ideal fuel. Feed-and-fuel stores in farming communities usually carry blacksmith coal. Although coke is a good fuel, too, I have never found a source for it.

Professional metal-working shops often have gas-fired forges, but for a beginning toolmaker, such a forge has several drawbacks. It is very noisy, since a forced-air blast is used to reach high heat; the large opening of the fire grate (about 1 by 3 inches) lets small forgings drop through accidentally; and steel oxidizes at a prodigious rate in such fires. In contrast, the regular coal-fired forge has a cast-iron grating (needing only six to ten ⅜-inch air holes), and creates very little steel oxidization. Also, the "carbon" character of coal tends to improve the steel rather than deteriorate it.

Other acceptable means of heating steel include acetylene torches, outdoor barbecue braziers, wood stoves, and fireplaces.

knuckles
anvil face
h
floor

correct height of anvil face
from floor level

MATERIALS

High-Carbon Steel

Most tools are made of high-carbon steel. This is *temperable* steel. It can be bought cheaply at steel scrapyards and automobile junkyards. And, once you develop an eye for it, great amounts are found strewn along highways and in vacant lots to add to your own scrap pile. No matter how beat-up or rusty a piece of discarded scrap may be, add it to your supply. Scrap is cheap, and as rusty, corroded surfaces are usually only skin-deep, they can easily be ground clean.

Your first trip for material should be to nearby dealers in scrap steel. If they permit you to roam over their yard, search particularly for all kinds of *spring steel:* leaf springs and coil springs of cars, garage-door springs, springs from garden swings, some heavy-gauge truck engine valve springs, broken starter springs, torsion bars, and stick-shift arms — any discarded machine parts that you suspect may be made of high-carbon steel — all contribute to your toolmaking supply. (Chapter 2 explains how to "test" steel for high-carbon quality.)

Note that spring steel has a sufficiently high carbon content that it can be tempered for hard cutting edges such as cold chisels. But the temper hardness used in car springs, though not hard enough for a cold chisel, often is hard enough for wood lathe-turning tools and wood chisels (if the wood to be worked is not too hard):

While many high-carbon steels have varying degrees of metals mixed through them, such as molybdenum, vanadium, tungsten, or other alloy additives, all such high-carbon steels are temperable, which is your main concern.

The question now is: Are all those different steels to be tempered differently? The more experience you gain in toolmaking, the more you come to realize that an "average" tempering method is practical for most of these steels, regardless of any scientifically prescribed tempering charts which list exact procedures for heat-treating (tempering) various alloy steels in industry.

If you decide at the outset to "make do" with the endless, unknown high-carbon steel varieties on scrap piles, it is the empirical (trial-and-error) methods set forth in this book that are to be followed. If you decide to make fine tools from scrap steel of unknown quality, then it is this type of steel that you must speculate about and deal with.

I can assure you that during the fifty years that I have made tools from questionable types of high-carbon steels, a very large percentage of these tools turned out to be excellent, and all of them have stood up under hard use — and often abuse as well. So don't worry if you feel that you know nothing about steel; you will learn enough as you work and practice.

Mild Steel

Such steel is of a low-carbon content and is not temperable, although its surface layer can be hardened through a process called *case hardening,* which applies a skin-deep hardness (see Chapter 2). Some tools can be made of mild steel, such as those that do not require hard-cutting or long-wearing parts (some garden tools, for instance).

From the same sources that produced high-carbon steel scrap can be found mild-steel plates, rods, and bars. All can be made into useful articles. In short, any and all steel parts you believe might be valuable in your metalworking activities are welcome, and tests in the shop will distinguish one steel from the other.

THE QUENCHING BATH

This bath may consist of water, brine, oil, or fat. Plain water for steel-tempering has been my choice all these years, but others find a salt brine, or other blacksmith-recommended quenching liquids, more desirable. I simply offer my own experience, recognizing, however, that the other liquids work as well. The merits of each different one are explained further when we discuss tempering (see Chapter 2).

Water

Water should be kept next to the forge. A water bucket will do, but a rectangular metal container is better. It should not be smaller than 10 by 24 inches and 15 to 20 inches deep, in order to quench long workpieces.

A tin can nailed to the end of a branch or stick makes a douser, and twenty-five or more little holes punched through its bottom with a small, sharp nail turns the dipper into a water sprinkler. This sprinkler keeps the coal surrounding the fire wet, and holds it to just the size of fire needed. Also, it is used occasionally to cool the steel workpiece that extends *outside* the fire by dousing it while the rest remains in the fire undisturbed. It is especially useful, then, to douse handheld steel rods that are rather short and thus may heat up through conductivity.

Because water is always nearby, fires are seldom started by accident in a blacksmith shop. The smith is always there when the forge is fired up, and can dip into the water trough in case a piece of hot steel or coal should be dropped on sawdust, wood, or fabric.

Brine

Brine is made by saturating water with common rock salt. The right amount of salt is determined by experimentation. The more salt-saturated the water, the higher the temperature at its boiling point. Some people have quenched a hot tool tip in powdered salt, with good results; such pure salt melts at a temperature much higher than the boiling point of brine or oil.

Oil or Lamb's Fat

Have some old motor oil on hand in a five-gallon can (covered, to avoid flash fires), for general use.

Rendered lamb's fat should be used to harden all very small, delicate tools. If possible, have as much as one gallon handy, kept in a lidded can to keep rodents away. This type of fat leaves the steel surface remarkably clean, while old motor oil often deposits some black carbon scale on the quenched steel; though it requires extra work to remove it, such deposits are otherwise harmless.

Historically, blacksmiths have always had this choice of quenching liquids: I believe it has caused many varied recommendations on how to temper steel, with as many varied results.

In due time, each smith naturally becomes convinced that, at long last, he has found the "best" way. Still, he often suspects that another smith may just have a secret recipe in *his* technique that he is not sharing. What could it be? It opens the door to mystery for the romantic craftsman. And there *is* romance in forming hot and malleable steel, then changing it into very hard steel. As makers of tools, we are bound to enter into endless exploration: using all sorts of steel; forming it into many shapes; tempering it in a variety of ways.

WOOD FOR HANDLES

A straight-grained, hard (not too brittle), tough wood is best for tool handles. Black and English walnut, hickory, ash, eucalyptus — all respond favorably. And don't hesitate to test wood that grows locally which you suspect might be suitable.

Many other materials will be needed in your toolmaking, such as grinding compounds, abrasive papers and grinding stones, steel tubing for ferrules, etc. They can all be added from time to time as they are needed.

2. Tempering Steel

Only *high-carbon* steels are temperable and can be hardened in the process called tempering. Mild, or low-carbon, steel cannot be tempered but may be case-hardened. Hardened tool edges must cut, shear, punch, emboss, and do many other tasks demanding both hardness and toughness. Such tools must not break or bend in normal use. The aim should be that tool edges, in well-designed and well-tempered tools, should stand up, even under momentary extra strain.

The methods for tempering recommended in this book are based on what I believe are the most reliable and practical when you work with scrap steel. Scrap piles are bound to contain a great variety of high-carbon steels from different machine and car parts. Such varieties, as a rule, react well to my tempering methods because of their high-carbon content, *not* because other metals have been added to produce special alloy steels.

Manufacturers' precise tempering procedures vary somewhat from mine because of the specialized industrial uses of alloy steels. Yet, in learning by trial and error, occasionally some high-carbon steels respond less favorably to my recommended methods.

An example of this would occur when making a cutting tool out of steel originally designed to remain fairly hard even under red-hot heat (such as automobile exhaust valves). Generally, the hardness of exhaust-valve steel remains *below* the hardness required for the cutting edges of our tools. Therefore, for any piece of steel, a test for temperability is always advised before you begin making a tool that must have a hardened and tough cutting edge.

TESTS FOR TEMPERABILITY

Test 1. From the scrap pile choose a steel rod that you suspect may be of a high-carbon quality. Hold it on the power grinder and examine the sparks; compare them with the sparks shown in the illustration, which distinguishes high- from low-carbon steel by spark characteristics. The rule of thumb in the shop is that a dull spark is mild steel, and a brilliant, sharply exploding spark is high-carbon steel. Most libraries have books you can examine showing the spark characteristics of various steels. If you still are not certain that the steel sample is of high-carbon quality, you can resort to the following sure and final test.

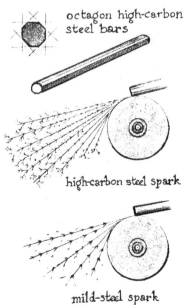

octagon high-carbon steel bars

high-carbon steel spark

mild-steel spark

heat to light cherry

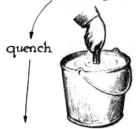

quench

test for hardness with file tip

heat range for tempering

heat to light cherry-red

Test 2. Build a medium-hot "clean" fire. (A "dirty" fire is any fire which emits smoke and yellow flame. It is an indication that combustible gases remain which may damage hot steel.) It should resemble a glowing charcoal fire.

Place the first inch of the rod horizontally in this fire, making sure that hot coals are always underneath the steel, as well as around and above it, to prevent fresh air from hitting it directly and thus oxidizing its surface. As soon as it heats up to a light cherry red glow (as judged in a semi-dark room), pull the rod out and immediately quench it completely in water at room temperature. It should emerge pearl gray in color.

Next, clamp the steel rod in the vise and, using the tip of a sharp file, pick on the gray quenched end. If the file tip slides off, like a needle on glass, it means that this steel is of high-carbon quality and thus temperable.

Various hardnesses of steel will be described separately and become more meaningful as we proceed in the practice of tool-tempering.

TEMPERING A TOOL BLADE

Once a tool blank made from tested steel is finished (as is the wood-carving gouge illustrated here), the blade can be tempered.

As described earlier, place the blade in a clean fire, gently fanning it to maintain an even, medium heat. Keep your eye on the tool, which is partly visible through the fire, but do not disturb the coals or the tool until you see that the blade has become light cherry red in color (again, as judged in a semi-dark room). My experience has been that, once the tool becomes invisible in the identical cherry-red glow of the fire, it is ready to be quenched.

When you are satisfied with the tool's color, withdraw it and immediately quench, holding it fully submerged and motionless in one of the following coolants:

Water, if the blade is fairly *thick*

Brine, if the blade is fairly *thin*

Oil (or rendered fat), if the blade is *very* thin

Rendered lamb's fat is ideally suited for tempering small, light-gauge tools.

COOLANT TEMPERATURE

All coolants should be kept at room temperature, to ensure that sufficient shock impact takes place here between the light cherry red heat glow of the blade and the room temperature of the quenching liquid. This shock impact produces the same outer hardness in the steel, no matter which coolant we choose.

While the boiling point of the liquid mantle that envelops the hot steel differs with each individual coolant, all coolants remain approximately at room temperature outside the hot mantle. Remember, therefore, that:

If the coolant is *water,* this mantle boils at approximately 100°C (212°F), cooling the steel toward the core *fastest.*

If the coolant is *brine,* this mantle boils at approximately 107°C (226°F) or thereabouts, depending on the concentration of salt in the brine, cooling the steel toward the core a little *slower.*

If the coolant is *oil or fat,* this mantle boils at approximately 150°C (290° F), cooling the steel toward the core *slowest.*

The importance of the choice of quenching liquid becomes clear once you know that the slower the hot steel cools, the "softer" its core becomes. This inside "softness" creates the tool toughness needed to keep the tool from breaking. At the same time, it should be noted that outer hardness penetrates deeper in fastest-cooling water than in slowest-cooling oil. Steel cools, through conductivity, at a pace dictated by the boiling temperature of the liquid mantle as well as by the steel's coefficient of conductivity. The result is that:

In *water,* the core becomes only a little less hard (fast cooling).
In *brine,* the core is a little softer (slower cooling).
In *oil,* the core is softer still (slowest cooling).

You can now see that the kind of tool you are making will dictate your choice of quenching liquid. For very thin and delicate tools, therefore, an oil quench is advised. Its slowest cooling toward the core makes the tool as tough as possible, reducing the chance of the steel cracking because of too-fast cooling and shrinkage. Moreover, in thin, delicate tool blades, the outer hardness penetrates deep enough during slower cooling to make the steel uniformly hard all the way through.

Such steel emerges with the clean pearl-gray color of a new file, and with a file's brittle hardness, as shown here.

Put a sheen on the surface of the brittle area with the fine-grit abrasives in your shop. A final polish on the cotton buffer, using tripoli compound, will make the blade shine like a mirror. This is desirable because it allows you to see the slightest change in *oxidation colors* during the heating.

DRAWING TEMPER COLORS

To "draw" color is the term used for the process of reheating a brittle, hard steel in order to temper it for a specific hardness. As steel heats, its shiny surface changes color, and each color change indicates a change in steel hardness. Specific color thus "matches" specific hardness.

Use a standard commercial propane torch or a plumber's gasoline blowtorch. The blue flame of a kitchen gas range functions well, provided you fit a bent coffee can over the burner to concentrate the flame, as with a Bunsen burner. Another method is to place two firebricks, spaced ½ to 1 inch apart, on top of the forge fire, so that heat escaping between the bricks becomes a localized heat column.

Now begin to heat the tool shank, holding it in the blue flame. Keep the mirror-smooth, brittle *blade* safely *outside* the flame. Soon, as the gradually increasing heat of the shank is conducted toward the blade, the first oxidation color appears — a *faint straw yellow* (see color chart).

As you heat the steel further, you will see the whole color spectrum appear as a full color band: *blue,* nearest the flame, followed by the full oxidation color spectrum as shown, to, finally, the original sheen. When this full color band moves down the shank to the beginning of the blade part, now hold that part quite high above the visible flame, trying to visualize the invisible heat column rising from it.

It is in this less-hot region that you should now hold the blade — quite high at first, playing it safe. Within about a minute, in that position, an even change of oxidation (temper) color is drawn over the total area of the shiny blade. It will be a faint straw color.

If this change of color appears at the extreme outer edge first, it means heat is entering that thin area too fast. Either move the blade slightly off-center in the visualized heat column, or hold the tool higher

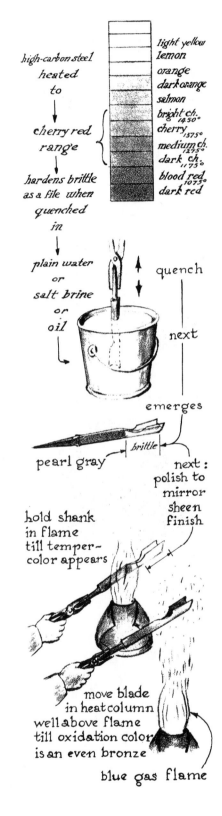

high-carbon steel heated to

cherry red range

hardens brittle as a file when quenched in

plain water or salt brine or oil

light yellow
lemon
orange
dark orange
salmon
bright ch. 1450°
cherry 1375°
medium ch. 1275°
dark ch. 1175°
blood red 1075°
dark red

quench

next

emerges

pearl gray — brittle

next: polish to mirror sheen finish

hold shank in flame till temper-color appears

move blade in heat column well above flame till oxidation color is an even bronze

blue gas flame

still, to reduce heat as well as speed of heat transfer. Continue thus to manipulate the tool, holding it now here, now there. This controls the speed of heat transfer, the exact area to be tempered, and the precise moment for quenching at correct temper color.

The following guidelines will help you select correct temper colors during the hardening of various tools:

Peacock. For thin delicate tools, and tools with spring-action parts that are hand-pushed.

Bronze. For heavier-gauge woodcarving gouges that are struck with hammers.

Dark straw. For center punches, small cold chisels, etc.

Light straw. For blunt-edged sturdy tools, such as large cold chisels, star drills, bush tools, etc.

Unexpected things, over which you may have little control, will probably happen. For example, after a water quench, some steels emerge brittle-hard with such great tension that, if not tempered within a short time, they may crack lengthwise all by themselves. Such steels probably were designed to be hardened in oil, and manufacturers' manuals *do* refer to water-hardening and oil-hardening steels. Therefore, the unknown (and unknowable) elements in your scrap pile can leave you guessing somewhat.

To summarize and underscore the tempering steps, reexamine the oxidation color spectrum and the annealing of a steel bar as shown in the drawings, as well as the color chart. (See also the cover illustrations.)

From my experience in dealing with high-carbon scrap steel, much that is hardened in oil shrinks less than that quenched in water or brine. And, as the tension that weakens steel is still further reduced during tempering in oil, it is wisest for delicate, thin tools to be hardened in oil or lamb's fat.

The many variables involved here in fact allow you an endless number of combinations to experiment with in the tempering process. That experimentation is what led many old-time, experienced blacksmiths to the art of hardening and tempering steel. Often they claimed to have supposedly secret methods. In fact, they simply varied their procedures, always seeking ways to improve their results; through trial and error, they often succeeded.

Thus, only through each individual's experience can he find for himself — through exploration, inventiveness, discovery, and judgment — what works best for him.

The breakthrough comes when all guessing is replaced by an intuition that is a blend between feeling and knowing. As maker of tools, you are bound to experience that most gratifying of sensations: the knowledge that, while making a good tool, the probability is that the next one will be the best one yet!

CASE HARDENING STEEL

Mild steel can be hardened by heating and applying to its yellow-hot surface the purest carbon powder or a proven commercial compound. By allowing carbon to penetrate the steel at its surface only, the skin, or "casing," is hardened and thus becomes temperable. Commercial compounds can be found at machine-supply stores.

An old-fashioned case-hardening method was to pack around the steel a layer of horn-shavings (hooves of horses, cows, goats, deer — all a pure form of carbon). The whole was then wrapped in cloth. To keep oxygen out, plaster or cement, mixed with firebrick grit and reinforced with chicken wire, was cast three inches thick around the cloth-swathed bundle and left to dry thoroughly.

The whole package was then buried overnight in the glowing coals of a boiler fire in a steam-engine plant. The next day, the plaster was knocked off and the clean, dark-yellow-hot piece quenched in a drum of water. Its surface would then be brittle hard, but that hardness was only skin-deep. The advantage was that the soft core made the interior tough and resilient, while the hard surface ensured durability.

Years ago, as a student, I helped a blacksmith friend of mine case harden just such a tool in his village shop: a thirty-pound hammer head, with a "skin" that must have been all of $1/16$ of an inch deep!

THE OXIDATION COLOR SPECTRUM

The illustrations here demonstrate what happens when a bar of highly polished steel is held over a hot flame and heated gradually. The colors that appear on the surface occur during the tempering of high-carbon steels. The resulting oxidation color spectrum is a kind of temperature-color equivalency chart. In high-carbon steels, each specific color represents a specific hardness when a brittle-hard steel is being tempered over an annealing flame. (Annealing is a process whereby heated steel is cooled very slowly rather than instantly as in quenching. It "softens" most steels.)

The first color that appears in the spectrum is *faint straw*. It represents the coloration that steel takes on when it is hardest and farthest removed from the flame; *light blue* is the last color that appears in the spectrum, indicative of steel when it is softest and closest to the flame, and hottest. This color band now travels, through heat conductivity, outward until it reaches the end of the bar and disappears, leaving the previously brittle-hard bar annealed (soft). When such a high-carbon steel bar is allowed to cool slowly after heating, it leaves the steel as soft as it can become.

The six-step demonstration of what happens to a steel bar held over an annealing flame should now make clear that if you choose to quench the steel when the specific color of your choice reaches the end of the bar, that color then matches a specific hardness of the steel at the end of the bar. Thus it has been tempered, just as if you had tempered the cutting edge of a tool. Once these sequences are understood and this procedure applied to temperable scrap steels, you can proceed with confidence to temper your own tool edges.

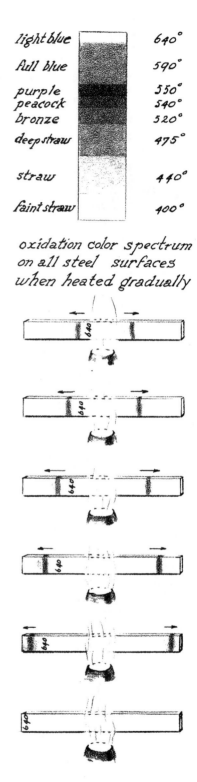

light blue	640°
full blue	590°
purple	550°
peacock	540°
bronze	520°
deep straw	475°
straw	440°
faint straw	400°

oxidation color spectrum
on all steel surfaces
when heated gradually

oxidation colors
during
heat-treating of
high-carbon steel

25

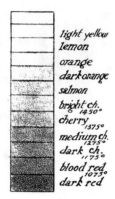

light yellow
lemon
orange
dark orange
salmon
bright ch.
  1450°
cherry
  1375°
medium ch.
  1275°
dark ch.
  1175°
blood red
  1075°
dark red

You might note that only in *brittle-hardened high-carbon steel* is the full oxidation color spectrum a graduated *hardness* indicator. Mild steel, though not temperable, will, however, show an identical oxidation color spectrum if its polished surface is heated gradually. In that situation all mild steels, regardless of a color spectrum appearance, remain soft enough to file easily.

THE COLOR RANGE OF HOT STEEL

Heat-Glow Colors

Heat-glow colors have a specific range, which the toolmaker should learn thoroughly and then determine only in the semi-dark. Much as a blacksmith constantly has to judge how hot he wants to heat steel before forging or tempering it, so must you do the same in toolmaking.

If bending the steel is all that is required, use a *yellow heat glow*. This makes steel malleable enough to bend easily. It is also a forging heat.

A *dark yellow heat glow* makes the steel a little less malleable.

A *light cherry red heat glow* (not malleable enough for forging or bending easily) is needed only when temperable steel is quenched to harden it.

A *dark cherry red heat glow* (in some temperable steels) is not quite hot enough for hardening in a direct quench.

In sum, visible heat glow relates to color and temperature, and the ability to judge these correctly is required when forging, bending, or tempering steel.

3. Making the First Tool: A Screw Driver

No matter how clearly the first steps of the very first exercises may be described — making the fire, heating the steel, hammering on the steel to shape it — the beginner must actually perform these steps himself before they become familiar. Only through practice, after repeating the steps again and again in routine sequence, will you gain the confidence that leads to competence.

This book's written and illustrated guidelines will become easier to follow because of your understanding as you read and experiment. Fewer detailed explanations will be needed since you can fill in the more obvious minute detail with which you have already become familiar.

The first lesson for the beginner is to make a screwdriver, because in this exercise, nearly every major element of toolmaking is employed with a minimum of complications.

If you are a complete novice you should first read the rules for grinding steel on motor grinders, and note all the rules on safety before beginning your first tool (see pages 292–3). Then, you may proceed with the following steps:

(1) Grinding a tapered point on the end of a rod.

From your scrap pile take a 10-inch-long engine push rod or other piece of high-carbon steel, about $^5/_{16}$ of an inch in diameter.

Grind a point, as shown, using the dry motor grinder. This point is to be the end of the screwdriver *tang,* which later will be driven into the wooden handle.

Make certain the tool rest is barely touching the rotating grinding wheel. Slide the tool back and forth as you grind, taking care not to cut grooves in the wheel. (This can happen if the sharp end of the rod is pushed hard against a single spot on the wheel face.

(2) Grinding four flat sides on the rod end.

While moving the tang back and forth, press it against the outer face of the wheel, using the tool rest to hold and steady it. Grind, without interruption, four flat surfaces to make a square cross-section, as shown.

Let the rod become as hot during sustained grinding as it can, so that the finished tang end can be forced hot into the handle. Use visegrip pliers if the rod gets too hot to hold by hand.

(3) Burning the tang into the wooden handle.

As the next step must be taken without pause, always keep a supply of standard commercial wooden file handles ready. They should have steel ferrules and be predrilled with ⅛- to $^3/_{16}$-inch diameter, 3-inch-deep holes to guide tangs into. Note that, at a $^5/_{16}$-inch diameter, the tang is larger than the predrilled handle hole; this ensures a snug fit once tang is burned into the handle.

Quickly, while it is at its hottest, clamp the rod in the vise, tang end up. Fit the handle hole over the tang point; then tap the handle rapidly with the hammer, driving it down on the tang to its full 3-inch depth. Smoke will indicate its "burning" heat.

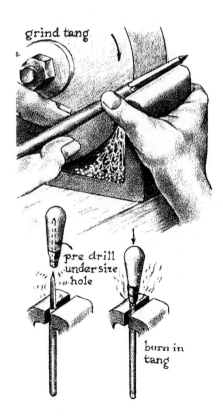

making a screwdriver

grind tang

pre drill under size hole

burn in tang

If you do these steps without interruption, the hot tang burns in evenly, without charring the wood excessively. Heated wood momentarily softens, yields, and (finally cooled) becomes hard once more. Thus, the handle will not split, is seated to perfection, and is locked against turning.

(4) Grinding and hardening the screwdriver bit.

The screwdriver bit should now be ground into a flat tapered end against the side of the wheel, as shown. As you grind the taper to knife-sharpness, you will (in a semi-dark room) begin to see a yellow heat glow appear — which is what you are after. The heat glow will "cool" to dark cherry red within one or two seconds after the tool is withdrawn from the grinding wheel. At that very moment, instantly quench it in the can of water kept alongside for that purpose.

If the resulting temper should prove to be too hard (the edge of the tool would chip in use) it means the quench has made the steel too brittle.

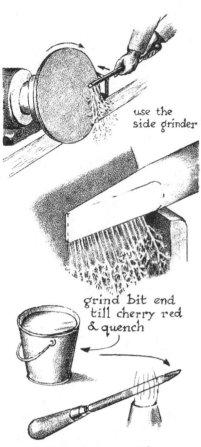

use the side grinder

grind bit end till cherry red & quench

polish sheen on bit end & over blue gas flame draw to bronze color & quench.

To repair, proceed as follows. Regrind the bit to an accurate taper (keeping it cool while grinding). Put a sheen on the bit end and draw to a bronze color over a gas flame, as shown.

And there you have it: a well-tempered screwdriver in about fifteen minutes! At the same time, the beginner has been introduced to the basic principles of toolmaking in the shop

Keep in mind, however, that while this simple tempering method can be applied to any small piece of steel (like a screwdriver), bigger workpieces must be heated in a forge fire to reach the necessary temperature.

4. Making a Cold Chisel and Other Simple Tools

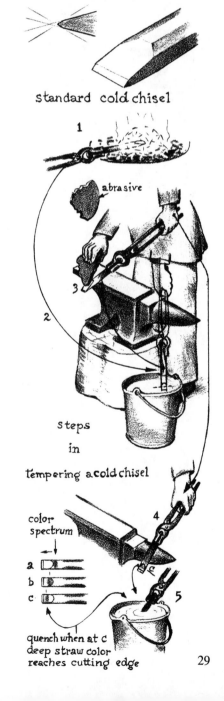

A COLD CHISEL

Select from your scrap pile a ⅝- to ⅞-inch-diameter round or octagonal bar, about 15 inches long. (As a general rule, the steel industry seems to mill much high-carbon steel into octagonal profile, though of course there are exceptions to this.)

Check, by holding the steel against the motor grinder, whether it is of high-carbon quality, as described in Chapter 2 (page 21). If in doubt, you can always resort to temperability test number 2, described in detail in Chapter 2.

SHAPING AND TEMPERING THE COLD CHISEL

(1) Shape the rod on the motor grinder, as shown, and cut off the length you want for your chisel.

(2) Build a medium-sized, clean, steady, smokeless fire.

(3) Place on the anvil, or nearby, a scrap of abrasive stone with which to put a sheen on the end of the chisel in order to judge the oxidation colors in the tempering process.

The quickest but least controlled tempering method is to hold the rod in tongs and heat about ⅜ of an inch of its end in the forge fire to a cherry red. Quench it immediately in water and test for hardness.

If you are not sure about judging the right moment at which to quench the steel, and if, in testing for hardness, you find the edge buckles or cracks, try the following, more controlled method.

Heat 1½ inches of the rod end to a dark yellow. Now, quench only ¾ of an inch of the end, as shown, and withdraw it when the visible heat glow of the rest of the steel has disappeared. This immersion step takes about ten to twenty seconds. Quickly transfer the still-hot rod to the vise or anvil and hold it firmly slanted downward over the edge. Immediately rub the end with the broken abrasive stone to make the steel shine enough to observe the oxidation colors travel through heat conductivity. Note when a dark yellow to bronze color reaches the cutting end, and at that moment quench the whole tool. This is then the right hardness for a cold chisel. (See also Chapter 2, on steel-tempering.) Whether you now harden the other end of the rod or not is optional; this is sometimes done in order to keep the steel from "cauliflowering" after long use.

In any case, cut it to the length you desire. (The average cold chisel is about 6 to 8 inches but special ones often have to be longer to cut steel in areas that are hard to get at.)

Finally, grind the bevels as shown, frequently dipping the tool in water to keep it cool during grinding so it will not lose its hardness.

standard cold chisel

1

abrasive

3

2

steps in tempering a cold chisel

color spectrum

a

b

c

4

5

quench when at c deep straw color reaches cutting edge

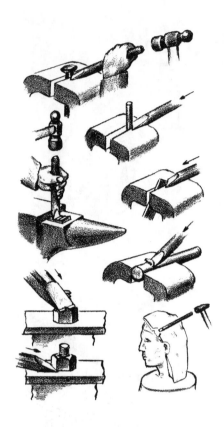

A simple cold chisel is one of the most useful tools in the shop. There are endless tasks that cannot be accomplished without this multi-purpose tool, among them:

To cut off the head of a nail clamped in the vise (to make pins, rivets, nail drills, etc.).

To score a hammer-stem wedge.

To hack off a corner of a steel part.

To cut a flat section on a round rod before trueing it up with a file.

To free a nut so rust-frozen on to a bolt that no other tool will budge it. A cold chisel can notch a toehold on the edge of the nut and, with a few hard hammer blows, loosen it. A wrench can then finish the job. If all else fails to loosen a nut, then the cold chisel, placed as shown, can be driven in as a wedge, creating such enormous strain on the bolt that it will break off. By using a shearing action, in time the chisel will cut the bolt off. No matter how frustrating it often is to loosen badly rusted nuts and bolts, the cold chisel will generally solve your problem.

A small, delicate cold chisel can be used to chip plaster from waste molds in the clay-modeling and plaster-casting arts.

A CENTER PUNCH

Make a center punch out of a ½-inch-diameter high-carbon steel rod. Grind as shown. Harden just as with the cold chisel. Cut off the length desired.

A NAIL SET

Use another length of a high-carbon steel rod, previously annealed, to make a nail set. Clamp it in the vise, as shown, and use a center punch to drive a crater-type depression in its center. Now grind, as for a punch, leaving a sharp little crater in the end, and temper like a cold chisel.

CHASING TOOLS

These tools are used to texture metal surfaces (such as the surface of bronze-cast statuary) in a punching action that leaves tiny craters as well as raised portions. Chasing tools may be made with a great variety of texture patterns to suit the user. Each "business" end is tempered the same way as is a cold chisel.

PUNCHES

Sheet-metal, leather, and paper punches are made to cut through material (or indent it, as in embossing and repoussé) and to create decorative designs. These punches can also vary greatly in shape of cutting edges, according to the user's need and inventiveness. Again, they are tempered just like a cold chisel.

A PAINT MIXER

Sometimes called an agitator, this tool can be cold-bent if it is mild steel, and filed as shown. It will function well if the rod is not less than $^5/_{16}$ inch in diameter and is used at the slower rpm (on the medium pulley) of a drill press. However, if greater speed is needed or more resistant paint is to be mixed, the rod should be high-carbon steel. Such steel is inherently both stronger and more resilient in the annealed state than is mild steel.

DRILL BITS

Indispensable as accessories, drill bits can be made from leaf springs and shaped with the abrasive cutoff wheel. They can be heated, flattened, annealed, filed, and ground into their final shapes.

Temper each cutting bit a peacock or purple color to prevent it from breaking under torsion strain.

Now that you have made these simple tools, you are ready to progress to more and more complicated ones. Your confidence will increase as you work.

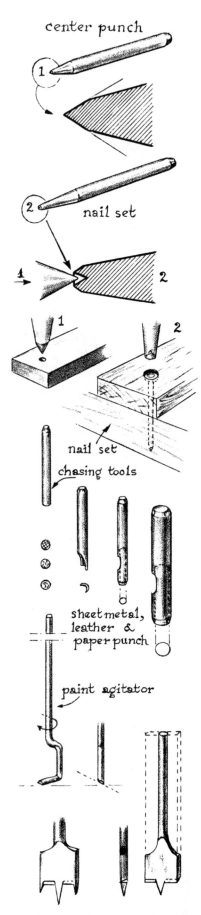

center punch

nail set

nail set

chasing tools

sheet metal, leather & paper punch

paint agitator

drill bits for wood

31

5. Making Stonecarving Tools

THE ONE-POINT TOOL

This stonecarving tool is simple for the beginner to make and does not need elaborate equipment. It looks like an elongated center punch and is the major tool used by sculptors working in stone. Other craftsmen use it too, for carving stone birdbaths, for chipping and fitting garden stones, and for carving other roughhewn pieces (in any stone softer than very hard granite).

Choose a ⅜- to ½-inch-diameter high-carbon steel rod (round, square, or octagonal). Test the steel with a file tip to make certain it is not too hard to cut with a hacksaw blade. Use the file *tip* only in order not to dull the file. If the file tip slides off, the stock is too hard to saw and should be annealed. (To anneal steel is to soften it. It is done simply by heating the steel to a red glow and cooling it slowly in ashes.)

Cut off an 8-inch length of the rod and grind or file its end into a point, as shown.

Heat ½ inch of this pointed end in the forge fire. As soon as a cherry red heat glow has been reached, quickly quench it in water.

Clamp the piece in the vise and test the pointed end for hardness with the file tip. If it slides off, the point is hardened. But it could be too hard: if the hardness has penetrated beyond the ½ inch at the end, the point may be too brittle. Therefore, probe with the file tip farther and farther in from the end until the file begins to grab the steel. Ideally, this should happen at ½ inch from the tip, which makes your tool just right for stonecutting. Should it occur as far in as ¾ to 1 inch, the point would be too brittle and might break off in use.

In this case, repeat every step since you first heated the tool. This time, however, the heat glow should be a little darker than cherry red before quenching. The tool tip will now be hardened correctly. The other end of the tool should be hardened in exactly the same way. (Various high-carbon steels in your scrap pile respond variably.)

Having finished this simple one-point tool, you are now well on your way to developing good judgment over the sequential steps always used in toolmaking: shaping the tool from raw stock into the tool blank; refining the blank into the final shape; hardening the tool's cutting end.

At this point, it is recommended that the beginner carefully reread the sections on tempering until all steps become thoroughly familiar. You will find that when the basic procedures are fully understood, you can work with greater ease and flexibility as you proceed to make the many tools covered in this book.

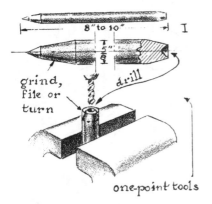

grind, file or turn

drill

8" to 10"

one-point tools

THE CLAW TOOL

The stock used for a stonecarving claw could be a ⅜- to ¾-inch-diameter high-carbon steel bar. But you may also find other usable stock: bars not less than ⁵/₁₆ of an inch thick, about 1 inch wide, and 10 inches long. Leaf springs from cars are often ⁵/₁₆ of an inch thick and of excellent steel; cut into smaller sections, one such leaf-spring blade may yield stock for a dozen or more claw blanks. Even though you find a type of scrap steel not mentioned here, it can be just as good or better, so don't hesitate to experiment with it.

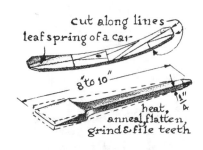

cut along lines
leaf spring of a car
8" to 10"
heat, anneal flatten, grind & file teeth
1¼"

Although an abrasive cutoff wheel can be used to cut through the leaf spring, you may prefer to have a welder do it for you with his acetylene cutting torch. In this case, all you need do is mark off with chalk the pattern of the various claws you want, just as a tailor marks off clothing patterns.

Remember, however, that any piece of high-carbon steel cut with the welder's torch is "burned" along these cuts. All burned edges have to be ground back on the motor grinder (a file would quickly dull on such hard edges) until unmarred steel is reached. The alternative to the welder's torch is the abrasive cutoff wheel (see Chapter 1, on cutting steel with an abrasive cutoff wheel), but it takes longer and assumes that you either own, or have access to, a table saw.

One end of the bar is now to be heated and flattened on the anvil with a hammer to form the shape you have chosen. The cutting end of the tool can now be annealed and the bevel either filed or ground on the motor grinder.

Clamp the beveled blank in the vise. With a triangular file (I often salvage some from a professional saw-filing shop), nick dividing marks on the edge, spacing the number of teeth you want — 5, 6, 7, 8, or more. Once you are satisfied that the division is accurate enough, cut the spaces by bearing down firmly on the file in the stroke that cuts *away* from you, idling lightly on the return stroke (so as not to dull the file any more than necessary).

File one side of the tool until the triangular grooves meet in sharp points on the tool edge; then do the other side. It is important to note that since the two outer teeth must remain as strong as those in between, that is why the blank has its two outer side facets ground as shown.

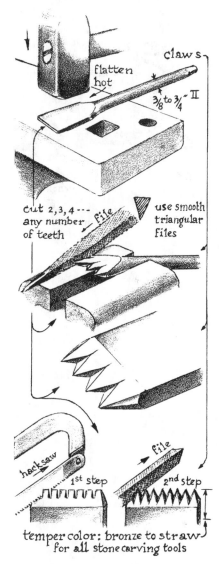

flatten hot

claws

⅜ to ¾" II

cut 2, 3, 4 ... any number of teeth

file

use smooth triangular files

hacksaw

1st step

file

2nd step

temper color: bronze to straw for all stone carving tools

Method 1

$1\frac{1}{2}$ to $\frac{3}{4}$"
remains annealed
to be tempered

3"

$2\frac{1}{4}$"

$\frac{3}{4}$"

heat to light cherry

next, hold 15 sec.
submerged

next, with abrasive,
put sheen on 2",
watch for bronze
color to reach end

2"

then

quench fully.

temper other end
the same way

over
blue
gas
flame

Method 2
quench whole end
after heating; polish,
draw to bronze color,
and quench full tool
at that instant

The tool end that is struck with the hammer should be hardened just like the one-point tool. But the delicate teeth of the claw require a more cautious tempering method, which can be accomplished in one of two ways:

(1) In the forge fire, heat 2¼ inches of the claw end to a light cherry glow. Quench by immersing only the final ¾ inch in the water, just long enough (about fifteen seconds) to see the heat glow disappear in the unsubmerged part. At that moment, quickly withdraw the tool and place it over the edge of the bucket. Rub it with a piece of sandstone or scrap of abrasive stone, creating a bright metallic sheen on the ¾-inch quenched end. The heat stored in the tool will now reheat the file-hard quenched end through conductivity. First, a light straw color will appear, but as the stored heat travels on, the teeth become hotter and hotter and light straw yellow turns to dark straw yellow, and next to bronze.

At this moment, quench the whole tool, to arrest the tempering process; this ensures that the teeth are now of a hardness to cut fairly hard stone, yet not so brittle that they might break off.

(2) This tempering method is similar to the first but gives you more control. This time, when 1 inch of the cutting end is heated to light cherry red, *quench the whole tool.*

Once the tool has cooled, polish the claw end with an abrasive stone or carborundum paper to a mirror sheen. Now use a blue flame (from a propane gas torch) to reheat the steel, thus drawing the temper colors. Watch the oxidation colors run their sequence as heat is conducted toward the claw end. Just as this color band can be arrested by full quenching, so can it be retarded by moving the tool farther from the heat source. When the teeth are drawn from a dark straw yellow to bronze, quench the tool once more.

This controlled method is always preferable when tempering an important or delicate tool.

THE BUSH TOOL

This tool with a serrated face is often referred to as a "nine-point." The bush tool actually crushes the stone, pulverizing its outer surface, whereas one-points and claws *chip* stone. The bush tool requires a steel stock about 1½ inches in diameter or square cross-section. Such heavy stock may be cut from a salvaged car axle, as shown.

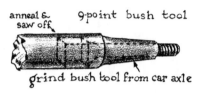

anneal & saw off   9-point bush tool

grind bush tool from car axle

When filing the teeth of a bush tool, make certain, as with the claw, that the steel is annealed by burying the heated tool end in ashes to cool the steel very slowly. This makes high-carbon steel as soft as it will ever get — sufficiently softer than files used on it, which otherwise would become dull quickly.

Once you are satisfied that the steel is properly annealed, clamp the tool blank vertically in the vise, as shown. First, prepare the teeth locations with a hacksaw. Follow up with a triangular file; and, finally, file the side facets on each tooth by slanting the file, as demonstrated in enlarged detail.

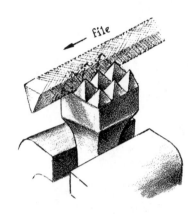

file

Once ground and filed, the points should be tempered as suggested in the first method for claw-tempering. This time, however, draw temper color to light straw yellow (harder than dark straw yellow). The reason is that the bush points will not be unduly strained by side tension, and thus can take a harder temper for longer wear.

Bush tools of a smaller caliber may be made from heavy-gauge coil springs. These are often to be found as scrap steel from heavy trucks. When a section of such a coil is cut on the abrasive cutoff wheel and heated and straightened, it will be long enough to make many sturdy stonecarving tools. The diameter of this steel is rarely over ¾ inch, just enough to file into its end-face a series of small points for a bush tool used to texture the delicate detail of stone sculptures.

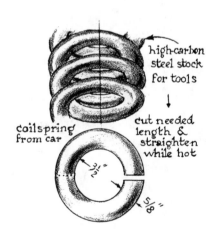

high-carbon steel stock for tools

cut needed length & straighten while hot

coilspring from car

$3\frac{1}{2}$"

$\frac{5}{8}$"

6. Sharpening Tools

Sharpening tools should be done on machine grinders if your shop is equipped for it, as power grinders not only save time and effort but result in a more perfect cutting edge as well. Sharpening tools on hand stones, although described at the end of this chapter, will prove your second choice once both methods have become familiar. You may then join me in saying, "Throw all hand stones out of the window" if machine grinders and buffers are available.

MAKING A SIDE GRINDER

The materials needed to build this power grinder are: a ¼ hp electric motor, scrap plywood, a salvaged abrasive cutoff wheel disc, setscrews, and some wood glue.

Cut out one ½-inch-thick plywood circle, about 10 inches in diameter (approximately the size of the salvaged abrasive cutoff wheel), and one ½-inch thick about 6 inches in diameter.

Also cut two that are 3 inches in diameter. Glue all four plywood pieces together, concentrically, between wood clamps.

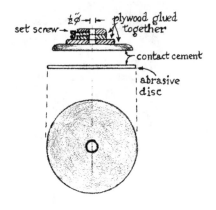

set screw → plywood glued together
contact cement
abrasive disc

Next, drill a ½-inch-diameter hole in the exact center of this assembly to receive the ½-inch-diameter shaft of a standard ¼ hp electric motor.

To secure this assembly on the shaft, the wooden hub part should accommodate a ⅜-inch-diameter wood screw as a "setscrew." Drill a ¼-inch-diameter hole in the hub, as shown (you can make the drill from a ¼-inch-diameter mild-steel rod 7 inches long, its end filed to resemble a drill tip). Then, forcibly screw a ⅜-inch wood screw, its threads rubbed in candle wax, into this undersized hole.

The motor shaft, as a rule, will have a key slot or flattened facet on to which the setscrew can lock. If not, drill a shallow hole into the shaft at a measured distance from its end, to seat the setscrew firmly.

Thus anchored to the motor shaft, the wood disc assembly can now operate as a wood lathe. An improvised tool rest, as shown, will assist your lathe-cutting.

When you are satisfied that the disc is perfectly flat and round (check with a straightedge), remove the assembly and apply contact cement to both the face of the wood disc and one side of the abrasive cutoff wheel. Once this cement has become dry enough (in a minute or so), press the two together. (A good way is to place the assembly flat on the floor, cutoff wheel down, and step on the hub, using your body weight to press the two cemented areas together.)

temporary toolrest nailed to bench

The next step is to remove the plastic coating from the face of the cutoff wheel so that the grit is entirely exposed. For this purpose, a piece of flat concrete, such as broken pieces of sidewalk or cement waste from building projects, is useful.

Place the cement slab horizontally, as shown, so that the abrasive disc can be moved over it in a grinding action. Sprinkle the slab constantly with coarse sand and a plentiful supply of water. The downward pressure will grind off the plastic, exposing the grit on the cutoff disc.

Finally, lock the finished assembly on to the motor shaft: you now have made one of the most useful machine tools in your shop — a side grinder.

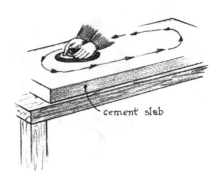

cement slab

USE OF THE SIDE GRINDER

The advantages of a single-purpose side grinder over standard multi-purpose grinding wheels are obvious. Because of the accessibility of the disc surface, you have a full, uninterrupted view when grinding specific bevel angles, cone shapes on blades, flat surfaces on long chisels and knives, as well as the small, flat facets required in V-shaped gouge uprights.

In short, all flat tool surfaces can now be safely and conveniently ground. The composition of this cutoff wheel is so hard and durable that I have found it to outlast every other stone for grinding purposes.

use of the side grinder

no interference by wheelguard

plywood back

thin wheel

Salvaging Worn Stones

When grinding wheels become worn down and thin, and side pressure would endanger their brittle structure to the point of breakage (a break would make them then fly apart), they must be reinforced. Fit a standard extension shaft and cup washer precisely against the shoulder of the motor extension. Between this washer and the thin stone, place a disc of plywood that is 1 inch thick and 1 inch larger in diameter than the worn stone. The side pressure will thus be absorbed by the plywood backing during grinding.

Anchor your now salvaged grinding unit to the shaft with washer and nut; the stone will serve well, though the washer and nut somewhat reduce the total usable surface of the stone.

As a double precaution, a protective wood ring can be nailed or glued on to the backing disc; or, the stone itself can be glued to the disc.

The Dressing Tool

A dressing tool is used to repair worn grinding wheels by smoothing out grooves and cleaning off dirt. The toothed wheels of the dresser are strung on a shaft smaller than the wheels' holes, causing them to spin independently and "rattle" against the revolving grinding wheel. This "rattling" action results in hammer-like blows that break down high spots on the stone's spinning surface.

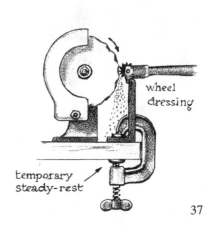

wheel dressing

temporary steady-rest

Dressing Worn Grinding Wheels

Dress a worn stone by adjusting the tool rest (which normally almost touches the stone) so that it is parallel to the stone. It must be far enough away to allow the dresser toe to hook over the edge of the tool rest, yet close enough so that the dresser wheels barely touch the stone's high spots. Then start the motor and move the dresser along the tool rest, wearing down the wheel. An important precaution to take during this operation is to use a nose mask and goggles. Goggles are essential because of the danger of flying sparks and stone particles. But the mask will not be necessary if you breathe properly: inhale deeply before starting the action, hold your breath while the dresser is cutting; exhale when the dresser stops cutting. Let dust settle, and repeat.

If the stone's surface is not sufficiently cleaned in the first operation, tap the tool rest lightly with a hammer to bring it a little closer to the stone. Each such readjustment permits the stone to be better dressed and, in time, the surface will be dressed with great accuracy.

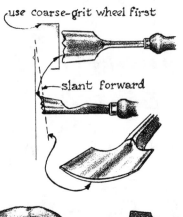

use coarse-grit wheel first

slant forward

Dressing a Broken Gouge Edge

In dressing a wood gouge, be sure to cool its edge frequently with water to prevent the hard steel from losing its temper on the coarse, hard stone of the grinder.

Slant the blade against the stone at the angle originally designed for it. (See Chapter 12, on wood-gouge design.) This slanted edge can be accurately aligned on the side grinder. The slight burr that forms must be scraped off the edge.

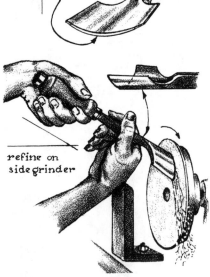

refine on
side grinder

You are now ready to grind the beveled edge of the tool on the various grinders. Take extra care against loss of temper, because the thinner the steel becomes, the sooner heat accumulates. To prevent it, more frequent cooling and slower grinding with less pressure are necessary.

The mirror-like reflection seen on the re-dressed edge must be watched during bevel grinding. That reflection guides you in maintaining an even and gradual approach of the outside bevel plane to the inside blade plane.

With continued grinding, the last little "sheen" will eventually disappear. This is the sign that the outside bevel and the inside plane of the blade have met, creating a minute burr, indicating that no further grinding is needed.

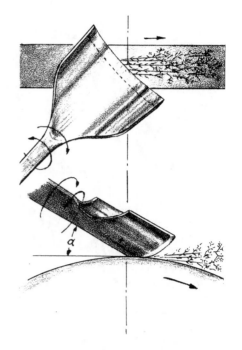

Refine the texture of the ground surface of the bevel on the rubber abrasive wheel and, finally, finish on the tripoli-treated buffer. This will sharpen your wood gouge to perfection.

THE RUBBER ABRASIVE WHEEL

The rubber abrasive wheel is used for honing tools after they have been sharpened on stones. Such wheels are very tricky to work with, but their advantages are great. Because rubber is inherently softer than stone, sharpened steel has a tendency to bite into — sometimes actually bite a hunk *out* of — the spinning rubber grinder. Such an accident can give one an expensive scare, since rubber abrasive wheels are costly. Nevertheless, it is worthwhile learning to use this tool *safely* so that you may enjoy its many benefits. It works much faster than honing by hand. To hand-hone a tool edge to a refined texture takes at least ten to twenty minutes; an abrasive rubber wheel produces an identical, or even smoother, result in ten to twenty seconds.

No wonder this modern honing method is a blessing. For instance, I can quickly prepare a dozen or more woodcarving gouges to start the day's carving by honing their edges on the rubber abrasive wheel, then adding a final finish on the buffing wheel. Such tool edges need no further honing or stropping on the buffer for days on end, even when one is using the tools uninterruptedly on clean wood of various hardnesses.

These two wheels — the rubber abrasive wheel and the cotton buffing wheel (treated with tripoli compound) — will prove indispensable in your shop.

If you have never used a rubber wheel before, I suggest you take a sharp-pointed thin wire or a common pin and touch its point to the rotating wheel lightly, as shown, to experience which angle is the safest between wheel and thin probe. In position 1, it grabs somewhat; in position 3, the probe is grabbed strongly; in position 2, the probe seems to stay on the spinning surface as if sliding over it (which is safe). Now, try to imagine how a razor-sharp tool would behave if pressed on the rubber wheel in any save one of these three tool positions:

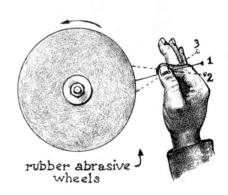

rubber abrasive wheels

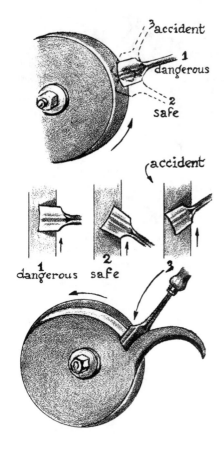

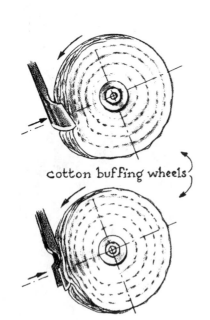

cotton buffing wheels

Position 1 is dangerous because any slight change of angle between tool edge and oncoming rubber surface will cause the tool to bite into the wheel. The soft rubber must slide under the steel edge unimpeded, to avoid accident.

Position 3 shows the sharp edge actually biting into the wheel. A strip of rubber is sliced off and the wheel thus badly damaged.

Position 2 is safe, as shown, because the whole tool bevel, in rocking motion, is pressed on the surface in the same direction that the wheel revolves. This ensures that the tool will not be "going against the grain."

Once in a while, the surface of the rubber becomes shiny with metal pulp, which reduces the effectiveness of its honing action. To clean the rubber, hold the sharp end of a broad flat file very firmly at a right angle to the surface of the wheel rim. This breaks through the outer rubber layer slightly, exposing a fresh, unused surface. As with all abrasive wheels, avoid breathing the pulp dust that is thrown off.

Rubber abrasive wheels are expensive, but should you accidentally slice a chunk out of one, it can be salvaged. Dismount the wheel and lay it on your workbench or the end of a wood stump. Cut off the uneven surface on the outside of the wheel with a chisel. After this new, smaller diameter has been cut, remount the wheel and use a file to refine it, with a scraping-cutting action, as was done in cleaning the metal residue from the wheel rim.

If more serious slicing occurs (perhaps by a heavy tool such as an ax), don't throw away the wheel remnants. Salvage any little portions, and cut them into discs of various diameters. These pieces can now be drilled through their centers (with a sharpened little tube) to receive stove-bolt arbors and washers. Such units will then serve excellently to refine the insides of your various woodcarving gouges. (Such gouges require smooth and even surfaces so that wood chips will slide off easily.)

The rubber abrasive wheel may give you a scare if there is an accident. But, handled with care, it will prove irresistibly useful daily for honing cutting tools.

THE BUFFING WHEEL

After honing, tools are finally refined by polishing on the buffing wheel. Buffing a steel surface is actually polishing it to an even finer texture than that left by the finest-grit grinding wheels and rubber abrasive wheels.

Besides putting on a sheen for appearance's sake, buffing is the absolutely essential final step in mechanical tool sharpening. The cutting edge is mechanically stropped on the buffer much as a razor blade is hand-stropped on leather. The tool, held tangentially to the buffer rim and pressed into it (first the tool's front, then its back), thus receives its finished edge.

Press a block of buffing compound on the rim of the fast-spinning cotton buffing wheel. This abrasive will momentarily melt, with the heat created by the spinning wheel, and adhere to it, transforming the cotton into the finest abrasive surface. I find that, for the buffing of steel, tripoli compound in its coarsest grit works quickly, yet is fine enough. During buffing action the wax-held granules are gradually flung off the wheel, so more compound must be added repeatedly.

Big industries, such as industrial plating companies, naturally have leftover pieces of buffing compound for which they have no further use. As such waste generally ends in the scrap bin, such shops are usually quite generous once their interest and sympathy are aroused by the work of the artist-hobbyist.

Some years ago, I stepped into the office of one of those plating companies. Showing the foreman some of my fine woodcarving tools, and some photographs of my stone and wood sculptures, I explained my need for buffing compound remnants. The boss took one look at my work and led me directly to a thirty-gallon scrap barrel full of fist-sized blocks of tripoli compound!

In industry, these pieces begin as large rectangular blocks, 10 inches long and 2½ by 1½ inches in cross-section. When they are hand-held against fast-spinning, 20-inch-diameter buffing wheels, the blocks wear rapidly and soon become too small to handle with ease.

The foreman urged me to help myself to these discards (especially since I had given him a picture of one of my sculptures!). He filled a large paper bag with about twenty pounds of those remnants, and for many years I was supplied — free — with the very best of buffing compounds.

USE OF HAND STONES

Tool-sharpening by hand requires the use, in sequence, of: one double-grit carborundum stone, one honing stone, one leather strop with emery powder, and thin oil or kerosene.

If an extremely dull chisel (or one that has been badly nicked) is to be sharpened, a coarse-grit stone must be used first, for quantity steel-removal. But if an only slightly dull chisel is to be sharpened, then begin with a fine-grit stone, and follow up by honing and stropping.

Submerge a double-grit (one side coarse, the other, fine) carborundum stone in thin oil until it is saturated. (The standard household brand, 3-in-One oil, will do fine and is available everywhere.) Carborundum stones absorb oil like a sponge, and still more oil must be added to act as a flushing agent to carry off the ground metal during tool-sharpening. The excess oil thus keeps the stone's pores open and its cutting granules exposed to the steel. (Kerosene, instead of thin oil, will also do.) Cradle the stone in some sheet steel or aluminum foil, leaving the top exposed, to keep the oil from draining out. Then, in order to have both hands free to handle the tool, clamp the stone to your workbench (or on a block of wood in your vise) with four little wood cleats, as shown.

Keep in mind that a dry, unsoaked stone will only cause the ground steel particles to clog up the stone's open pores, rendering it, in due time, completely ineffective. Therefore, keep flushing the steel pulp with oil, from time to time. Use additional oil and a small rag to wipe the steel-pulp-laden oil off before adding fresh oil. These are good habits to practice when hand-sharpening tools. (Incidentally, if the stone is of a type where water is recommended as a cutting liquid, the same procedure holds true.)

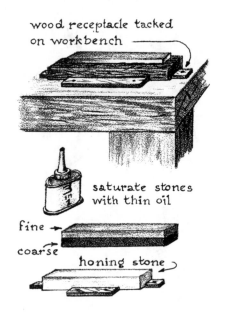

wood receptacle tacked on workbench

saturate stones with thin oil

fine

coarse

honing stone

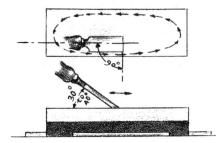

SHARPENING A CARPENTER'S CHISEL

Grinding Motions

A carpenter's chisel, or wood chisel, has two angles, both of which must concern you: the 30° to 40° angle of the cutting bevel; and the 90° angle, at which the chisel itself is held (see illustration). Therefore, you must sharpen the bevel without changing its correct angle, while at the same time maintaining the 90° angle of the edge to the length of the chisel. Only practice will give you the necessary skill to move the chisel over the stone evenly while maintaining both angles.

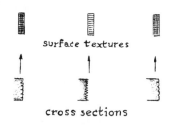

surface textures

cross sections

If you rock the chisel up and down while sharpening it, the bevel will become rounded instead of flat: a rounded bevel tends to make the tool jump out of any wood you try to cut, whereas a flat bevel keeps the tool steady as it is manipulated.

If you bear down on the chisel more on one side or the other, you will in time affect the 90° angle of the edge to the length of the chisel.

Assuming, now, that you are holding the chisel steady, and correctly (both angles unvarying during the grinding), the movement ought to be elliptical, as shown.

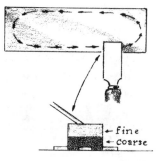

fine
coarse

An elliptical movement continually "crosscuts" the tiny grooves ground into the steel surface, whereas a straight back-and-forth movement grinds the grooves straight. The crosscut texture produced by elliptical grinding cuts the steel faster, while, at the same time, leaving a finer texture on the tool. A straight back-and-forth movement, however, cuts the steel more slowly, since grit slides continually in the same grooves, leaving a coarser texture.

In any case, whether you use elliptical or straight movements, try to utilize as much of the stone's surface as possible, to avoid wearing deep grooves in it. Should it become worn or grooved, resurface it with the same method you use in re-dressing the side grinder (see page 37).

To summarize: Press the tool steadily downward, move it elliptically, and always at the correct angle. Use as much of the stone's surface as possible, so that no local wear hollows out the stone too fast; remember to add oil from time to time; and wipe off steel pulp before each fresh oiling.

Honing the Burr

Grind the tool bevel down on the coarse-grit side of the stone until it meets the other side of the blade at its edge; then turn the stone and continue grinding on the fine-grit side until an almost imperceptible, fine burr forms. Do *not* attempt to remove the burr yet. The burr is so thin that it will bend away from the stone, instead of being ground off. A correctly tempered chisel produces a springlike burr. (Never grind the *flat* side of the chisel in an attempt to grind off the burr.) If the tool edge of a delicate wood gouge or chisel should not form a burr, it simply means that the steel was tempered too hard, making the burr crack off during grinding, instead of bending. This leaves a microscopic jagged edge which will scratch the surface of soft wood, and break on hard wood.

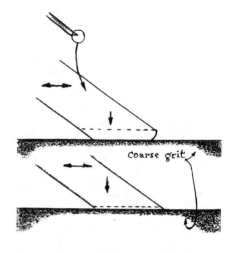

Any further grinding of the burr will only lengthen it, while reducing the life span of the tool needlessly. So turn, now, to the honing stone to refine the texture of the bevel surface. Use thin oil or water as a flushing agent, and hone until the bevel becomes mirror-smooth. When the entire bevel surface shows the honing stone's texture, the burr should still be there.

Stropping the Burr

It is now time to remove the burr. Fix a strip of leather on a piece of flat wood, then clamp it in your vise or to your workbench. This leather strop should now be oiled a little and sprinkled with some extra-fine emery dust.

Just as a barber strops an old-fashioned razor on a leather strop, draw the tool's edge backward over the leather as shown — first one side, then the other — pressing the tool flush with each stroke.

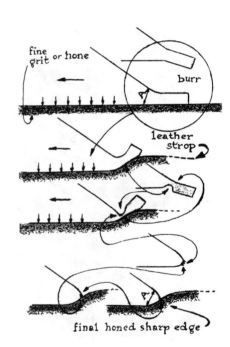

In due time, the emery dust held by the leather and oil will wear the burr so thin that metal fatigue finally makes it crack off, like a tin-can lid bent back and forth.

Now, only a minute, saw-toothed edge remains. Continue stropping *after* the burr has come off, so as to wear this jagged edge down to a microscopically smooth, rounded edge. This achieves an ideal cutting edge, and one that, because it is too tough to bend, is also extremely durable.

Since the grinding effect of the strop deals with a microscopic thickness of steel, it ensures an edge that will remain sharp enough to need only occasional stropping. The life of the tool is thus prolonged, since the harsher grinding on stones will not be needed for some time.

You may now recognize that the foregoing explanation is a welcome answer to that persistent, nagging question: What is the best way to sharpen a tool? Judging from my own long-time experience, the method I have developed and offer here is the best, since it results in the ideal cutting edge and is accomplished in seconds, instead of hours, thanks to power tools and a better understanding of what is involved.

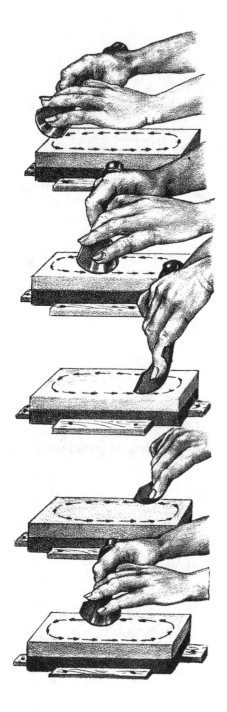

SHARPENING ROUND GOUGES

Like wood chisels, round gouges should be ground in elliptical movements, in a continuous cycle of tool and hand over stone. As shown, the left hand guides the blade end while the right hand rotates the tool around its lengthwise axis. This rotation creates the rocking movement of the blade, while both hands aim to maintain the angle of the tool bevel.

Remember that while some tools may be moved in a rotary action during grinding, their cutting facets must remain flat, in order not to "jump out" of the wood. The tool then would fight your hand instead of obeying it under easy guidance.

USE OF THE DRY GRINDER

⅓ to ½ hp electric dry grinder

Most tools in your shop can be ground on a standard ¼ hp 1750 rpm motor grinder with two wheels (one coarse-grit, the other fine.)

The tool rest, preferably bolted to your workbench, is adjusted simply by tapping it with a hammer to close the gap between tool rest and stone. A can of water to cool the tool should hang between the two wheels, so that instant quenching is possible should the tool edge become too hot to touch.

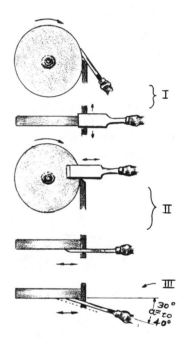

In sharpening a flat carpenter's chisel, for example, hold the blade at the correct angle to the wheel; then move the chisel back and forth over the wheel-rim surface, while maintaining the angle. This back-and-forth movement is corrective: it tends to keep an initially accurately dressed wheel aligned, and makes a *less* accurate wheel *more* accurate. If the wheel should be badly grooved, first re-dress it.

Should a badly abused tool come into your possession, first refine the *flat side* of the chisel on the side of the wheel, or on the side grinder. Do not grind the bevel until a flat surface has been restored.

Grinding new bevels can be tricky, and if, somehow, they turn out less than perfect, a gentle and careful touch-up of the bevel facet on the side of the wheel may save the day. Avoid, however, using the wheel's side if there is a reasonable chance that the wheel's rim can do the job. It is far more difficult to repair a groove in a wheel's side than on its rim.

testing for sharpness

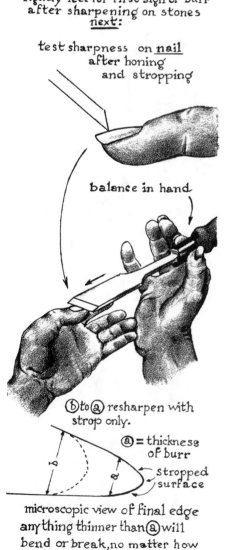

right way to detect burr

wrong way

lightly feel for first sign of burr after sharpening on stones
next:

test sharpness on **nail** after honing and stropping

balance in hand

ⓑ to ⓐ resharpen with strop only.

ⓐ = thickness of burr

stropped surface

microscopic view of final edge
anything thinner than ⓐ will bend or break, no matter how perfect the temper

TESTING FOR SHARPNESS

Lightly touch the flat side of a tool blade with your finger, drawing your finger over the edge, as shown. If a burr has formed at the cutting edge, you will feel it instantly, even if you cannot see it without a magnifying glass.

You cannot tell if a burr has formed if you scrape or rub your finger *across* the tool edge.

Running your finger *along* the edge also is not satisfactory because you are likely to cut yourself. Burrs are razor sharp, with jagged edges.

Once you have established that a slight burr has formed evenly over the whole tool edge, strop it off, as described earlier. The only test now left is that for sharpness.

If the tool edge, gently placed on the fingernail, as shown, seems to "pick" at the nail, rather than slide off, that particular spot of the tool edge is sharp. Any part of the edge that slides off the nail needs further sharpening. Grind, hone, and strop until the edge "picks" at your nail at every local spot along its entire width (while the tool is balanced in your other hand at a barely "cutting" slant). Your tool is now truly sharp.

Note from the enlarged microscopic diagram of an ideally sharp tool that the edge is actually *rounded*.

7. Making Carpenter's Chisels

THE NARROW CHISEL

From your scrap pile, select a leaf spring of a car as stock for a flat, one-inch chisel. Heat one foot of it to a dark yellow in the forge, straighten the curve on the anvil, then let it cool slowly in ashes. This anneals the steel so that it can be cut with a hacksaw and filed without damaging saw or file.

Suppose the leaf spring is 2½ inches wide by ¼ inch thick, and that you want a 1-inch-wide chisel. Saw the spring lengthwise. (One blank will be 1 inch wide, the other 1⅜ inches, with a ⅛-inch loss in saw cut; the wider piece can be saved for future use.)

Scribe the exact outline of the chisel on the blank, allowing for a 3½-inch tang. Cut out the blank by filing or grinding along this outline.

If the blade is thicker than you wish, file it down gradually, tapering toward the cutting edge. Use the side grinder if you prefer. Refine the surface with progressively finer grits on small rubber-backed abrasive discs. Finish up on the buffer for a final mirror-smoothness.

Next, grind the cutting-edge bevel at an angle of approximately 30°. Finally, temper the tool and burn the tang into the handle as described in Chapter 10.

THE BROAD-BLADE CHISEL

A wide chisel is made in the same way as a narrow chisel, with one exception: any blade broader than 2 inches must be brittle-quenched in *oil* instead of water, to avoid or minimize the chance of warping. You will doubtless have to experiment (holding the blade on the vertical, horizontal, or diagonal during quenching) before you find the best way to maintain its straight edge.

The next step is to clean and polish the blade mirror-smooth in order to draw its temper color. Here, especially, with a wide blade, you must be extremely careful to prevent warping during reheating over the annealing flame. Aim for a slow, even drawing of oxidation color over the full blade.

Once the desired temper color has been reached, immediately withdraw the blade from the flame and cool it slowly at room temperature. The tension developed during the initial brittle-quenching will be released by this slow cooling process, thus strengthening the blade overall.

Caution must also be exercised when sharpening broad-bladed chisels. It is best to let them cool slowly from time to time, as you sharpen. Or, if you grind slowly enough, the blade will be cooled by the air that is centrifugally blown along the wheel rim as you work. Don't attempt to cool the blade during grinding by quenching it in water, for that, too, can warp it.

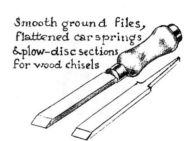

Smooth ground files, flattened car springs & plow-disc sections for wood chisels

Files must be
ground smooth
before shaping them
into tools

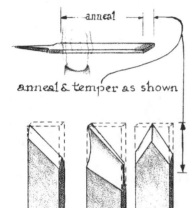

anneal

anneal & temper as shown

wood lathe-turning tools
made from old files

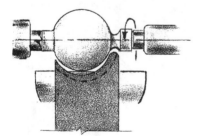

leafsprings ground into
special profiles for turning

FILES AS STOCK FOR CHISELS AND LATHE-TURNING TOOLS

Old flat files lend themselves excellently to making flat chisels (and lathe-turning tools) because the carbon content of files is much higher than that of spring steel.

First, be sure to grind the old file clean of all traces of corrugations, because any remaining grooves can cause the steel to "crack" during quenching in the tempering process (the same principle as cracking glass with a glass cutter).

During grinding, do not lose the file's brittle hardness through overheating (the color *blue* would appear). Assuming that the steel has not lost its hardness, the cutting end can now be drawn a dark bronze to peacock color for a hard cutting edge. The remainder of the tool should be annealed, and the tang heated and burned into the handle.

Finish the chisel by grinding and sharpening a beveled edge, taking great care not to overheat and so anneal it.

The variously shaped cutting ends of lathe-turning tools, which can also be made from old files, ground clean, may be drawn to a straw yellow, if the steel is not thinner than ¼ inch. Anneal the remainder.

If leaf springs of cars are used, choose a section 2 inches wide and cut it 18 inches long. The cutting ends may be ground (be careful not to lose hardness by overheating during grinding) and shaped into various profiles for special wood-turning projects. No handle is needed on such 18-inch-long tools. Turning a wood sphere, as shown, is a one-movement operation.

I have many sections of car leaf springs (most retain the original slight spring curve) whose ends have been ground into various profiles useful for cutting lathe-turned curves in woods.

8. Making Cutting Tools

KNIFE FOR CARDBOARD AND PAPER

A small cutting knife is invaluable when you are working with cardboard and paper. It can be used to make all sorts of articles: boxes, mobiles, art constructions, commercial displays and decorations, mats for watercolors, prints, photographs, and the like.

This tool's design grew out of a Swedish system of manual training called *sloyd*. Taught in many Scandinavian countries, the sloyd-knife art extends to a variety of cardboard and woodcarving crafts. The knife shown here is designed especially to fit the hand as you bear down on the tool during cutting.

There are many kinds of scrap steel suitable as stock for this sloyd knife (as well as for many other cutting tools). The following are suggested: heavy-gauge industrial hacksaw blades, discarded lawn-mower blades, industrial band and circular-saw blades, old butcher knives or cleavers, and — since the advent of the power chain saw — old handsaws, which are often found either secondhand or abandoned around farmyards or barns. All such steel is high-carbon, hard-tempered, and therefore highly suitable for cutting tools.

Step 1. Cut the blank, as shown, on your abrasive cutoff wheel. *Keep the steel cool*, in order to preserve its hardened state, frequently quenching in water, particularly the part intended for the knife blade.

Step 2. The wooden pieces of the handle will be attached to the steel with rivets, but first the two holes in the steel must be carefully prepared. On hard-tempered steel, your drill will be ruined in short order, so the spots where the holes will be drilled must be annealed.

Mark off the two locations on the blade. Now, make a simple mild-steel drill from a nail by cutting off its head and grinding its end flat, as shown. Put it in the drill press, and at *high* speed, press this drill on your hole mark until the friction heats the spot. When it turns blue, it is annealed. Do the same for the second hole mark.

Step 3. Next, use a high-speed-steel twist drill (the size of the rivets you plan to use), but run it at *slow* speed. It should have no difficulty in cutting through the annealed area to make the holes.

Step 4. Choose a piece of hardwood for the handle, selecting whichever grain or color suits your taste. Cut two pieces, a little larger than the desired handle size.

The two holes in the steel can be marked off on each wood piece and drilled. To seat the washer in the handle, drill a slight depression with a blunt-ended bit, the diameter of the washer, or rout-countersink in the wood.

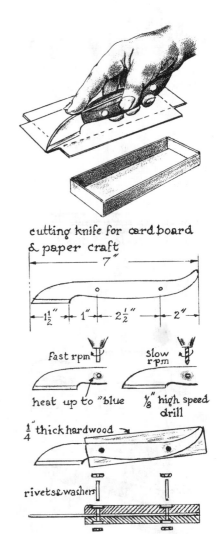

cutting knife for cardboard & paper craft

7″

1½″ · 1″ · 2½″ · 2″

fast rpm · slow rpm

heat up to "blue" · ⅛″ high speed drill

¼″ thick hardwood

rivets & washers

Next, rivet the wood sections on to the metal handle with small brass or steel washers that have countersunk holes. You can countersink the washers on the drill press (using a regular large drill) by holding them in visegrip pliers. A quicker way is to put the washer on a hard-wood stump and hammer it once with a blunt center punch. This will spread the washer hole into a little conical depression wide enough to fit the rivet head.

If you have no flatheaded store rivets on hand, improvise by clipping off nails to a size that will fit the holes snugly. Clipped sections of brass welding rod will do as well. Allow a little excess length, to form the rivet heads.

Step 5. After inserting rivet sections into washers, and washers into the wood-steel-wood handle assembly, tape it all together with masking tape and place it on your anvil.

Strike the rivet first with the flat side of a lightweight ball-peen hammer. The mild steel spreads outward as you hammer, forming a little burr edge, which now begins to act as a rivet head so that you can turn the knife handle over without the washer dropping out. The other end of the rivet, backed by the hard anvil, will also have been compacted and spread somewhat.

Hammering with ball-peen and flat face alternately on both sides of the handle, the forming rivet heads soon fill the washers' counter-sinks completely, leaving little to be trimmed off.

Step 6. Using either a rasp, a flexible rubber-backed sanding disc, or a chisel, trim all excess wood and shape the handle, as shown. Hold it against your side grinder so that the rivet heads can be ground flush with the wood. Take care not to grind the heads more than necessary, leaving enough to secure the handle assembly.

Refine and polish the wood finish as described in Chapter 10.

A CHERRY PITTER

This all-purpose kitchen tool can also be used to pick out the eyes of potatoes and pineapples after they have been peeled, to remove small seeds and bad spots in almost any fruit or vegetable, and to form fruits, such as melon balls, for fancy desserts. This tool cuts by rotating around its axis in a scooping motion.

A ⅜-inch-diameter straightened coil spring can be used for stock. Cut a 14-inch length so that it can be comfortably hand-held while its cutting end is heated to a yellow glow in the fire. Place the hot tip of the rod on the polished, rounded, and tempered end of a ¾-inch-diameter bar that you have previously clamped in the vise to serve as a saddle. Pound the hot tip over this saddle with a 1½-pound hammer until you judge that the "spoon" that has formed is only $1/16$ inch thick in the center.

Shape the spoon on the motor grinder so that the blade gradually thins out to razor-sharpness at its edge.

Grind a round shank, $3/16$ of an inch in diameter and 2 inches long, followed by a 1½-inch tang, which should be square and tapered.

Temper the blade to the hardness indicated by a straw yellow oxidation color, and burn the tang into its handle as described in Chapter 12. Finally, buff the entire blade to a smooth and sharp finish on your cotton buffing wheel.

wrap with masking tape

trim excess wood & polish

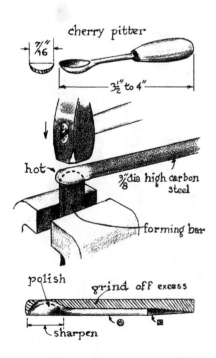
cherry pitter
$\frac{7}{16}$"
3½" to 4"
hot
⅜" dia high carbon steel
forming bar
polish
grind off excess
sharpen

GARDEN TOOLS

The fishhook-like tool is designed to cut plant roots. By forcing it down into the earth next to the root, the hook, which is sharpened on the inside, cuts the root on the up stroke.

Use a piece of a car bumper as stock. This steel is of high-carbon quality and should be about ⅛ to ³/₁₆ of an inch thick.

Scribe the pattern of the blade on a properly annealed and flattened section. Drill the two holes where the rivets will attach the two wooden pieces to the handle.

Next, sharpen and temper the whole blade a peacock color and rivet the wood pieces, as with the sloyd knife handle.

If you prefer, you can seat the tool in a one-piece wooden handle. In that case, the tang should be ground to a slight taper, then heated and burned into the handle, as is done with woodcarving tools (see Chapter 10). Leave a ⅛-inch space between handle and tool-shank shoulder and let the whole unit cool somewhat; then drive the handle down flush with the shoulder. This still warm, yet resilient, wood grasps the tapered tang and, once fully cooled, locks in securely.

In tools such as these, which exert a pulling rather than pushing action, tangs should extend a little beyond the wood handles; the ⅛ to ¼ inch of steel that protrudes can then be hooked over the end, locking handle and blade together permanently.

Try to avoid accidental overheating during grinding since it might require retempering the knife blade. If this does become necessary, be careful, if the handle is already attached, that the wood does not overheat and scorch. To prevent this, bind the handle with a soaking-wet cloth and thin wire.

Tempering procedures for garden tools will vary according to the amount of stress you expect a given tool to undergo. For example, to ensure resilience at a location apt to break under heavy strain (where handle and blade meet), anneal such areas locally or else temper the steel to purple to avoid brittleness.

The *harvesting tool,* which cuts with a downward stroke, is useful in harvesting thick-rooted vegetables like lettuce, cabbage, and asparagus. It is easily made from a scrap industrial hacksaw blade. Make the wooden handle as for the fishhook tool, and grind the steel blade cautiously so that it does not lose its temper.

A small, narrow *scoop shovel,* useful in transplanting seedlings, can be made from an old handsaw. Cut the section desired with your abrasive cutoff wheel. Next, heat the shank and tang portion to a yellow glow and transfer it immediately to your vise. Fold the hot steel between the vise-jaws (as described in Chapter 12 on making woodcarving gouges).

Taper the tang and burn it into the wooden handle. If the handle has a reinforcement ferrule and a slightly smaller predrilled hole, the tang can probably be driven in cold without splitting the handle.

A small *garden hand hoe* can be cut out of a scrap plow disc. Heat the blank and fold it "hot" over the edge of the anvil. Then, while the tool's cutting edge is still dark red hot, quench ¼ inch of it in water. That part of it is now hardened, while the unsubmerged portion, which will not suffer strain in use, remains soft.

The little *hand rake* is identical to the hand hoe except for its teeth, which should first be slotted on the abrasive cutoff wheel, then bent hot over the anvil. Harden the teeth tips in the same way as the hoe edge.

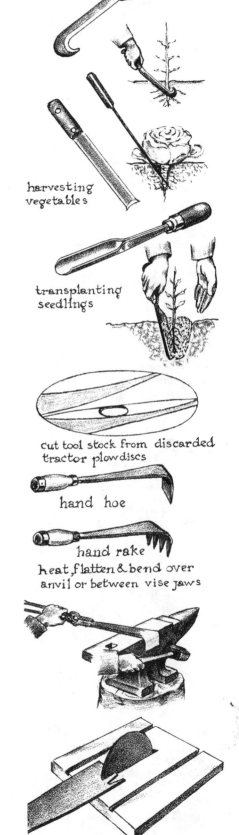

cutting tough roots in garden

harvesting vegetables

transplanting seedlings

cut tool stock from discarded tractor plow discs

hand hoe

hand rake
heat, flatten & bend over anvil or between vise jaws

cut teeth on abrasive wheel before or after bending

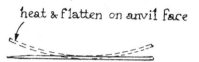

stock for cleaver

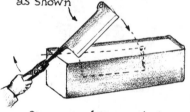

Salvaged worn steel disc from tractor plow is cut as shown to make stock for tools

heat & flatten on anvil face

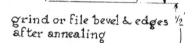

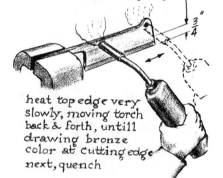

grind or file bevel & edges ½" after annealing

heat in elongated fire ½" to cherry red, and quench as shown

after quenching, polish a sheen for ¾" along edge

heat top edge very slowly, moving torch back & forth, untill drawing bronze color at cutting edge next, quench

fire clay   fire brick

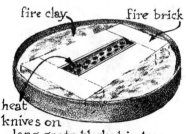

heat knives on long grate blocked in by fire brick & clay placed over forge-grate

KITCHEN CLEAVER

Often called a butcher's cleaver, this axlike instrument is invaluable in cutting up meat, bones, and frozen foods of all sorts.

Discarded plow discs are useful, both in their shape and high-carbon quality, as stock for cleavers. (These plow discs come in various sizes and are good to have on hand in your scrap pile for other tools, as well.)

Scribe the blade design you have in mind on the disc. Cut out the blank on your abrasive cutoff wheel, or use a welder's torch. In this event, grind off every trace of steel residue left by the melting action of the welder's torch. You will now have a "clean" disc section, whose thinner outer edge will form the cleaver's blade, and whose thicker center portion becomes the cleaver's spine and tang.

Heat the blank in the forge fire, and hammer out the disc's curve on the anvil. Let it cool slowly, then file, drill, or grind this annealed blank into its final shape.

Now heat the blade, in an elongated fire, as shown. Temper the cleaver just like a broad-blade wood chisel (see Chapter 8) but use a long quenching trough and a propane torch for local annealing. Either burn the tang into the handle, as described in Chapter 10, or rivet it between two wood pieces, as with the sloyd knife.

9. Making Eyebolts and Hooks

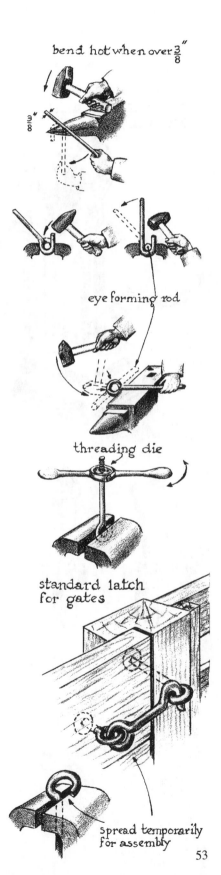

bend hot when over $\frac{3}{8}$"

$\frac{3}{8}$"

eye forming rod

threading die

standard latch for gates

spread temporarily for assembly

Eyebolts and hooks are basic devices for fastening one element to another: gate and door latches; anchors for rope or chain, etc.

Use mild-steel stock, which, if ⅜ of an inch in diameter or less, can be cold-bent in the vise. If thicker stock is used, heat it in the forge first (unless you happen to own a very sturdy vise of fifty pounds or better).

To make the eyebolt, hammer 3 inches of a ⅜-inch rod over the anvil horn until it bends into a 1-inch curve, as shown.

Now seat a 1-inch-diameter steel bar in the curve of the bent rod and clamp the two firmly in the vise. The bolt is formed by hammering first the upright shank end 45° off the vertical. Next, close the eye end by hammering it down flush with the forming bar.

Then remove the whole assembly (forming bar included) from the vise and place the eye over the anvil edge. Hammer out any misalignment, then knock out the forming bar and clamp the eye in the vise. Use a ⅜-inch threading die to thread about 1½ inches of the eyebolt shank-end.

The hook starts out as an eyebolt, but instead of threading the shank, bend it into a hook, just as you did when forming the eye — only leave the end open, as shown.

To link the eyes of hook and bolt, spread one of the eyes open, hook it into the closed eye, then clamp it closed again, either with a hammer or the vise.

A standard latching assembly requires two eyebolts and one hook. The eyebolt through which the hook falls may have to be slightly longer than the eyebolt to which the hook is linked.

53

10. Making Tool Handles

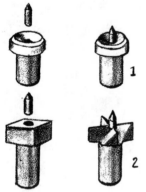

1

2

A standard wood lathe is always horizontal, and more convenient, but not having one perhaps you can adjust yourself easily to a "vertical" lathe to cut the wood for your handles. The drill press can be converted into such a vertical lathe on which to turn wooden tool handles. Once you get used to this vertical, instead of horizontal, lathe-turning, you will find it satisfactory. (For lathe tools, follow the method described in Chapter 7.)

CONVERTING THE DRILL PRESS INTO A LATHE

The first step in making a vertical wood lathe is to improvise a tool rest to guide your lathe tool. Rivet together a section of heavy-gauge angle iron and a piece of ¼- to ⅜-inch-thick angle iron, as shown. Then drill a ½-inch hole at the base angle iron to receive the bolt that anchors this tool rest on the drill base.

To make the lathe centers, two ½-inch cap screws (or tap bolts) are used. With a ³/₁₆-inch drill, center a ½-inch-deep hole in the head of each cap screw. These holes will later receive snug-fitting pointed pins, which impale the wooden handle stock at either end for stable lathe-turning. You can make these pins by cutting off the pointed ends of two ³/₁₆-inch-diameter nails.

File the butterfly headstock, as shown, and tamp in its ½-inch center pin. Then clamp this assembly in the drill-press chuck.

Now countersink the tailstock, using a ½-inch drill; then, either with a file or power grinder, grind the square head round until it forms a knifelike edge with the countersunk depression. Tamp in the second ½-inch center pin and seat this assembly in the base of the drill press. Your converted wood lathe is now prepared to turn whatever handle stock you choose.

It is perhaps worth noting that, for this very operation — grinding the corners off a square-headed cap screw — this vertical lathe would come in handy, since at the drill press's lowest speed, even steel can be ground successfully. You simply brace a file between vertical tool rest and rotating cap screw; this "filing" action quickly and accurately rounds off the screw head — or any other piece of metal that must be turned, provided the part has first been previously ground by hand the approximate diameter needed. Once I realized how well this works, I acquired a carbide-tipped tool for just such improvised metal-turning by hand.

STOCK FOR TOOL HANDLES

Now that your converted wood lathe is ready, it is time to select handle stock to turn in it. Hard, fibrous wood, such as black or English walnut, eucalyptus, ash, hickory, acacia, maple, and comparable woods stand up very well. The wood fibers should always be *straight,* in order to transfer the force of the hammer blows effectively. When the butt end of a wood handle has been plied by hammer extensively, it should begin to fray, like the end of a rope; if, instead, the butt end pulverizes, that type of wood should be avoided in the future. Never use burled wood for handle stock; such wood is "rubbery" and tends to disperse force and thus waste it.

SHAPING THE HANDLE

First, cut the stock to the desired length and thickness. Then drill, in each end, a centered $3/16$-inch-diameter hole to a depth of ¼ inch.

The end that is to be anchored by the butterfly headstock should have a ⅛-inch-deep X cut in it in which to set the four wing edges. This is made with a saw bisecting the leading hole at right angles.

Adjust your drill head and tool rest to accommodate the handle blank. Rub candle wax on the bottom of the blank to lubricate it, then center it on the drill press. Next, set drill speed at about 2000 rpm (somewhat faster than the standard 1750 rpm electric motor speed). Place wood between centers. Before switching on the motor, carefully align butterfly headstock with the prepared center hole and saw slots in the wood blank. Press the tail end firmly onto the tail center. Release the pressure slightly, and switch on the motor.

Once the wood is spinning, gently lower the drill-press head enough so it countersinks the wood on the tail center; then lock the drill head at that position.

You are now ready to turn a wood handle, using the lathe tools as shown in the standard horizontal wood-lathe setup. Whether you use the latter, or the vertical drill-press conversion, the principles of wood-lathe-turning apply equally and the steps are the same.

The proper angle, or stance, of wood-lathe tool to handle stock is one you must gauge for yourself; only experience, after your first tries, can tell you which tool position works best for you. Obviously, the sharper the tool edge, the smoother the wood surface becomes. However, a major rule of thumb is to keep the gap between tool rest and spinning wood to a minimum. And, whether the tool rest is horizontal or vertical, press the lathe tool down onto the tool rest firmly and hold it steadily as you move your hands along the tool rest.

The shape of a tool handle is a matter of personal preference, and can vary in length, slenderness, or stubbiness, depending on the tool's size and intended use. The handle shown here (some 5 to 6 inches long) has proved to be both practical and all-purpose. But once you have chosen a design and turned the handle accordingly, you are ready to tackle the other tool parts.

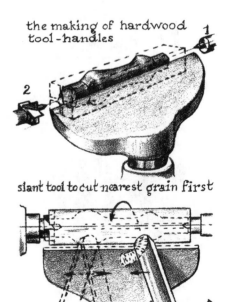

the making of hardwood tool-handles

slant tool to cut nearest grain first

horizontal standard wood lathe setup

FERRULES

Ferrules are metal reinforcement collars, banded around the ends of wooden handles to keep them from splitting. Ferrules can be made from electric conduit pipe of various diameters, as well as metal tubing of many kinds — rifle cartridges, CO_2 cartridges, lipstick tubes. All make effective ferrules when cut into sections.

Use a hacksaw, a pipe cutter, or your abrasive cutoff wheel to cut the length appropriate for each tool. Remove any burrs caused by this cutting with a file or grinding wheel. A rotary file mounted in the drill chuck and run at medium speed is ideal — both for removing burrs and for beveling the inside of ferrules. (An old round hand file, broken into several pieces, makes a fine rotary file. Saw shops often discard ones that you can probably use for this purpose.)

Use one square-holed washer between tool shoulder and the handle. Since tangs are best when square in cross-section (to keep them from loosening and turning in the handle), the washers must fit the square at the point where the tool's shoulder meets it. The main purpose of the washer is to prevent a small shoulder from entering the wood handle under forceful hammer blows when the tool is in use. If the tool's shoulder should be large, a washer is not needed. There are three possible methods of "squaring" a round-holed washer, any one of which will work:

(1) Choose a standard steel bolt washer whose hole is a size smaller than the tool's tang. Clamp the washer lip in your vise so that the hole is free and clear. Use a small square file to square the hole to fit the tang.

(2) Instead of a file, you can use a narrow cold chisel (same width as the tang), shearing the steel flush along the vise jaw, as shown.

(3) A third method is to clamp the tool shank tightly in the vise and slip the undersized washer over the upright tang (it will get hung up about halfway down). Then, take a 5-inch section of ⅜- to ½-inch plumbing pipe and drop it down on the tang so that it sits on the washer. Grease the tang (the part exposed below) a little, and hammer the pipe down until the washer has been forced flush with the tool shoulder. The mild-steel washer yields to the harder steel of the tang. This partly cuts the washer steel and partly compresses it, forcing it tightly onto the shoulder ledges.

It is often difficult to true up shoulder ledges accurately. A simple method is to clamp the tool shoulder (tang upright) in your vise, leaving about ⅛ of an inch of shoulder protruding above the jaws. Slip a fairly large washer (thinner than ⅛ of an inch, but larger than the tool shoulder) over tang and shoulder so that it rests loosely on the vise jaws. With a flat file, cut away any shoulder "excess" around the tang, using the washer as a trueing jig. The washer also acts as a tool rest, keeping the file clear of the vise so that neither it nor the file can be harmed. To remain strong, the tang must be thickest at its shoulder location. Make sure, therefore, that the file's edge does not groove the tang accidentally while you are filing the shoulder flush with the washer.

square-holed washer

tube forces washer onto tool shoulder

washer is jig to help file shoulder accurately

file shoulder

FITTING HANDLES TO FERRULES

Each ferrule should have one inside-bevel end. The bevel, when pressed over a slightly oversized handle end, will not cut the wood but squeeze it in a tight fit with the ferrule. You can measure the precise fit by pressing the sharp edge of the ferrule onto a slightly tapered handle end. This pressure leaves a circular dent on the wood, indicating how much you still need to cut off, yet leaving enough so that the beveled ferrule end will squeeze onto the handle under great pressure when it is hammered, or pressed on, as shown.

One more step may be needed. If the wood chosen for the handle was not adequately seasoned (dried out), shrinkage would eventually loosen the end ferrule. To prevent that, hammer a few depressions in the ferrule (once it's seated) with a center punch. This locks the ferrule to the wood with little dowel-like points. Subsequent hammering, while using the tool, will eventually expand the wood enough to "fill" the ferrule, and thus anchor it permanently.

FINISHING THE HANDLE

Once the ferrules are pressed and locked on, center the assembled handle once again in the lathe and trim off any excess wood.

At this point, either dismount the improvised vertical tool rest (if you are using the converted drill press) or move the standard horizontal tool rest well away from the handle. This clears the work area for the hand sanding and polishing which you now begin.

Refine the handle with progressively finer abrasive paper, until it is extremely smooth. Then, with the handle still spinning, hold a little wad of shellac-soaked cloth against it to seal the wood. Now, before the shellac dries, hold a piece of beeswax against the spinning handle, to act as a lubricant while polishing the surface. The last step is to hold a dry, clean wad of cloth firmly against the spinning handle, moving it back and forth as a final buffing.

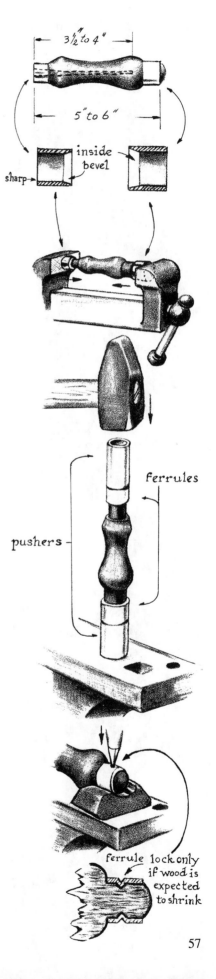

burning tool tang into a wooden handle

heat ¾" tang tip

I

II

III

↓ $\frac{1}{8}$"

IV

heat light cherry red

tempering tool blade after assembly

↓ quench

draw color over gas flame

ASSEMBLING HANDLE AND TOOL

Assuming that your tool blank has already been tempered, and fitted snugly in the square-holed washer, you should now drill a ⅛-inch tang-hole in the 5- to 6-inch handle deep enough to hold a 3½- to 4-inch tool tang. Heat the tang in your forge fire until ¾ of an inch of its tip shows a dark heat glow. This indicates, because of heat conductivity, that the whole tang is hot enough to burn into wood, without risking temper loss to tool shank or blade (should they have been previously tempered).

Clamp the tool shank (tang upward) in the vise and quickly slip the handle down over the hot tang. Let both flame and smoke escape while you rapidly, but lightly, hammer the handle down until it is ⅛ of an inch short of the shoulder washer. Let the whole assembly cool somewhat; then "cinch" the handle down on the tool flush with the washer, thereby seating and holding the burned-in tang permanently. There is now little chance that the handle will split, even with maximum use, since the wood, which became temporarily soft where the hot tang contacted the wood, once again becomes resiliently hard after final cinching and complete cooling.

TEMPERING THE TOOL BLADE

You may find that a tool blade occasionally has to be tempered (or retempered) after it has been burned into its handle. In such an instance, wind a wet strip of cloth around the handle to keep the wood from scorching. Then heat the tool blade to a light cherry glow, quench, and finally, after polishing the blade mirror-smooth, draw the temper color over a blue gas flame.

58

11. Making Hammers

HAMMER DESIGN

In toolmaking, the hammer is an all-important tool. In order to design one that meets its function perfectly, you must first understand the principle of hammering.

A hammer's prime purpose is to release stored energy, on impact. A hammer that strikes a chisel to cut wood must deliver the right amount of energy to make the cut without overloading the chisel so that it bends or breaks. The size of the chisel, the hardness of the wood, and the weight of the hammer must all be correlated. The craftsman should learn to "feel" the ideal relationship, or harmony, between the three elements. Choosing the right weight of hammer (whether it is made of steel, plastic, wood, or rawhide) for the job at hand can prevent frustration, and save time and energy when work is to continue hour after hour.

This same "feel" applies in choosing *any* tool for a given task, even if you simply want to hammer a nail into wood.

Hammer Weight

If you try to drive a thin nail into hard wood, many light taps with a lightweight hammer works best; a heavy hammer would collapse such a thin nail.

To drive a sturdy nail into hard wood, deliver several well-directed medium blows with a medium-weight hammer; too heavy a hammer might collapse such a nail if the wood is very hard.

To drive a heavy spike into hard wood, a heavy hammer, delivering many well-directed blows, will be needed.

Of course, there is a point when even the sturdiest nail or spike will collapse under heavy blows with a heavy hammer if the wood is too hard. Then, only predrilling a slightly undersized hole to receive the nail or spike will work.

relating hammer weight to size of tool & hardness of wood

relating hammer weight to size of nail, hardness of wood

drill when wood is too hard

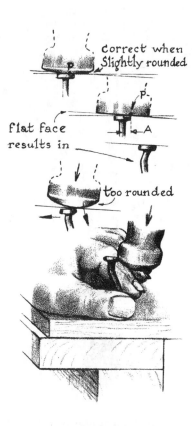

correct when
slightly rounded

flat face
results in

too rounded

carpenter's hammer

The Face

Most hand-held hammers that have hard faces have one design feature in common: a slightly rounded face.

The reason becomes clear once you try to use a flat-faced hammer on a standard flat-headed nail. As shown, a slightly unaligned blow (on a slant, instead of parallel to the face of the nailhead) contacts the nailhead at point P; the force of the blow, delivered from distance A toward the center of the nail stem, will either bend the nail, as shown, or cause the hammer to glance off.

This same unaligned blow, delivered by a slightly rounded hammer face, contacts the nail head at P so close to the center of the nail stem that it will not bend.

But, note that too rounded a face spells trouble — in the form of a bloody thumb and finger!

Since hand-hammering is necessarily less precise than machine-hammering, this slightly rounded hammer face design allows the craftsmen a margin of human error, but near-machine precision.

The Claw

When nails are not driven accurately — or permanently — they must be extracted. Thus, a *claw* was added to the hammer's design, giving us the present-day carpenter's hammer.

The curve of the claw varies to meet the needs of pulling out different nails. A steeply curved claw works best to pull out a large nail that is firmly held in the wood; but it is slow work and requires the aid of progressively bigger props. A flatter claw will pull out a small nail that is not held too firmly in a couple of tugs, with or without a prop.

The Stem

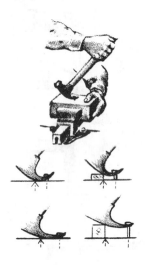

shoemaker's hammer

The length of the hammer stem (handle) depends on the craftman's needs. For example, a long stem can be a great impediment when you are doing close work hour after hour, in cramped quarters. Conversely, if a great deal of heavy hammering must be done — large spikes, hard wood — a short stem would soon exhaust you. Generally speaking, stem length is determined by hammer head weight, which in turn is determined by the magnitude of the task at hand. A tack hammer is short-stemmed and lightweight; a sledgehammer is long-stemmed and heavyweight.

A shoemaker's hammer is specifically designed so that the cobbler can drive in short-stemmed tacks with broad flat heads, without denting the surrounding leather. Such a hammer has a large, flat face, a slightly rounded rim, and is fairly light in weight. In one stroke, the tack automatically aligns itself with the hammer face. Such tack hammers are also used by carpet-layers and upholsterers, who have to tack carpeting or material tight without crushing fiber or fabric. These hammers often have magnetized faces, to which tacks "stick." The upholsterer thus has both hands free, one to wield the hammer and the other to stretch the material tight. In short, analyzing your needs correctly will lead to logical design of the hammers you make.

THE CROSS PEEN HAMMER

Shown here is a simple, all-purpose, lightweight hammer with a double-duty head: one end has a slightly rounded face; the other, a cross peen. It is useful in many shop activities — driving nails and forming heads on small rivets, bending light-gauge metal parts in the vise, etc.

Begin with a rod of high-carbon steel that is ¾ to 1 inch in diameter. (You can use a square cross-section bar, a torsion bar, or a car axle, should your scrap pile yield these.)

Clamp the end of the rod into the drill vise and make two center-punch marks ⅜ of an inch apart; then drill two holes, ⅜ of an inch in diameter, through the bar at these marks. If the steel is too hard to drill, heat it in the forge and cool it slowly in ashes to anneal it.

The holes, made by a ⅜-inch drill, should be close together but just missing one another. Be sure that they are exactly centered (not lopsided) in the bar.

Now select a ⅜-inch-diameter mild-steel rod and saw off two sections 1-inch long to be used as plugs. Dent them with a few hammer blows on the anvil so that they will "grab" when pounded into the two prepared holes. Once the plugs are in place, grind off any excess so their heads are flush with the rod surface.

Drill a third hole, midway between the first two, then drive out the two plug remnants. You will end up with a roughly oval hole, as shown. Smooth out the inside with a hand file.

Now fashion the cross peen, either by grinding it on a motor grinder, or cutting it with a hacksaw. Cut off the last 3 inches of the rod with a hacksaw to form the hammer head.

blacksmith's hammer

1" DIA.

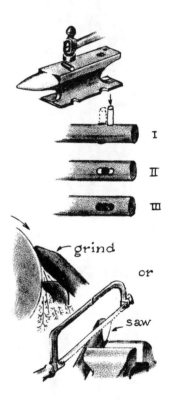

I

II

III

grind

or

saw

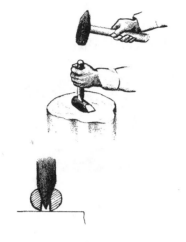

Place this blank on your freestanding wood stump and insert a ¾-inch cold chisel into the oval hole, as a wedge. Holding the chisel firmly in vise-grip pliers or tongs, use your heaviest bench hammer to drive the chisel in until it spreads the sides of the steel blank slightly. This gives you a roughly conical hole, into which the wooden stem will later be locked. Knock out the cold chisel (easy, because of its taper) and your hammer blank is now ready for tempering.

String a piece of baling wire through the hammer hole so that you can submerge the whole blank in the forge fire, as shown. When it glows a dark cherry red, quench it at once deep in oil, moving the hammer head up and down a little to speed the cooling somewhat.

Any piece of hot steel as large as this blank will require at least five gallons of oil for adequate quenching. Old crankcase oil will work very well, but make sure your oil container has a hinged lid so that you can close it quickly to snuff out accidental flash fires.

Since the hammer head has become thin on both sides of the hole, quenching in water is too risky; the steel may well crack at such vulnerable points.

Test the cooled steel for hardness with the file tip, pressing down firmly. (Never apply full file strokes on hardened steel for you will ruin your file in short order.) Recall that if the file tip slides off like a needle on glass, your steel is sufficiently hard.

Now anneal the sides of the hole to prevent the hammer head from breaking there during severe use.

This local annealing can be accomplished while simultaneously tempering the hammer head as a whole. Grind the sides of the hammer on the rubber-backed sanding disc until a mirror-smooth sheen appears from end to end.

Now take a ¾-inch-diameter mild-steel rod and grind its end flat and tapered, like a cold chisel, so that it will slip loosely into the hammer hole.

Heat the tapered rod end in the fire. When it becomes yellow hot, hold it upright and slip the hardened hammer blank on it. Hold it above the water bucket and watch for a yellow oxidation color to spread to both ends of the blank. At this point, the sides of the hole will be purple, whereupon you should immediately knock the blank off the rod and into the water bucket.

Once it is quenched and cool, hold the blank at right angles to your rubber-backed abrasive disc to true up the hammer face (check with your square for accuracy, as shown).

As you proceed to grind, rotating the blank over the spinning disc, you will find that the natural limitations of hand work result in concentric inaccuracies. Instead of a perfectly flat surface, the hammer face will become just slightly rounded — but symmetrically so, and therefore exactly what you need!

Now prepare a 10- to 12-inch hammer stem made of ash, hickory, eucalyptus, or any hard-fiber wood of your choice. Drive it into the smaller opening of the conical oval hole.

Make a steel wedge out of a thick nail by cold-hammering its end into a taper on the anvil. Score it with a cold chisel to make sure it holds tightly when driven into the wood. Drive the nail wedge in diagonally, as far as it will go, thereby spreading the wooden stem to fill the cone-shaped hole. Cut the nail excess off with a hacksaw and grind it flush with the hammer head.

Check to make sure that the stem is correctly aligned with the hammer head, from all angles. If stem alignment is off, but the wood is thick enough, reshape it with a wood rasp or sanding disc, as shown. If the stem is askew (as shown in profile), nothing can be done but to cut it off where it enters the hammer and punch out the wasted piece. You will lose that length when refitting the stem, but enough may remain to be refitted accurately.

You have now completed a well-tempered, all-purpose hammer for your collection, and, at the same time, learned the principles to apply in making future hammers of different design.

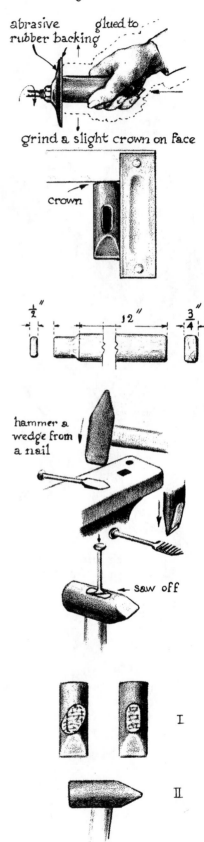

making a hammer

abrasive glued to
rubber backing

grind a slight crown on face

crown

hammer a wedge from a nail

saw off

I

II

12. Making Sculptors' Woodcarving Gouges

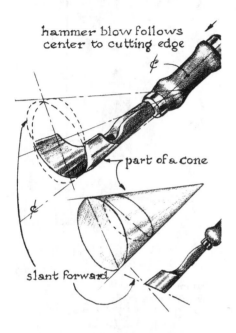

hammer blow follows
center to cutting edge

part of a cone

slant forward

Follows curves
without binding

because blade is
part of a cone

DESIGN OF THE SCULPTORS' WOODCARVING GOUGE

This chapter is concerned mainly with sculptors' woodcarving gouges, yet the design principles are basic to almost any wood gouge or chisel.

Other crafts may require particular features in woodcarving tools that are not present in sculptors' gouges. For instance, certain crafts use chisels to *pry* with (as in making mortise-joint seats); to *scrape-cut* (deep wooden bowls); or to *plane* (a millwright's slick).

The function of the sculptor's woodcarving gouge is simply to remove chips of wood, much as a beaver does when felling a tree. The tool design, as shown, becomes self-explanatory. If a chisel has the conventional "cylindrical" type blade, it cannot help but bind as it cuts a deep groove, especially in a curve. Note, then, the difference of the "cone" type blade, and how, without resisting sideways guidance, the cone design allows the blade to follow the carver's manipulation.

The illustration shows how the cutting edge should be tailored so that the *upper* part of the edge slants forward. This blade design ensures that the outer grain (fiber) of the wood is cut first, thus freeing and releasing the woodchip without binding.

Conventional chisels have the lower part of their cutting edge ground to protrude forward. This results in a "wedge" action, in which chips can be released only after forced driving. That same force will often tear the outer fibers, instead of cutting them, especially in a hard or brittle wood. Soft wood can be cut well enough if the conventional chisel is thin, since the wedge action of the tool hardly compacts the wood. But the hard mallet or hammer blows required to free chips tend to bend or break such a thin tool, making it a liability at best.

All woodcarving gouges that are hand-held can profit from the conical design. When modified by the inventive craftsman, they can serve the most unexpected woodcarving needs (bowl- and scroll-carving, cutting letters in signboards, marking measuring scales, for instance).

It may interest the reader to know that this cone blade design, though probably not "new" in a historical sense, came to me many years ago when I began to design woodcarving tools for my own use. I had found that commercial chisels were not adequate for wood sculpture, and so began to design my own. The cone blade quickly caught on with other sculptors, who saw it as the ideal answer to their particular needs.

basic design principle for woodcarving gouges

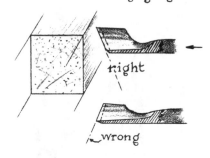

right

wrong

upper grain is cut first

therefore all lower grain is cut next, freeing chip to come loose easily

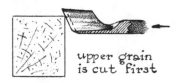

SMALL WOODCARVING GOUGES

The hand-pushed engraver's-style small woodcarving gouge, as shown, is adapted from the wood-engraver's burin. Thus, the woodcarver can utilize the engraver's flexible hand and tool manipulation when he works with a small gouge.

The most distinctive feature of the engraver's-style gouge is its flat upright shank, traditionally designed so that the thumb, braced firmly against it, acts as an anchor as well as a guide along which the tool slides as it cuts. The tool is held by the fingers, while the handle is cushioned against the heel of the hand and is pushed by contraction of the hand muscles. For increased stability and reduced danger of the tool shooting forward, which so often happens in conventional handling of carving tools, the thumb is securely anchored down on the wood by the other thumb (see photographs, pages 90–91). Applying the traditional wood-engraving tool design and manipulation to small sculpture carving gouges gives magnificent control as well as flexibility in straightforward or curved-form wood cutting. In addition, the flat upright shank sliding against the thumb cannot cut into it, whereas the standard commercial tool, with its small sharp-edged square shank, can.

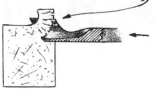

hand-pushed engraver style small woodcarving gouges

4½" to 5"

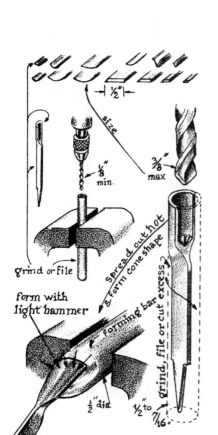

grind or file

form with
light hammer

Wood-engraving burins, small sculptors' wood gouges, as well as wood-block and linoleum-block gouges, can all be made without the advanced skills of the blacksmith, though you will need a forge for minor shaping and, of course, for tempering.

A 1/8-inch-deep Gouge

Cut a 6-inch length of a ¼- to ⁵/₁₆-inch-diameter high-carbon steel rod. Spring steel from a straightened coil spring is excellent for this purpose.

Place this annealed rod upright in your drill vise and, with a ⅛-inch drill, center a hole ¾ of an inch deep. Be sure that you drill in the exact center of the rod. An eccentric hole is dangerous because the drill may accidentally break when it comes through a thin part, and become lodged there.

Keep small drills razor-sharp and symmetrically ground to avoid wandering action that causes such strains and breaks. Progress a little at a time, withdrawing the drill often to remove steel pulp. Use lard as a cutting agent.

Next, grind or file the tool blank, as shown. Refine its texture using rubber-backed abrasive discs, and grind an outside bevel to form the tool's cutting edge. Finally, temper the tool blade.

If all has gone well, the drill track should have a smooth final and finished texture, and the blank is now ready to be joined to a handle.

Note that from now on the razor-sharp and hard-tempered cutting edge of the gouge is a present danger at each step during the assembly of tool and handle. Therefore, *be careful* of the gouge when it is clamped in the vise for further work, and *never* leave its blade exposed when you are away from it, even temporarily. You, or someone else in the shop, could be severely injured by bumping into that lethal edge.

Somewhat wider deep gouges are made in the same way.

A Wide-Bladed Engraver's-Style Gouge

When finished, this gouge will have a 1-inch-radius blade curve. Begin as you did in making the small gouge. Select a ¾-inch-diameter high-carbon steel rod and drill it with a ⅜-inch drill. File or grind off the excess to arrive at the blank's blade, as shown by dotted lines.

Clamp a ½-inch-diameter forming bar in your vise to act as a "saddle." Excellent "saddles" can be filed or ground from car bumpers, and plumbing pipes and fittings; halves of pulleys and ball-bearing races. These can be filed or ground into smooth-surfaced forming bars or saddles. Shaping saddles can be devised from most any concave or convex steel implement, provided it is sufficiently resistant and of the proper diameter. Plumbing pipes (as shown), old pillow blocks, halves of steel collars, pulleys, and gear hubs — all make good forming saddles with varying diameters, giving you plenty of choice for whatever gouge curves you want to make.

It is best to preheat the saddle a little with a propane torch before placing the gouge blade on it. This way the thin hot blade does not cool so fast and stays malleable longer, and the work in reshaping it may be done in one heating period, after some experience. (If the saddle is icy cold, a thin hot blade placed on it will be "hardened" as if quenched. Preheating the saddle to about 300° C. (572° F.) is a good habit.)

Heat the tool blade in the forge to a yellow glow and place it over the preheated saddle. Using a ½- to ¾-pound hammer, tap quickly but lightly on the blade blank, starting with the thicker curved portion (where the material resists reshaping most) and working progressively outward toward the thinner end (cutting edge) of the blade. Rapid, even, and gentle strokes over the whole surface will shape the blade accurately over the saddle curve. Reheat the blade periodically if it becomes too cool.

Try to keep the tool's shank and blade well aligned while hammering; if the blank "twists," make sure it is hot, then wrench it back with tongs or pliers.

Complete the blank as with the other small gouges.

Finishing Gouges

Unlike the engraver's-style and V-shaped gouges we have discussed, small finishing gouges are sometimes so shallow they are almost flat; thus, they can be made from flat steel stock. A broken starter spring, a heavy-gauge clock spring, or a section of a handsaw are all good stock.

Heat the shank section of the stock to a yellow glow and clamp its lower part in the vise. Then, with tongs, twist the exposed upright portion 90°, as shown. Align the shank and blade on the anvil face with a hammer.

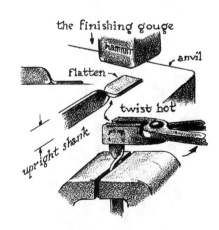

This shaping method applies to all *flat* small gouges; the desired width of blade can be ground or filed later on, at the time you fashion the tang. Making these flat gouge blanks can be the first steps in making deeper curved roughing out gouges as well as shallow curved finishing gouges.

Once gouge blanks have been shaped, aligned, filed, ground, and polished, they are finished by tempering and sharpening. Finally, the tangs can be burned into wooden handles as described on page 68.

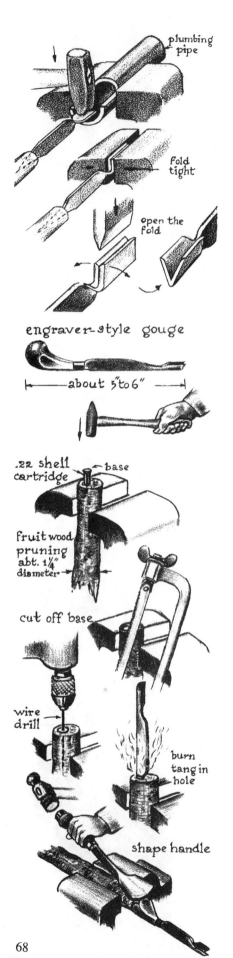

plumbing pipe

Fold tight

open the fold

engraver-style gouge

about 5" to 6"

.22 shell cartridge   base

fruit wood pruning abt. 1¼" diameter

cut off base

wire drill

burn tang in hole

shape handle

A V-shaped Gouge Made by Folding

If a flat blade is hammered on to a curved saddle, as shown, it can be reheated and folded in the vise as the first step in making a V gouge.

To open the hot folded blade, use a flat cold chisel with double-beveled edge as a wedge, as shown. If no one is available to help you hold the blank on the flat anvil face, you can put the cold chisel in the vise (cutting edge up). Hold the heated blank with tongs, and place the folded blade on the chisel. Hammer the blade open with a flat wooden stick (so as not to dent the outside of the V-shaped blade).

Reheat the blank once more, and again place it over the cold chisel in the vise. Now refine the blade alignment and the final angle of the V.

With practice, you can use this method for making V-shaped tools with very small angles. By filing down the excess of the blade uprights, the tool will become fine enough to cut lines of hairline widths on wood blocks.

An important feature here is that the strains involved in cutting minute lines are so small that you can afford an almost "brittle" hardness in tempering the very end of the blade, and if not abused, such tools never need to be resharpened.

Small-Tool Handles

Fruitwood makes excellent handles for small tools, but the wood need not be too hard. From your garden, or orchard (or even a friend's firewood supply), select straight, dead branches about 1 to 1¼ inches thick and at least 10 inches long. Fruitwood prunings salvaged from local orchards are a fine source of supply.

Preferably, the handle should have a small ferrule reinforcing it where it joins the tang (see pages 56–57, on ferrules). Suitable stock for such ferrules includes small brass tubing, or empty .22 pistol or rifle cartridges.

Clamp the branch vertically in the vise and lightly tap (with a ¼-pound hammer) the .22 cartridge into the branch, as shown, leaving $^1/_{16}$ of an inch of its base exposed. Cut off this base with a hacksaw so that the ferrule is flush with the wood.

Taper-file a $^1/_{16}$-inch-diameter nail stem to make a wire drill. Then drill this wire down into the branch, in the dead center of the .22 cartridge. The improvised mild-steel drill will get hot and burn itself in, preparing a hole for the gouge tang.

Heat the tool tang somewhat, then gently drive it straight into the hole with a wooden mallet until it seems secure.

If you prefer, reverse the procedure by clamping the tool shank in the vise, hot tang upright. Then slip the prepared branch over the hot tang and tap it down as far as you think necessary. Whichever method you choose, keep both tool and branch aligned throughout.

Now, with the tool well secured, reclamp the branch horizontally in the vise, allowing a 2½- to 3-inch overhang, as shown.

Use a flat carpenter's chisel (or a larger wood gouge, if you've made one) with a half-pound mallet or hammer, to carve the branch as shown. Once the handle has been shaped, use a hacksaw or coping saw to incise the branch where it meets the handle to ⅜ of an inch all around. Use a 1-inch cone-blade gouge, as shown, to round off the handle; when only a core of wood remains, cut the handle off with a hacksaw. Now refine the whole handle, with rasp, abrasive paper, or on your rubber-backed abrasive wheel.

The experience gained in making the small woodcarving gouge should enable you now to make a whole set of small gouges, as simple or elaborate as your talent allows and your needs dictate. They will prove ideal for the carving of small sculptures, low-relief work, cameos, linoleum blocks, side-grain wood blocks, and many other jobs requiring only small tools.

A Small V-shaped Gouge

The V-shaped cutting edge of a veining tool will prove more difficult to make than the previous gouge design. The very bottom of the V seems to resist most efforts to create a truly sharp meeting place of the two upright sides of the tool. I believe the most direct way to make this tool by hand is to use a triangular file and a steel cutting chisel ground to the same angle as the file.

Stock for this gouge can be a straightened ⅜-inch-diameter coil spring. Heat it in the forge and bend 1 inch of the hot end to a 45° angle, as shown. Then let the hot metal anneal by placing it in ashes for twenty minutes. If still too hot to handle, quench the blank in water and clamp it firmly in the vise. Use the small triangular file and in the bent section file a groove as even as your skill allows. Reheat in fire to yellow glow; then straighten, align, and anneal.

The natural limitation in sharpness of any file will prevent you from making the bottom of the groove really sharp. Thus it is time to turn to a triangular cold-chisel-type tool.

Grind the chisel to the identical angle of the triangular file, but sharpen only its very bottom cutting edge, as shown. Leave the chisel's upright edges dull but polished: this ensures a perfect alignment of chisel in groove, and protects the gouge's filed uprights against scoring by chisel and hammer action.

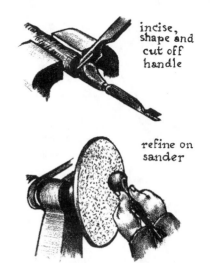

incise, shape and cut off handle

refine on sander

making a small V gouge

bend hot a ⅜ dia. high-carbon steel rod

file groove

straighten while hot

rounded groove sharpened

grind to fit groove as shown

sharp polished & dull

Align the chisel in the tool's groove, and cut as sharp a groove as possible, with many light hammer taps (instead of a few hard blows) to cut the very bottom of the groove sharp.

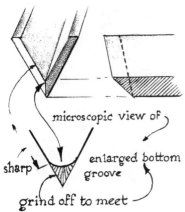

microscopic view of

enlarged bottom groove

sharp

grind off to meet

finished tool

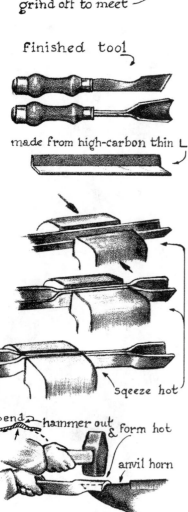

made from high-carbon thin L

sqeeze hot

bend hammer out & form hot

anvil horn

LARGE WOODCARVING GOUGES

The design features of large woodcarving gouges are the same as for small gouges which are not struck with a mallet or hammer.

High-carbon angle-iron bars salvaged from steel bed frames, garden swings, or dishwashers all make good tool material for large gouges because these articles as a rule are light in weight but very strong. Equally suitable stock for large gouges is the lightweight, yet rigid, steel tubing often used in high-grade bicycle frames.

To make your first large gouge, select an angle bar from a bed frame. Cut off a 20-inch length — long enough to hold by hand during heating in the forge. Heat a 7-inch section (for the shank) to a yellow heat glow in a slightly elongated fire. The heat should be spread as evenly as possible and 1½ inches longer than the vise jaws in which the piece will be clamped later on. Have everything ready so that not a second is wasted when the bar has become hot. Immediately transfer the hot bar to your vise, and tighten the jaws quickly to squeeze the angle sides together until they meet. Repeat this same process for the tang end.

It will take several reheatings to accomplish this forming, but aim for as few as possible since excessive reheating will damage the steel's composition. *Never overheat steel to white hot* (sparks will fly from the fire at such heat), for this is close to the melting point of steel and may ruin it for good.

Now heat the 1½-inch blade end to a yellow heat glow. Place it over the anvil horn, as shown, and flatten out the angle bend with hammer blows, leaving the middle section of the curved blade as thick as possible.

The next heating must be spread evenly over the entire, now slightly curved, blade. Place the hot blade over the anvil horn again, and, with a 1- to 2-pound hammer, tap rapidly down each side, alternately, to bend the blade edges into a gradual curve from shank to blade edge while following the cone shape of the horn as well. This gradual curve between shank and blade allows the hammer blows during woodcarving to be transferred more evenly over the gouge's cutting edge. An abrupt meeting of blade with shank tends to cause breakage where the blade meets the shank, particularly if the tool is somewhat thin at that point.

The final shaping of the blade on the anvil horn should be done with a light hammer. Keep the blade malleable (yellow hot) and in constant alignment with shank and horn as you tap it, now here, now there, in very rapid progression.

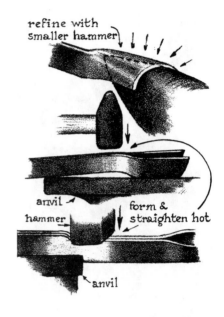

refine with smaller hammer

anvil & hammer

form & straighten hot

anvil

Remember, if your first attempt at this shaping of a tool blank is something of a struggle, each time you make another you will benefit from the previous experience. Practice will steadily improve your results.

What you have actually been doing here is what in coppersmithing and sheet metal work is the main skill and activity: it is bending already flattened steel, much as a tailor works with paper. The blacksmith may engage in this type of activity on occasion, but the specialty of his craft is in hammering one given solid volume of hot steel into another shape of the same volume. This *forming* includes "upsetting" (making long pieces of steel shorter) and "drawing-out" (making short pieces longer). Forming steel this way is true forging, whereas *bending* steel may be done without true forging (blacksmith) skill. (Some modern thin mild-steel plates can often be bent and stretched cold, as is done in the automobile industry when whole car body parts are shaped cold in giant dies. No true forging enters into this.)

Your hot-bent work should now have produced a suitable large-gouge blank, as shown. If there is a slight jog between shank and blade, it can be eliminated, though it concerns more the appearance than the effectiveness of the finished gouge. If you wish to bring the bottom of the blade into alignment with the shank, reheat the shank at the jog location and lay the whole tool blank on the anvil face. Deliver one or more exact blows with a medium-heavy hammer, as shown. This will bend out the jog and correct the shank alignment. Any localized spots forced out of true alignment by this hammering can be corrected with light taps. Reheat the blank if necessary.

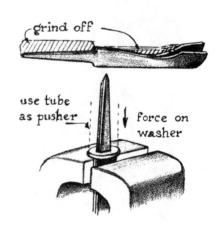

grind off

use tube as pusher

force on washer

Finally, grind the blank as shown, removing any steel not needed for tool strength. Where shank and tang meet, leave a notch broad enough to "shoulder" the ferrule of the handle.

If you prefer having a washer between shoulder and handle, it must rest on two small shoulder ledges: heat the area yellow hot and hold the blank over the anvil edge, crease down. A few precise heavy blows will force the shank down, creating a bottom shoulder where it binds at the anvil edge, as shown. Any slight distortion this causes in the blank must be corrected now, while the steel is still hot, and only with the flat of the hammer. Grind the blank at the point where the tang begins, to form the upper shoulder ledge.

Next, clamp the blank firmly in your vise (tang up) and slip a heated, undersized washer over the tang. Have ready a tube a size smaller than the washer to use as a pusher. Quickly hammer the washer down until it is flush with the shoulder ledges, thus forcing a snug fit.

Now wedge the tang into a temporary wooden handle, so that you can hold the tool blank comfortably during its final and precise grinding and sharpening (see page 37).

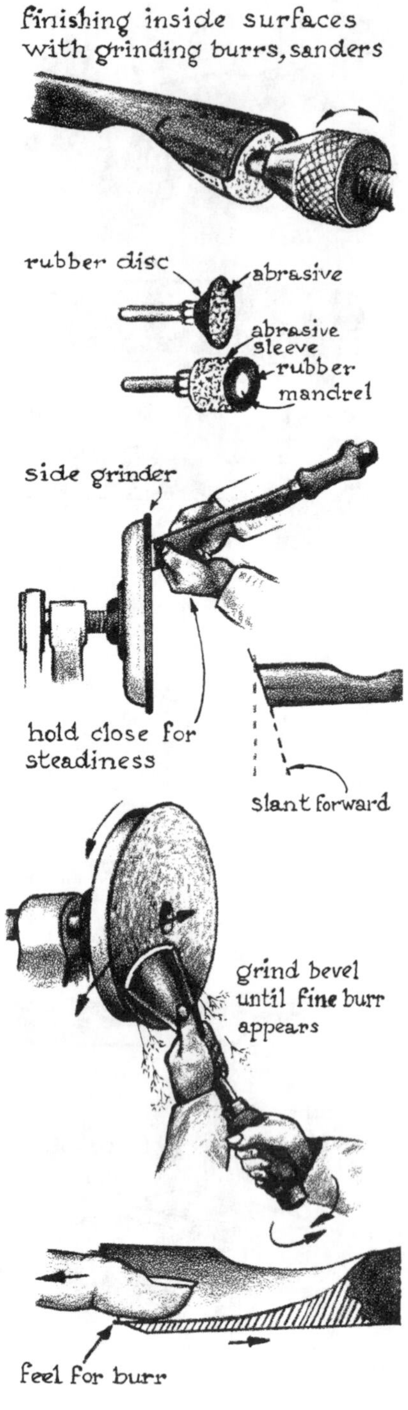

finishing inside surfaces
with grinding burrs, sanders

rubber disc
abrasive
abrasive
sleeve
rubber
mandrel

side grinder

hold close for
steadiness

slant forward

grind bevel
until fine burr
appears

feel for burr

Grinding the Bevel

Refine the inside of the blade with chuck-insert burr grinders and rubber-backed abrasive sleeves of a diameter to match the curve of the blade. Grind the outside of the blade evenly and toward its shoulders, following the conical curves as closely as possible. Then refine the surface with progressively more fine-grit abrasives, finishing up on the buffing wheel to give the blade a mirror sheen.

Begin the grinding of the outside (or cutting edge) bevel by first making a clean, mirror-shiny, cross-sectional cut at the blade-end: take the tool by its blade (both hands as close to the abrasive disc as you dare) and hold it, at a constant angle, to the side-grinder wheel. As soon as the blade-end is ground evenly, scrape off the burr that has formed on its inside with the sharp steel edge of a file tip. You will now see a mirror-shiny cross section, revealing clearly the exact thickness of the blade at the cutting end.

Slanting the tool against the grinding wheel at such an angle that the mirror surface reflects the condition of the edge during grinding, you can guide your movement so as to avoid overgrinding. Only practice will enable you to become skilled at this. The tiniest remaining glimmer of the cross section must gradually disappear, leaving only a faintly noticeable burr. When this burr is felt, as shown, it means that the inside and outside surfaces of the blade's cutting edges have met. (See also the illustration on page 39.) Any further grinding is useless and will only shorten its lifetime.

Removing the Burr

Refine the texture that the grinding has left on the bevel surface with the rubber abrasive wheel, or (in view of what this razor-sharp tool can do to rubber), with an extremely fine-grit abrasive stone wheel.

Next, refine the bevel on your cotton buffing wheel, with tripoli compound rubbed into its spinning buffer face. First, hold the bevel-edge plane tangentially against the rotating wheel rim, so that the cotton fabric slides off the steel as the tool is pressed against it. Then reverse the blade and "grind" the inside of the blade, tangentially. The burr thus is "ground" off (as described in the chapter on tool sharpening).

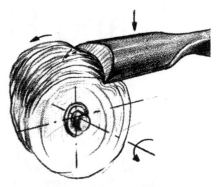

grind off burr on buffer with Tripoli compound

Tempering and Assembly

You may temper the finished tool blank either now, or after its tang has been burned into a permanent handle. In the latter case, the wooden handle should first be wrapped in a wet rag to prevent scorching (see page 58).

When the blade is tempered and blank and handle assembled, test the tool on the wood you plan to carve, allowing the tool its full bite, but never burying the blade so that its whole edge is out of sight. If it does not buckle, crack, or chip, yet scratches the wood surface somewhat, it suggests that a microscopic sawtooth cutting edge still remains. This can be remedied by buffing the bevel a little more on both sides. With that, you will have finished your first large sculptor's gouge and given it an ideally sharp edge.

13. Making a Seating Cutter and Hinge Joints

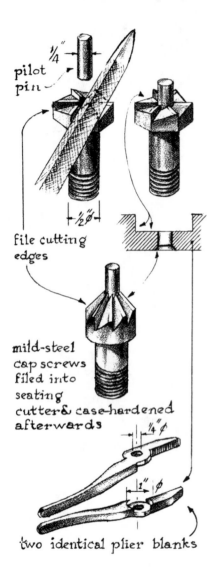

pilot pin

¼"

file cutting edges

½⌀

mild-steel cap screws filed into seating cutter & case-hardened afterwards

¼⌀

1"⌀

two identical plier blanks

THE SEATING CUTTER

A seating cutter, which sometimes acts as a milling cutter and drill combined, is made here with six cutting teeth and a pilot pin. This ensures precise bearing surfaces for the tools, which, when joined with a hinge pin, achieve the hinging action essential to pliers, shears, tongs, tinsnips, and other hinging tools.

The seating cutter shown here is filed from a standard hexagon-headed ½-inch-diameter mild-steel cap screw, which later is case hardened.

First, drill a $^3/_{16}$-inch pilot-pin hole ¾ of an inch deep into the exact center of the hexagon head. By clamping the cap screw in the drill press vise, you can be sure of drilling the hole dead center, and in true alignment with the cap-screw shank.

File the teeth in the hexagonal pattern, as shown, first with a triangular bastard (medium-coarse) file, then refine with a triangular smooth file.

Lock the cap-screw shank in the drill chuck and place under it, on the drill press table, an abrasive cutoff disc (or salvaged remnant). Set the drill press at medium speed, and lower the cap screw gently onto the abrasive, barely touching it. This action will grind shiny little horizontal facets on the six teeth of the head.

Now clamp the cap screw in the vise, with its teeth up, and with the smooth file sharpen the six cutting edges exactly. All the teeth have the same length so as to mill a perfect plane and absolute circle in your tool blank. Use outside calipers to measure the length of each tooth.

The mild-steel cutter teeth are now ready to be case hardened. A commercial case-hardening compound powder is available in any machinists' supply house. Ask for a non-toxic brand and follow the instructions given on the can.

Build a small, but even and clean, forge fire, and make a little depression with the poker in the fire mound center. Place the cutter blank in it, teeth down. As soon as the end of the blank is medium-yellow hot, lift the cutter out gently, with tongs, without disturbing the fire. Dip the toothed part only in the case-hardening powder for one or two seconds; then carefully withdraw it, so as not to disturb the jacket of powder which has caked around the dipped part. Replace the cutter, teeth down, in the little depression in the fire, which should have remained intact.

This pocket now should hold the cutter blank, with its layer of caked powder, in upright position. As the powder begins to bubble and melt, gently fan the fire to keep the steel dark-yellow hot (even if you do not see it behind that bubbling jacket).

Once the powder has burned off (20 to 30 seconds), immediately quench the hot blank in water (it will sound like a firecracker). Finally, use the regular file-tip test and observe how glassy-hard the teeth have become as a result of this case hardening procedure.

Into the prepared pilot-hole, press a tight-fitting pilot pin made from a $3/16$-inch spring-steel rod (better for this pin than mild steel). Leave $1/2$ inch of the pin protruding. This strong pin will hold the cutter firmly centered during milling.

Similarly, another cutter can be made to countersink holes. A bolt with a round head may be used with six or more teeth, and a pilot pin for exact concentric cutting. The hinge pin riveted during assembly of pliers, for instance, then fills up the cone-shaped space with precise contact surface. This ensures an even bearing as well as holding action for the pliers.

You have now added to your shop a tool accessory essential to the making of precise bearing surfaces for all hinged tools.

Rotary cutters for milling are various and sundry; the design of each one is dictated by the milling job it must perform. You can, for instance, grind a two-bladed cutter with lead pilot out of a $3/8$-inch-thick spring-steel blade, as shown. While this is easier to make than the hexagonal cap-screw cutter described above, it is less satisfactory to use in milling seats for hinged tools, since it cuts more slowly and less precisely.

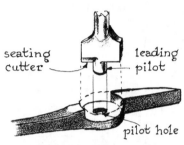

cutting hinge bearing surface for a plier blank

HINGE JOINTS

To make good tongs, pliers, tinsnips, and shears you must be able to make accurate, flat bearing surfaces. It is important to remember that the hinge joints must transfer the force of hand pressure to the blades or jaws of the tool. Thus, the following notes are essential to making the tools described in Chapters 14, 15, 16, and 17, as well as other hinging tools.

An undersized hinge pin tends to bend or break under stress. If the bearing surface between the two halves of the tool is too narrow, the slightest looseness will allow the jaws to flop around. Therefore, a good tool should have oversized (snug-fitting) rather than undersized hinge mechanisms, as well as precise bearing surfaces.

If the bearing surface is 1 inch in diameter and the tool half at that point is ¼ of an inch thick, the hinge pin diameter should not be less than $5/16$ inch. These dimension relationships are recommended for all such hinged tools, of whatever caliber.

grind flat hinge&blade
surface on side grinder

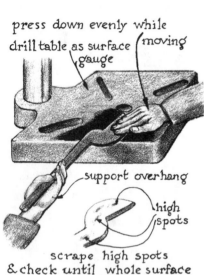

press down evenly while
drill table as surface
gauge (moving)

support overhang

high spots

scrape high spots
& check until whole surface
is accurate. Next, assemble
and drill hinge hole

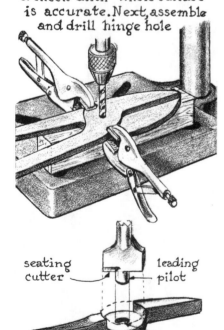

seating cutter

leading pilot

pilot hole

cutting hinge bearing
surface for a plier blank

Making A Strong and Accurate Hinge Joint

The hinging mechanism of a pair of tinsnips is representative of the hinge-joint in many other tools.

Assume both tool halves of your tinsnips are ready as rough blanks, now to be prepared for assembly (see page 74). Make the contact surfaces accurately flat on the side grinder. Test them as follows.

Smear a tiny bit of blackening paste (printer's ink or lampblack mixed with grease) on any surface that you know to be perfectly flat (your drill press table is ideal). Place the ground surface of the blank on it and rub the blank back and forth evenly. Examine it for blackened high spots. Scrape or file these off. (You can make a scraper from an old triangular file that has been ground smooth to make razor-sharp edges.) Scrape and test as many times as necessary, until an even contact indicates that the whole bearing surface is perfectly flat.

An alternative method of making the surface accurate is to place the annealed bearing surface of the blank on the flat face of a hard, unscarred, accurate anvil. Cover it with a piece of flat-surfaced ½-inch-thick steel plate. Strike this hard with a very heavy hammer; one well-centered blow will even out all inaccuracies. A well-annealed steel blank (if no thicker than ¼ of an inch) will respond very well to this treatment.

Refine the surface further either on your side grinder or with a file. The latter method is best done in the vise: make a jig by nailing two $3/16$-inch-thick strips of perfectly flat steel 2 inches apart (on either side of the blank) on a small piece of wood. These will guide the file strokes. Clamp the wood level in your vise, and move the file across the blank's bearing surface. As soon as the file begins to graze both strips simultaneously, it has cut away all the excess steel of the hinge-joint area and the filed surface has become accurate and flat as the jig itself.

A third method of making the bearing surface flat is to recess the blank accurately in a flat-planed hardwood jig, dispensing with the steel alignment guides altogether. This is illustrated on page 78.

Other inventive ways to ensure accuracy for flat bearing surfaces may occur to you; however, they will be increasingly less important once your freehand filing skill becomes adequate to do the job.

Assembling the Hinge Joint

When both tinsnip blanks have been accurately refined, clamp them together, as shown, in visegrip pliers. Punch-mark a dead-center location for drilling the hinge-pin hole.

Now drill a preliminary hole with a ¼-inch drill. As a precaution (to cushion the bit when it bores through the steel), clamp an oblong piece of hardwood in the drill vise. This wood "cushion" should be flat-topped and set at a right angle to the drill. Place the hinge joint (still clamped in the pliers) over the wood and drill the premarked hole through both tool halves.

Separate the two blanks and enlarge the ¼-inch hole in one of them with a ⁵/₁₆-inch drill. Thread the other hole with a ⁵/₁₆-inch threading tap to fit the ⁵/₁₆-inch-diameter hinge pin. (In threading holes with taps of ¼ inch diameter or less, I find that approximate sizes are perfectly adequate as long as one proceeds cautiously, little by little, clockwise and counter-clockwise.) If the taps are a little dull, use plenty of lard as a cutting agent and steady, aligned movements during the cutting. If a tap should break off, heat that part in the forge fire, anneal, and drill out the broken tap.

You are now ready to assemble the hinge mechanism. Insert the hinge pin and screw into the threaded half. It can now be locked by a nut if you are satisfied the fit is exact and there is no binding friction between the bearing surfaces of the hinge joint.

The locknut and cap-screw system may be improved still further by placing a smooth washer between the cap-screw head and the blank. Lubricated, the hinging action will prove more even, and more lasting in accuracy, with the washer.

Sometimes smooth, unthreaded hinge pins may be riveted on just enough to eliminate any play in the hinge joint. However, in time the rivet head will wear down as it rubs against the blank as the pin turns. Riveted-on countersunk washers will improve this system, as shown. This hinging method is treated in Chapter 15.

The above procedures apply to all tools that have similar hinge joints. More elaborate setups are found in carefully machined combination pliers and wire cutters. But all aim at hinging actions that remain smooth, strong, and without play. Play, or slack, causes wobbling and malfunction of the tool.

Depending upon your skill, you can now extend your projects to the making of various hinging tools.

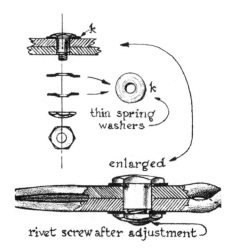

thin spring washers

enlarged

rivet screw after adjustment

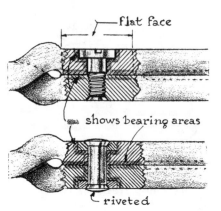

flat face

shows bearing areas

riveted

14. Making Tinsnips

Just as with shears, pliers, and tongs, the keystone of a superior pair of tinsnips is its hinge mechanism. From this, all else follows; and, in view of its significance, bear these points in mind as you make your first hinged tool: (1) The hinge joint must have smooth bearing surfaces that stay in contact without play or friction. (2) The hinge pin mechanism must stay in place, not work itself loose under stress or in time. (3) The shank of the hinge pin should be large enough to prevent rapid wear, and fit snugly but not bind. (4) The proper dimensional relationship between hinge-pin diameter and thickness of bearing surface should be observed. (5) From time to time, the whole mechanism should be oiled.

LIGHT-GAUGE TINSNIPS

The stock for such tinsnips can be cut from a salvaged automobile bumper. As a rule, bumper steel is about ⅛ of an inch thick (⅛-inch in small cars; ³/₁₆-inch in heavy trucks). Cut out the section you need with the abrasive cutoff wheel, then heat it evenly in your forge fire. While it is hot, pound it flat on the anvil face and bury it in ashes to anneal.

Cut a cardboard template (pattern), as shown. For symmetrical tinsnips, make a duplicate template. Turn the one over onto the other, and, using a thumbtack as hinge pin, move the cardboard blades in a simulated cutting action.

This allows you to examine the feasibility of your design, and alter it, if necessary. Once you are satisfied with it, scribe the templates on the annealed plate-steel stock. Use either a hacksaw or the abrasive cutoff wheel, to cut out the two blanks.

To make the openings in the handle grips, use a ¼-inch high-speed steel drill at *low* speed, cutting a series of holes around grip openings as shown. Then knock out the waste and file all jagged edges inside and out, so both blanks are completely smooth.

Next, create a jog between the hinge part and the handle by hot-bending the steel; this clears the file strokes during the filing of the hinge bearing surface. This hinge bearing surface and blade must be accurately filed flat. File freehand, if you have developed that skill, or use a jig, as shown, if it gives you more confidence.

Next, the file texture is removed by grinding gently on the side grinder or scraping with a triangular scraper. Use the test for flatness (on a blackened surface gauge, as described on page 76).

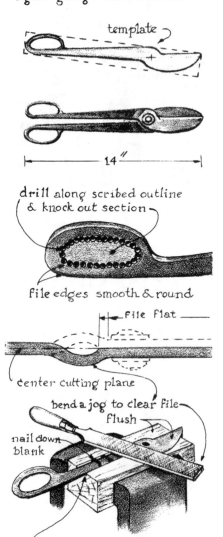

tinsnips for cutting light gauge sheet metal

template

14″

drill along scribed outline & knock out section

file edges smooth & round

file flat

center cutting plane

bend a jog to clear file flush

nail down blank

recess wood to hold blank for accurate filing.

Now clamp the blanks together with visegrip pliers, in mock assembly, and center them on the drill press table with its woodblock prop. Drill a ¼-inch-diameter hole through both blades and put in a temporary ¼-inch hinge pin that fits snugly. Remove the pliers (see also page 76). Heat the tool handles in the forge fire for final bending. Clamp the hot handles in the vise and force them into alignment between the vise jaws. (The blades are twisted similarly into that same alignment.) Then disassemble the blades and prepare for *hollow grinding*.

Hollow-Grinding the Cutting Blades

Begin the hollow grinding of the blade on your ½ hp grinder, using a large, extrafine-grit wheel. Before you begin grinding, make sure the tool rest almost touches the wheel. (If there is a gap, the blade tends to become wedged into the opening.) Place the blade gently against the wheel so that its spine is on the tool rest and its cutting edge is upright and just clear of the spinning wheel face. This position places the flat of the blade against the grindstone, so that a hollow impression is ground in the inner surface of the blade, all along it, but stopping $1/16$ of an inch short of the cutting edge, and just short of the hinge bearing surface (see illustrations). Similar results can be obtained by filing at a slight slant, as shown. Hollow grinding ensures that only the actual *cutting* edges (instead of the flats) of the blades are in contact once the tool is assembled.

Please note that extreme care should be taken while grinding, so you do not "run over," into either the hinge area or the $1/16$-inch cutting edge. If you overgrind in either direction, you will destroy the accuracy of the hinge bearing surface or the accuracy of the cutting edge — or both. In either case, the blades will not properly "meet," hence cannot cut or shear at that point.

Since the steel has been annealed, you can cold-bend a slight curve in each blade after hollow-grinding or filing. This curve dictates the amount of pressure exerted by the blade's cutting edge, but only experience can tell you just how *much* of a curve is needed. In general, however, the degree of curve is determined by thickness and hardness of the metal to be cut, as well as length and springiness of the tinsnip blade itself.

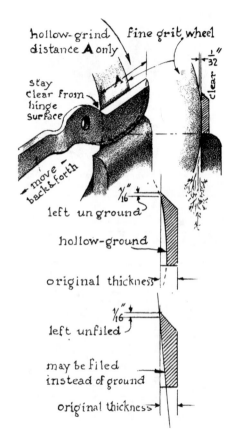

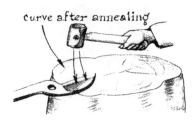

curve after annealing

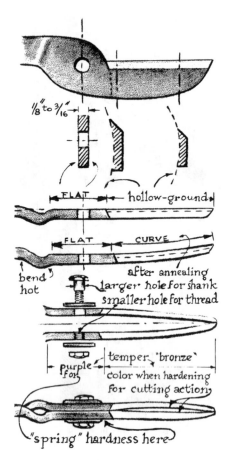

⅛" to 3/16"

FLAT — hollow-ground

FLAT — CURVE

bend hot — after annealing
larger hole for shank
smaller hole for thread

purple for — temper "bronze"
color when hardening
for cutting action

"spring" hardness here

Curving the Blades

While the blade is still annealed, place it on the wood stump, with the inside facing up and the hinge area free, as shown. Using a 1½-pound wooden mallet, pound fairly heavily on the blade, working from the hinge area toward the tip. Sight the gradually forming curve in the blade from time to time, until you believe it to be sufficient.

If the steel is too thin around the hinge point, it may produce too springy a blade. This can be partially corrected by use of a hinge screw and sturdy, large-diameter washers that fit snugly.

Tempering the Blade

Temper the blade and hinge area, heating evenly to a light-cherry-red heat glow, and quench.

It is at this point that we encounter the endless controversy on how best to quench a delicate, thin, and long steel part so that its original shape will be least affected by the uneven shrinkage that causes warping.

Should the curve in the blade warp radically after the quench, all you can do is reheat, reestablish the intended curve, and try slanting the blade at a different angle in the quenching bath. A lopsided slant *may* compensate for an unsymmetrical blade, so that the surfaces that shrink first can achieve the curve originally intended.

Here experience is not always reliable, since you are probably working with unknown types of steel from your scrap pile. One piece may be a typical water-hardening steel, another, an oil-hardening type. The best you can do is experiment, hoping that an undistorted brittle-hard blade will finally emerge from the quench. Only then can you proceed to carefully polish, clean, and temper the blade for specific hardness.

Drawing Temper Colors

Begin by holding the *hinge* part over the heat core, far enough from the flame to ensure slow heating. At the first sign of a faint straw color, move the blade part slowly back and forth through the heat core, until a light straw color indicates that the whole blade is uniformly drawn. Keep heating it very gradually until you see the color changes from light straw to dark straw to bronze. At this point withdraw the blade and let it cool at room temperature for about a minute, *without* quenching.

While the tool is still fairly hot, hold it by the blade and return it to the heat core, so that the handle part bordering the hinge area will heat to a peacock to dark purple. When this color spreads over the hinge area, that portion then has achieved a ''spring'' hardness, which will prevent the blade from breaking at the hinge, where strain is generally the greatest. The balance of the handle remains annealed, as before. All that is left is to assemble the tool, as detailed in Chapter 13 (see page 74), on making hinge joints.

The hinge pin used with washers and lock nut (see page 77) is especially suited for light-gauge tinsnips. A $^5/_{16}$-inch-diameter cap screw which has a smooth shank acts as the bearing for one blade in a $^5/_{16}$-inch hole; the threaded portion of the cap screw fits the threaded hole in the other blade. The hinge pin should be long enough to allow room for a smooth washer to be inserted between cap-screw head and one blade, and for a flat locknut to be added to the outside of the other blade.

HEAVY-DUTY TINSNIPS

The only difference between light and heavy tinsnips is the gauge of stock used to make them. More rigid blades and hinge joints that will stay in contact while cutting heavier sheet metal require heavier-gauge steel. Such snips can be used to cut small nails and thin wire, as well as heavier sheet metal. As stock, use car leaf springs no less than ¼ of an inch thick. If you do not use the abrasive cutoff wheel, the blank can be cut by a welder with a cutting torch.

The thicker the sheet metal to be cut, the sturdier and longer the handles will have to be, in order to provide enough leverage to cut comfortably by hand.

Curving such thick blades will be difficult unless you heat them first, but take great care not to distort the flatness of the hinge bearing surface.

Draw the cutting edges of heavy tinsnips to a light straw temper color (after brittle quenching), and the rest of the tool to peacock.

The Hinge

The thicker steel involved here may require a countersunk cap-screw type of hinge pin.

This flat, thin, screw-driver slotted head can be sunk flush in one blade, as in the illustration on page 77; but it may instead bear on the outer surface of the blade, without being countersunk, if you feel that the head of the cap-screw will not be in the way.

The second blade has a threaded hinge hole into which the threaded end of the cap screw fits, adjusted so that the blades will bear on one another but not bind. Enough threaded section must protrude to be riveted flush with the countersunk hole on the outer surface of the blade.

One advantage of this adjustable hinge pin is that future slack, through wear, may be taken up by tightening the screw a little, and riveting the protruding thread-end flatter. When properly adjusted, there should be just enough clearance between the bearing surfaces to prevent any hinge wobble or faulty alignment during cutting. And finally, remember that occasional lubrication will extend the life of your hinge mechanism.

Should you find that "naked" handle grips are rough when you are cutting great amounts of heavy metal, try winding them with soft leather thong.

Having successfully completed a pair of heavy-gauge tinsnips, you are thoroughly prepared for making the many other tools that employ a hinging, cutting, or gripping action. Now try your hand at making scissors, nail- and wire cutters, shears, pliers, or tongs.

leather thong winding

15. Making Wire and Nail Cutters

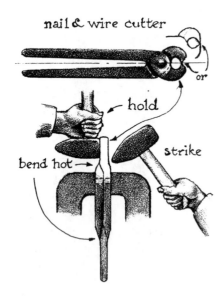

nail & wire cutter

or

hold

bend hot →

strike

NARROW-JAWED CUTTERS

Use as stock a leaf spring from a heavy-duty truck; such steel is apt to be ⅜ to ½ inch thick. After the leaf spring has been heated, flattened, and annealed in ashes, make a cardboard template (following the symmetrical design shown) and scribe it on the steel stock.

Use a ¼-inch high-speed drill, run at slow speed, to cut out the tool blank (a ⅛-inch drill breaks too easily should it bend). The waste pieces can simply be knocked off if only paper thin sections are left between the holes. For greater accuracy, mark off all hole locations beforehand with center-punch depressions. The drill then does not wander and interfere with the previous hole. If it should, withdraw it, turn the blank over, and punch-mark that side accurately. Then drill from that side, to meet the opposite hole halfway. (Another way of controlling a drill that tends to wander is to plug the adjacent hole with a tight-fitting pin, flush with the blank surface. Then readjust the punch mark and redrill accurately, using lard as a cutting agent: it works better than regular machine oil.)

If the drill holes were not spaced closely enough, the steel in between may have to be cut through part-way with a cold chisel. The waste can then be knocked off without difficulty.

If the drilling method seems too much work, try the abrasive cutoff wheel. With thick steel, however, this too may take more effort than you care to spend, and the welder's cutting torch would then be the answer.

Once the blank has been cut out, grind all edges smooth and drill the temporary hinge holes. These must match the diameter of the lead-pilot of the seating cutter which you will use later on instead of a file to mill a clean face on the hinge bearing surface.

Now assemble the two blanks with a snug-fitting temporary hinge pin, to hold them together while hot-bending their jaws. Clamp the hinge part in the vise, and use two hammers, as shown, to align the yellow-hot jaws. The jaws can be further adjusted to make the cutting edges meet.

Now heat the handles and clamp their hot ends in the vise. The visegrip pliers, holding the hinge assembly firmly clamped together, are used to twist it into alignment with the yellow-hot handles.

Next, knock out the temporary hinge pin so that the hinging surface of both blanks can now be milled with the milling cutter. Adjusted to the slowest drill speed, the milling cutter is used only to clean the surface, barely cutting the steel, to produce an accurate bearing surface.

Once more assemble the two blanks, with the holes aligned, and firmly clamp them in visegrip pliers (as shown with the light-gauge tinsnips, page 78). Place the assembly on the accurate wood prop on the drill press table, and drill the final hinge hole.

The remaining steps in finishing nail- and wire cutters follow the procedure described for making tinsnips: tempering, making a permanent hinge pin, fitting the hinge mechanism in exact adjustment with a locknut (see page 77).

WIDE-JAWED CUTTERS

For wide-jawed nail- and wire cutters use the same stock, but leave the blank end wider. This end of the blank, where the jaws are to be, can be twisted hot, as the illustration shows. Reheat the twisted jaw and hammer it over a forming rod held in the vise to make the proper curve.

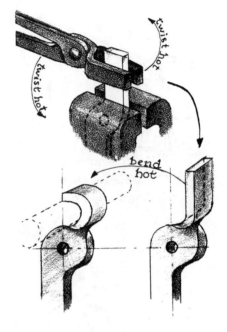

Once both blanks have been identically curved, the remaining steps duplicate those for making the narrow-jawed cutter: assemble with temporary hinge pin; heat, clamp in vise, align hinging part, jaws, and handles with the two-hammer and vise action: file cutting edges on the jaws where they must meet precisely; disassemble and mill clean each hinge bearing surface with the seating cutter.

Reassemble, between visegrip pliers, and drill final hinge hole to receive a rivet-type hinge pin, as shown on page 77.

With washer-size seating cutter, countersink each hinge hole about ⅛ inch to receive washers. Each washer should be countersunk to receive a sturdy rivet head. These heads can be cold-riveted with the hammer peen, then evened out flush with the blade surface with the hammer face. Using a few drops of kerosene as a flushing agent, open and close the hinge forcefully to wear down any unevenness that may cause binding. Flush out any metal pulp with more kerosene.

Continue this riveting and "wearing" until the hinge action is snug but easy-moving. All bearing surfaces should be seated without undue slack.

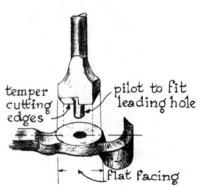

This hand-fitting will prove to be as good — or better — than elaborate machine-fitting, and it will approximate the accuracy of hand-seating engine valves with abrasive compounds. Do not, however, use abrasive compounds in these particular hinge mechanisms, since they would wear down the pin itself, and so ruin the fine fit between pin and hole.

16. Making Large Shears

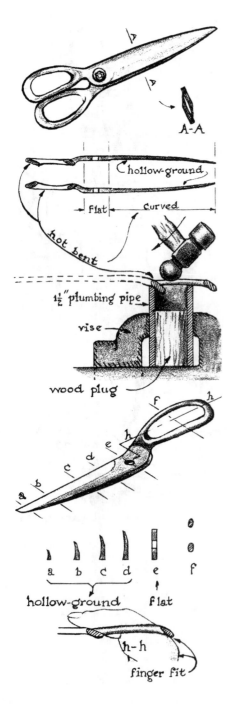

A-A

hollow-ground

Flat   curved

hot bent

1½" plumbing pipe

vise

wood plug

f   h

h

e
d
c
b
a

a   b   c   d   e   f

hollow-ground   flat

h-h

finger fit

Ordinary household scissors are made in the same way as large shears. Use a portion of a salvaged car bumper about ³/₁₆-inch-thick as stock, or a section cut from a plow disc. Heat the stock and straighten out its curve on the anvil, then anneal slowly, in ashes.

As these shear blades will be symmetrical, make two cardboard templates and assemble them with a thumbtack to see whether the design needs correction or improvement.

Scribe the outlines for the two blanks on the annealed, flattened stock. (A discarded dental tool or ice pick makes a good scribe.) Cut out the blanks on your abrasive cutoff wheel, or use a hacksaw. True up the blanks on the grinder, or with a file, until all edges are smooth and exact. Cut out the handle grips and smooth them as described on page 78.

Now heat the handle grip to a dark yellow heat glow and place it, as shown, over the tube (a plumbing-pipe forming bar, reinforced with a wood plug) clamped in the vise. Using the ball end of a one-pound hammer, bend the parts of the steel over the tube rim so that hand and fingers will not have to encounter sharp edges when you use the shears.

Hollow-grind and curve the blades as with the tinsnips (page 79). However, since shears have longer blades than tinsnips, they will require even greater observation and corrective manipulation during quenching for hardness to prevent warping. Distortion is probably best avoided by using an oil quench.

Once the blade emerges brittle-hard, and with its intended curve intact, buff it mirror-smooth, before temper colors are drawn. Draw color exactly as with the light-gauge tinsnips, but take extra care to draw the bronze color as evenly as possible over the whole blank.

Now draw the hinge bearing surface locally, to purple, to ensure adequate spring action in that area. For this localized drawing of color, the small, concentrated flame of a propane torch, held about 4 or 5 inches away, is most effective.

The only difference in procedure between the hinge design of these shears and that of light-gauge tinsnips is in the size of the hinge pin and the type of washer used. For the shears, the hinge pin shank acts as a bearing in *both* blanks instead of only the one. Just enough thread should protrude at its far end for a flat locknut to be added, as shown.

Instead of the straight, flat washers of the tinsnips, shears should use "spring-action" washers (see page 77). These exert an inherent spring pressure between the flat hinge surfaces. Once correct pressure is established, the protruding thread is riveted flush with the nut, locking the pin to prevent its working loose. As always, lubricate all hinge bearing surfaces.

Poorly made shearing tools often suffer from short-lived hinging mechanisms, generally because the cap screw is too small in diameter. This requires the threaded part of the screw, riveted into the blank, to be frequently adjusted to take up the slack between blades. Such blades tend to wobble and cut poorly at best, and their thin hinge pin often bends under strain.

So, at the risk of repetition, bear in mind that, *if* the steel is sound and not too thin, *if* the steel is tempered correctly, *if* the blades are curved and hollow-ground to the proper degree, then only the *hinge mechanism* can be at fault, if the tool malfunctions.

17. Making Pliers

There is a wide choice of possible designs for pliers, but as your first project make a simple pair of symmetrical ones.

You will need a speed-reducing transmission device for your drill press (available through some mail-order houses that feature power-tool accessories). Just as the lowest gear in a truck transmission exerts maximum pull, so will this reduction device slow down even the lowest speed in a standard ½ hp drill press. This slow, powerful action is needed when a hinge seating has to be milled in steel as thick as plier blanks. (It is also essential in other shop projects, such as when using large drills, which would quickly dull if run at high speed in heavy going.)

PLIER BLANKS

For pliers to be of high quality, the all-important consideration, of course, is a hinge mechanism that fits snugly, without wobble. The diameter of your seating cutter (one inch, for instance) determines the size of the hinge and, in turn, the size of the pliers.

Choose a piece of spring steel, about ⅜ of an inch thick and 2½ inches wide. Heat an 8-inch length, flatten it on the anvil, and anneal it as soft as possible (cooled slowly, while buried in ashes).

Scribe the two blanks from the cardboard template you designed beforehand, being sure to allow a $1^{1}/_{16}$- to 1⅛-inch diameter in the hinge seating area.

Cut out the blanks as described earlier, in the sections on tinsnips and shears.

Drill a ¼-inch-diameter hinge hole in the center of each blank to receive the ¼-inch-diameter milling cutter pilot (see page 75). The pilot anchors the seating cutter so that it remains centered while seating the hinge-bearing surface.

If you want to relieve your seating cutter of some of the work in cutting, pre-scribe the exact one-inch circle around the pilot hole. Within that circle, drill (with a ¼-inch drill) as many ⅛-inch-deep depressions as you can fit. By thus "pitting" the area, only a small amount of steel remains to be milled away by the seating cutter to reach a $^3/_{16}$-inch depth.

Search in your scrap pile for a 1-inch-diameter rod to match the 1-inch-diameter seating depression of the plier blank. Cut a 3-inch length of that rod and file or grind its end flat, to form a precise right angle with its length.

Seat the rod in the plier hinge depression, and clamp it in the vise. Then, using this jig as a file rest and guide, file off any steel in excess of the 1-inch hinge diameter.

Now assemble both blanks with a temporary ¼-inch bolt, and, with some abrasive valve-grinding compound (available in automobile-parts stores) smeared on the flat bearing surfaces only, grind in the hinge surfaces precisely. When you are satisfied with their seating, disassemble the plier blanks and clean all abrasive remnants away with a little cleaning solvent. Countersink the hinge holes slightly in their outer surface, to later receive riveted hinge-pin heads.

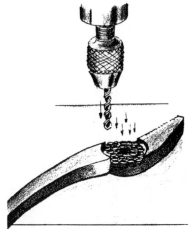

predrill ⅛ inch deep over bearing area for seating-cutter follow up

TEMPERING THE JAWS

Temper the plier jaws as with tinsnips: the jaws, *hard;* the hinge, *springy.* Assemble the plier halves once more and insert the permanent hinge pin; this pin should be a section of annealed, high-carbon ¼-inch-diameter rod, long enough so that its ends can be riveted into heads on the outside of the pliers.

Place the assembly on the anvil face and rivet the end of the pin with the ball end of a ball-peen hammer, striking in the center of the pin to spread it a little; then use the flat of the hammer to "stretch" the outer rim of the pin still further. Alternate hammering with the ball and the flat — now on one end of the pin, now on the other. In due time, this combined hammering and stretching of the pin's head will fill the counter sunk depression in the hinge. Any excess rivet-head steel can be filed off, flush with the plier surface. From time to time open and close the pliers to test the snugness of the hinge fit, and oil generously, to work in all bearing surfaces.

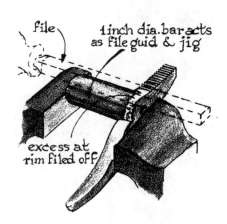

file

1 inch dia. bar acts as file guid & jig

excess at rim filed off

The area of the pin heads, where these touch the steel surfaces of the countersunk holes, also acts as an important bearing surface. But the side strain placed on plier jaws is minimal compared to the pincer strain placed on the hinge mechanism. Thus your major consideration remains: the precise fit of all hinge bearing surfaces, which is what makes one pair of pliers better than another.

ALIGNING JAWS AND HANDLES

Once you find that the hinge assembly functions correctly, but jaw and handle are slightly misaligned, correct this by cold bending. First, make certain that the hinge area is tempered "springy" (purple to blue). Only plier jaws are to be tempered hard (bronze).

Bear in mind that unannealed high-carbon steel that is tempered springy may break during bending. Therefore, if significant bending is required, it is better to anneal the tool first, bend it correctly, then retemper it afterward. Major realignment should only be done by clamping the hinge section in the vise; then use a large pipe as a lever to bend jaw or handle into line, if you are certain that the steel next to the hinge will "give" a little without breaking.

If the tool parts are only slightly off-line, or if you are afraid of breaking the tool during bending, try grinding the excess off on the side grinder when the pliers are permanently assembled.

If you have a saw table, an excellent way to align jaws is to substitute a thin plastic abrasive disc for the circular saw blade. Lay the assembled pliers, fastened with a temporary hinge pin, flat on the saw table. Check to see if the two plier elements are flush with one another. If they are not, it means that the hinge seating depressions were not cut to exactly half the thickness of the plier blanks.

True up the difference by grinding the temporarily assembled pliers on the side grinder, as shown.

When they are accurately aligned, place the pliers open and flat on the saw table. While the thin plastic abrasive disc turns at full speed, the opened plier jaws are now gently closed on the spinning abrasive disc. Any remaining inaccuracy of jaw alignment will in this manner be ground off both jaws simultaneously.

Once the whole tool is polished again, you can restore color and give it a beautiful finish as described in Chapter 18.

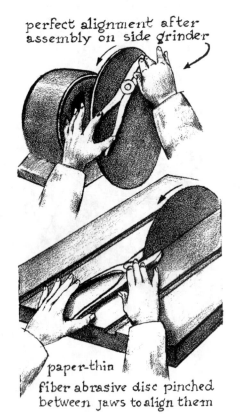

perfect alignment after assembly on side grinder

paper-thin fiber abrasive disc pinched between jaws to align them

18. Applying Color Patina to Steel Surfaces

Oxidation colors that appear on steel surfaces during the tempering process are, in effect, color patinas. Their beauty, no doubt, has already impressed you during the foregoing tempering exercises; it may have occurred to you that such colors can be restored to the finished tool. Mild-steel tools as well as high-carbon ones can be colored this way.

Suppose, for example, that a long-used, well-tempered tool has lost its surface color through wear, and become a bit rusty here and there. Simply rebuff the whole surface to the mirror sheen it had before tempering, making sure you do not overheat it so as to destroy its temper.

Now, holding it over a gas flame, you can actually "color" the whole tool just to please the eye. But here be warned that the *hard* part of the tool must be colored only within the *yellow* range (matching the critical temperature for that hardness), in order not to lose its correct temper. All other areas may be patinated to your taste, again and again, without harm to either tool or cutting edge.

The steel handles of pliers, shears, scissors, and tinsnips may, if you feel like it, have graduated colors: light yellow at the jaw or cutting ends; darker yellow at the hinge (the mild-steel pin will become blue); peacock at the midway point on the handles; purple, farther along the handles to, finally, blue at their very tip.

These patina colors, breathtakingly beautiful as they are, are only mildly functional: the patina is a slight protection against rust.

In olden days, the steel of gun barrels was often colored in this way. But nowadays color-patina chemicals in liquid form have become commercially available and can be applied "cold" by gun fanciers. I do not practice this, since I prefer the natural drawing of colors through heating the steel, with end results that have always proved very satisfying.

It is here that the good craftsman, the "artisan," blends craft with art; when pleasure in aesthetics is a bonus added to the pleasure of making things.

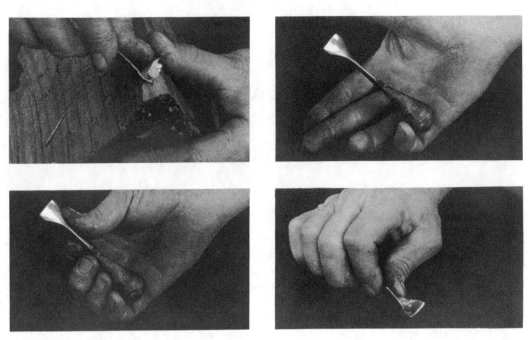

The small woodcarving gouge is manipulated in the same way as the engraver's burins. The flat upright slides along thumb during each stroke of the tool.

The engraving reproduced here was engraved on endgrain wood, using traditional burins. The burin's flat upright allows it to slide along the thumb during the tool's cutting strokes.

Bali Bella Donna, *endgrain wood. Engraving, 7½x12.*

Testing cone design of gouge, cutting an S-curve through splintery, soft, redwood board. The bevel edge is on the outside. The slant of the edge is an original design feature described in Chapter 12.

Mother and Child. *Java teakwood, lifesize. The woodcarving gouges described in Chapter 12 were used to carve this sculpture.*

Stonecarving tools made as described in Chapter 5 were used to carve this lifesize stone sculpture. PHOTO: Jim Ziegler

THE
MODERN
BLACKSMITH

CONTENTS

Introduction

The art of blacksmithing, beginning before recorded history, has changed very little over the centuries. In the recent past it reached such a peak of perfection that it will be difficult to attain that excellence again. But the very fact that our present society has entered into a renaissance of handcrafts now places the skill of working at the forge in a most promising light.

Modern equipment and trends have introduced new elements into this age-old art of hammering iron into various shapes on an anvil. For instance, that very useful new device, the visegrip pliers, is a very welcome additional tool in the modern blacksmith shop, not replacing traditional tongs, but supplementing them.

The utilization of salvaged steel material is a modern phenomenon. The present "economy of waste" has created an abundance of discarded, high-quality steel from scrapped automobiles and a variety of mechanical and industrial equipment. This gives the contemporary blacksmith excellent material at almost no cost. In addition, modern power-tool equipment (secondhand or new) can be repaired or converted to meet specific tasks. Soon, an ideally equipped workshop comes about. No matter how "junky" it may seem in the eyes of others, it enables the modern blacksmith to forge useful and beautiful things from a seemingly endless variety of salvaged scrap.

Some years ago I set up a good, but simple, blacksmith shop in my second-story sculpture studio in Berkeley, California. All I had was a 100-pound borrowed anvil mounted on a wood stump; a coal fire in an ordinary wood-burning stove; a standard household hammer; a pair of visegrip pliers; and a 1/4 hp motor, salvaged from a discarded washing machine, with which I drove my grinding and buffing wheel unit. Limited though it was, I had sufficient equipment to make my own sculpture tools for wood- and stonecarving.

The modern blacksmith must learn to do by himself that which the old-time blacksmith and his helper did as a team. So it is this I am proposing to teach here: how to resort to whatever we may invent, improvise, and construct in order to reduce the handicap of not having an apprentice helper.

In all of these activities, the machine plus modern hand tools become our assistants. Hand-filing, hand-drilling, and sledging can easily be replaced with the power grinder, table power drill, and power hammer. Nevertheless, the forgings do not necessarily need to assume a mechanized look. If the machine remains strictly the helper, the overall results remain hand-rendered in appearance. The modern blacksmith who truly loves his craft will scrupulously want to perfect his skills in pure forging. Machine "hypnosis" must always be held at bay so that the craftsman will remain in full control of the machine and not the other way around.

Although I was taught the several methods of welding, I have purposely eliminated this aspect of metal work because, as I see it, blacksmithing, without welding of any sort, is the "mother art" in forging. I believe that, like good wine, the pure flavor of the craft should remain undiluted.

Forge-welding, once mastered, becomes too readily the easy way out, like a short cut. It is used primarily when joining two or more separate elements together, and in the recent past was overdone by tacking on this or that endlessly. For those who want to learn forge-welding, there are ample sources to encourage and teach this craft. (Unfortunately for the forge-welding purist, the field has had recently to contend with the acetylene torch and electric arc, which have made it obsolete.)

The mechanical welding field is a thing apart, and I leave it to others.

Decorative wrought-iron work is simply an offshoot of blacksmithing. So is horseshoeing. Doing either one does not make a blacksmith. Ornamental work and sheet-metal practices are only touched upon in this book in order to round out the general activities in the shop and to acquaint the reader with their possibilities.

The illustrations in this book are meant to represent, as nearly as possible, live demonstrations in the shop. They are intended to show *how* something can be done: not the *only* way, but one of many possible ways. They therefore are not to be considered as inflexible blueprints; instead, the craftsman is encouraged to *improvise,* using these basic guidelines for his constantly expanding skill as he works at the forge. Above and beyond showing *how* something can be done, *why* we do what we do is stressed as of overriding importance.

The student of pure blacksmithing will find this to be a no-nonsense book. Practicing the progressive steps will result in an independent craftsman, able to make "things" out of "nothing" in his simple shop. The promise of success, then, is limited only by the talent and enterprise of the beginner, and not necessarily by the lack of expensive or elaborate equipment and materials.

TWO MAKESHIFT FORGES

Many years ago, in Lawang, Java, I had an opportunity to try out some rare wood for carving. The forestry department there had given me a freshly cut section of podocarpus wood (about 24 inches long by 20 inches in diameter) which I had selected from their samples for its very fine grain and ivory color. But I was warned that the wood was subject to severe checking during drying, and I hoped to meet the problem by first roughing out the composition and then hollowing out the core, leaving a uniform thickness so that the remaining layers would check insignificantly.

My problem, however, was how to make the tools for the carving and hollowing out. I was able to purchase, for a few pennies, some old, worn files as metal for my tools at the native marketplace in the nearby village of Porong. And, luckily, I located there a blacksmith shop where the smith allowed me to spend the day forging the tool blanks. The shop was primitive, but, after all, forging is mainly confined to the simple requirements of fire, hammer, and anvil.

Charcoal was the fuel for the fire. There were a few worn hammers. The anvil had no horn, but there were a few round, heavy, broken steel

Village blacksmith shop
in Java, Indonesia, 1935

little boy pumps 2 cocks feather
& reed pistons in bambu tubes

Fuel: teakwood charcoal
anvil: broken shaft of sugar mill
       crusher, set in wood stump
Forge: basin & air ducts in clay
quenching tub: hollowed out log
shop: bambu structure & clay floor

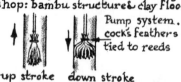

Pump system.
cock's feathers
tied to reeds

up stroke   down stroke

|———————— 15" ————————|

the tool was forged & tempered
in the above shop

98

rods to be used instead. And so, in the same amount of time it would have taken me in a modern shop, I had soon finished my tool blanks.

The illustration shows how the native blacksmith shop actually looked. Not shown is the feeling of kinship the Javanese smith, his little boy, and I had while working side by side as we shared the common language of this basic craft. The final result was the needed large tools and the carving of the lifesize wood sculpture shown on page 180.

Ten years later I again had to improvise a way to make some needed tools: a set of small woodcarving gouges and some engravers' burins.

I had settled in a small cabin in the woods in California where, as yet, I had no workshop and no electricity. I was limited to a few implements: a small machinist's vise; a pair of pliers as tongs; a table-clamped hand-grinder; a heavy bulldozer part on a stump as my anvil; a regular two-pound carpenter's hammer; and a bucket of water for quenching. Charcoal from the doused embers of the fireplace was my "blacksmith's" fuel.

I made my forge from a coffee can with an opening cut in the side and air holes punched in the bottom. This I attached to the lower end of a 12-foot-long, 5-inch-diameter steel irrigation pipe, given to me by a farmer friend. I hung it from a tree branch, and the draft up the pipe fanned the charcoal fire in the can, giving me my forging heat.

With this setup, I forged the little tools which made it possible for me to do the sculpture carving and the engraving on wood shown in the photographs on page 180.

outdoor arrangement to forge small artifacts

salvaged farmer's irrigation pipe used as chimney

can is filled with charcoal. & removable pin holds can to chimney

fire

water bucket

salvaged part of a bulldozer used as anvil

5"

charcoal made from fireplace hot coals snuffed out in airtight can

small tools easily forged in above setup

1. The Blacksmith Shop and its Equipment

semi-dark blacksmith shop

practical home-made forge

THE FORGE

Blacksmith forges, with hand-cranked air blowers, are still being made and can be ordered new through some hardware stores. If you are lucky, you may find a very useful old one secondhand. However, a simple forge is not difficult to make from available salvaged material. The examples shown here are only a few of a great variety. You may very well invent your own setup, just as long as it accomplishes its basic function.

The basin holds in its center a sufficient mound of glowing coal in which to heat the steel. The flow of centrifugally fanned air entering from below can be controlled for a fire of less, or more, heat.

A hand-cranked centrifugal fan is good. One driven by an electric motor, controlled by a foot-pedal rheostat switch, functions well. Even an old hair-dryer fan, without the heating element, can be adapted to do the job.

It is here that the question comes up: which is preferred, a hand-cranked blower or a machine-driven one? My preference has always been for the hand-cranked blower. It has the self-governing feature of the air flow stopping automatically when you stop cranking.

If you use an electric fan, make certain to install a foot-operated rheostat which shuts itself off once the foot is removed. Such a switch automatically stops the air flow, and the fire remains dormant while the smith is forging.

If you use machine-driven blowers without a foot-operated rheostat, use a damper in the air feed. It should have a spring action that automatically shuts off the air flow when the foot is removed from the damper pedal. A mechanically driven system without such a safety device requires remembering to shut off the air flow manually each time forging is to begin. To forget invites a dangerous situation: an ever-growing fire at your back while you forge. In your absence it may radiate so much heat that it could ignite nearby paper or cloth, or burn steel left unattended in the fire. *Make certain your forge is of a type to prevent overheating of a temporarily unattended fire.*

The modern improvement of a hinging smoke-catcher is an additional desirable feature for a forge. It is similar to the antipollutant device in some automobiles. In cars it catches oil smoke and leads it into the combustion chambers of the engine. In the forge, the smoke emission, mixed with surrounding oxygen when starting up the fire, is led by the catcher back into the fire by way of the air intake. Once the fire flames up, it consumes the smoke by itself and the smoke-catcher, no longer needed, is hinged out of the way. A relatively smokeless fire is maintained thereafter.

In my student days in Holland, we had the old-fashioned, overhead leather bellows. The lower chamber pumped air into the top chamber, which stored it under pressure. This still remains a good system, as are all types of blowers that accumulate air in storage bags or chambers, to be released under even pressure. If you plan to make or restore one, note that such bellows should not be less than four by five feet.

In arranging the shop, keep the forge, anvil, water trough and dipper, tong and hammer racks, etc. so spaced that simply by taking a single step you can pick up what you need while working between the forge and the anvil. No time is lost then, and forging can begin within one or two seconds after the hot steel has been pulled out of the fire.

The forge and tools should be located in the darkest part of the shop, so that the heat of the steel can be judged correctly, in a semi-dark space. In broad daylight too hot a steel can hardly be seen, and overheating may then ruin it structurally. This is why the outdoor rivet-heating forges, used in bridge building, always have a shading hood to shield the fire from sunlight. These little forges, if you can find them, are ideal in a hobby shop. They are completely adequate for the size of most forgings described in this book. Only when large pieces are to be forged will a more sturdy and roomy forge be needed. Then the whole setup will require a more specialized type of shop, with a more elaborate array of accessory tools.

THE BLACKSMITH'S FIRE

Keeping a medium-sized fire clean and effective all day long is one of the more difficult skills for the novice to acquire. Strange to say, what seems at first the least of our problems in the blacksmith's craft — starting and maintaining the fire — often proves to be a stumbling block for the beginner. But in due time it becomes routine, and experience will cancel out the initial frustration.

In the day-to-day work of blacksmithing, each previous coal fire leaves a remnant of coke, small cinders, and ashes. Smoke and yellow flame indicate the burning of unwanted elements that are not pure carbon.

Coke is what coal becomes after it is heated. Like charcoal, coke lights easily and gives off a blue flame. It makes a "clean" fire. A "dirty" fire (one which still smokes) should not be used in heating steel, since it harms the composition of the steel.

Slag is a melted mix of noncombustible matter in coal. It lumps together at the air grate below the fire, often plugging up the holes and obstructing the flow of air, thus lowering the fire's required heat. Each time, before starting a fresh fire, remove the slag from the air-grate holes with the point of the poker.

Ashes and small cinders will drop through the air grate into the ash pit below the forge. This built-in ash pit should have an easy-release door

smoke-catcher is hinged out of the way

after first flame shows

fire brick or fire clay

outdoor riveting forge

hood to shade against sunlight

centrifugal blower

poker

fresh coal
coke

air grate

ash release

so that its contents can be emptied into a metal bucket below, or fall in a heap on the dirt or stone floor. (In these accumulated ashes I often poke a piece of hot steel that needs to be very slowly annealed.)

To start the fire, use a handful of wood chips, or some coke left from a previous fire. Fill the surrounding spaces with ample fresh coal. (If it is in large lumps, break it up into pea-sized pieces.) This surrounding coal is the supply which will be raked into the going fire from time to time to replenish whatever has been consumed. It should be kept wet with the sprinkler can so that the fire will not spread out larger than is needed. Blacksmith coal is expensive, so be economical with it!

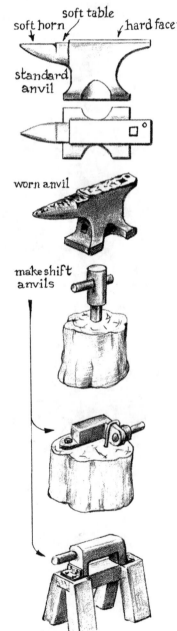

sprinkling can

soft table
soft horn
hard face
standard anvil

worn anvil

makeshift anvils

Crumple half a sheet of newspaper into a fairly tight ball. Light it, and place the burning wad on the cleared firegrate, holding it flame downward with the poker. *Gently* crank the air fan with the free hand until the flame is burning well, then crank vigorously. Now let go of the fan and, with the freed hand, rake the ready coke or wood chips (but not the fresh coal) over the flaming wad of paper. The fan meanwhile idles by its own momentum; when cranking is resumed, the paper wad is fanned into still more vigorous flame.

At this stage the greater amount of smoke can now be caught by the smoke-catcher hinged in position over the fire. Some smoke is bound to escape, but will soon change into flame. When the flame breaks through, hinge the smoke-catcher hood out of the way.

Now let the fire become evenly hot before raking in some of the surrounding fresh coal. It is best to rake in only a little of the fresh coal from time to time to avoid smoke emission; a yellow flame that combusts the smoke is preferable to having to use the smoke-catcher too often.

Throughout the many years that I have forged tools and artifacts, I have rarely needed a *big* fire. In old blacksmith shops, however, I have experienced entirely different situations, where the blacksmith had to weld and repair heavy equipment. His forge had a fire twice as large as we need. Therefore, if you should find, inherit, or purchase a large, old blacksmith layout, the first thing you should do is reduce the space around the air grate by about one-half. The economical way is to start with a small fire, enlarging it only when specially required.

THE WATER DIPPER

A water dipper can be made from a one-pound coffee can nailed onto a wooden handle (a small tree branch will do). The bottom is punctured, with a fine-pointed ice pick or 1/32-inch nail, with holes about 1/2 inch apart. The dipper is used like a sieve for sprinkling water *slowly*. As a water quencher it can wet down coal, or the part of the steel that sticks out of the fire, to keep it from overheating, especially when steel is held by hand instead of with tongs.

THE ANVIL

Some hardware stores will still take special orders for anvils, but prices run high.

During patriotic frenzy in past wars, anvils that had been neglected in farmyards and old shops were given to the government to be melted into weapons, making it almost impossible to find one secondhand today. (Already, however, the blacksmith had practically been replaced by mechanized industry.) The limited number of available used anvils are often quite worn, or severely damaged, and have to be refaced. Consequently, you may have to make do with any hunk of scrap steel (35 pounds or more) that is suitable to forge on. The illustrations show the shapes of makeshift anvils that I have used quite satisfactorily. It is also possible to make an anvil from a section of large-gauge railroad rail (how to do this is shown in Chapter 25). Therefore, the lack of a professional anvil is no reason to postpone your first experiences in blacksmithing.

Secure the anvil, or a substitute, with bolts or spikes, on a block of wood or tree stump. The height of the block plus the anvil should be such that the knuckles of your fist, with arm hanging freely, just touch the face of the anvil. At this height a hammer blow can strike the steel at the end of a full-length arm stroke.

Always keep in mind that whatever is used to ''pound'' on can never be too heavy. An anvil must never be so light that a hammer blow can move it.

anvil made from scrap railroad rail

correct height of anvil on stump

THE BLACKSMITH'S VISE

The gradual abandonment of equipment from old blacksmith shops has left fewer and fewer used post vises available. New ones can still be ordered, of course, but at great expense.

A heavy *machinist's bench vise* (35 pounds or over) will do very well for a simple shop. If you should be lucky and find a secondhand 100-pound monster (even a somewhat damaged but still functioning one), do not hesitate to acquire it. The heavy mass of a vise (as with the anvil) must be great enough that each blow on the hot steel held in it will be fully effective.

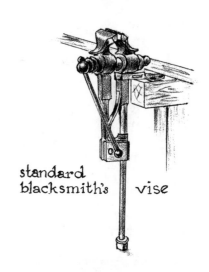

standard blacksmith's vise

standard machinist's bench vise

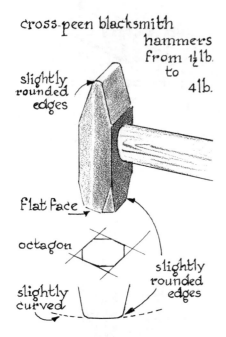

cross-peen blacksmith hammers from 1½ lb. to 4 lb.

slightly rounded edges

flat face

octagon

slightly rounded edges

slightly curved

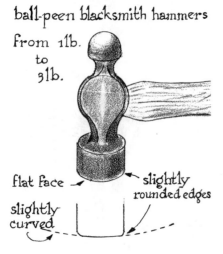

ball-peen blacksmith hammers from 1 lb. to 3 lb.

flat face

slightly rounded edges

slightly curved

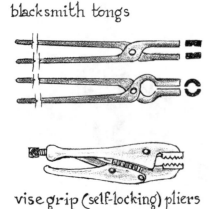

blacksmith tongs

visegrip (self-locking) pliers

HAMMERS

Blacksmiths' hammers are available through hardware stores and mail-order houses. They have changed little in design, although the steel composition may have improved through modern metallurgy.

The cross-peen-type hammer is the most useful. One end of the hammer has the cross peen, which is used to draw out steel, while the other end has a flat, octagonal face, slightly rounded off at its edges to leave an almost circular face. The face itself is very slightly convex, so that any minor inaccurate blows by the hammer's edge, not parallel to the anvil face, will not leave deep local markings in the steel.

Several hammer sizes (weights) and designs are used, but the beginner will not need more than a few at the outset.

The *flat and cross peen,* and the *flat and ball peen* are the all-round blacksmiths' hammers. Sizes of 1,2,3, and 4 pounds are preferred. Any specially shaped hammers you will learn to make later on (see chapter 21).

The heavier hammers, as a rule, are the *sledgehammers.* In olden times the smith had his apprentice helper to swing the sledge when heavy stock was to be forged. There is a limit, therefore, to the weight of hammer you can use by yourself. If you can acquire a small-caliber mechanical hammer, it will do the heavy sledging for you (see chapter 26).

TONGS

You can see old collections of tongs displayed in blacksmiths' shops in museum towns such as Williamsburg, Virginia, and Sutter's Fort in Sacramento, California. They are of every conceivable size and shape, revealing how the smith would make a special pair for each new forging problem. The jaws are made in infinite variety, to hold each particular steel workpiece firmly and easily during forging. And today, as in the past, there is no limit to the usefulness of having many designs at hand as you work. You will learn to make a pair of tongs in the same way that the early blacksmith did, and thus gradually be able to build up your own set exactly as you want it.

The old smith did not have our modern visegrip self-locking pliers, a very welcome, practical addition to the blacksmith's shop. It can be used most successfully in place of the old type of tongs which had a clamping ring that held the jaws firmly together. Visegrip pliers can hold the widest range of sizes and shapes of stock during forging, more easily and better than the old tongs. I recommend them highly.

STEEL FOR THE BLACKSMITH

This book stresses the fact that excellent steel can be found in the vast scrap piles across the country. The salvageable items in their varied sizes, shapes, and qualities seem endless; waste is one of the U.S. economy's natural by-products. Gathering such waste can become a great pleasure, because finding anything one has good use for, that others throw away, is like finding treasure.

High-carbon steels are the choice items. In cars, all springs are of sufficiently high-carbon content (over 0.2%) to make such steel temperable: this means of a hardness to cut wood and mild steel. In all the years that I have made tools of such steel, I have never had one that disappointed me.

A most important source of steel is the auto-wrecking yard where scrap car parts are to be found; the useful ones include leaf springs, coil springs, starter springs, axles, valve springs, push rods, valves, stick shifts, steering cross-arms, linkage rods, torsion bars, and bumpers, and any and all items you *suspect* may be of useful size and shape (see photographs, pages 182 and 183).

The local auto wrecker is generally selling salvaged replacement parts from cars. He earns a greater margin of profit per pound of steel than the dealer in scrap steel. If, therefore, economy is your aim, go to the dealer who sells scrap steel *only*. If he allows you to roam through his yard, you can gather an incredible variety of items.

Bars and rods in all sizes lend themselves to making many forged items. Coil springs can easily be straightened under heat (as described on page 147). Sometimes flat steel, even if it has been curved, also offers the promise of good stock. Heavy-gauge old saw blades (straight and circular saws and heavy industrial hacksaws), tractor plow discs, chain-saw bars, old files — all are excellent material for your collection.

Old rusted tools, such as cold chisels, carpenters' chisels, center punches, crowbars, and cleavers are useful, as are remnants of reinforcement steel and waste plugs that come from rivet holes punched out of plates at boiler factories and steel-construction plants. Abandoned farm machinery will prove rich in high-carbon steel.

Look also along highways and country roads. Big companies, as well as individuals, often seem to prefer leaving steel waste strewn around rather than carrying it off to the salvage yards. In this way, somewhat bent steel braces, bolts, bars, and plate from electric poles, overgrown with weeds, have found their way to my steel scrap pile.

Additional useful items to salvage are discarded ball bearings (barrels full are thrown out by electric-motor repair shops, garages, etc.). These lend themselves to making tools and jigs and makeshift instruments and a great variety of forgings. Old cast-iron pulleys, gears, and heavy items make good forming blocks on which to pound hot steel into curves.

Use your imagination in selecting items which might be useful for something or other. But be cautioned: store them inconspicuously. To your neighbor, your pile of scrap steel may be an eyesore. Keep it out of sight; and you will find that everyone will enjoy the finished treasures you make out of the things that once were just junk.

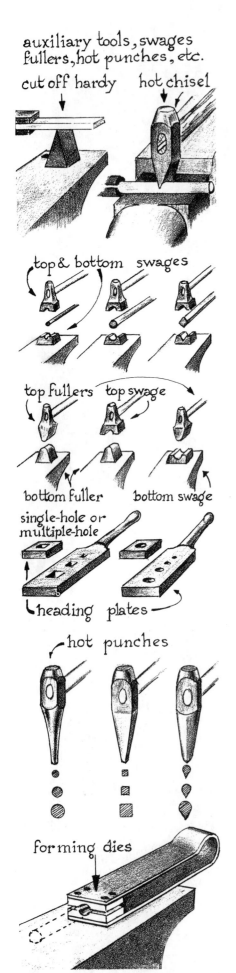

auxiliary tools, swages
fullers, hot punches, etc.

cut off hardy   hot chisel

top & bottom   swages

top fullers   top swage

bottom fuller   bottom swage

single-hole or
multiple-hole

heading   plates

hot punches

forming dies

BLACKSMITHS' COAL

The coal we use is always called "blacksmiths' coal" to distinguish it from household coal. It is more expensive, but with correct use it lasts longer. It also burns hotter. It leaves clinkers (slag) instead of ashes and is therefore cleaner, releasing less ash dust into the air. Whatever its composition in scientific terms, the farmer's feed-and-fuel stores throughout rural areas that sell this coal always call it "blacksmiths' coal." As a rule it is sold in 100-pound sacks. It is this coal that is used by horseshoers (farriers).

CONTAINERS FOR QUENCHING LIQUIDS

The Water Trough or Bucket. This is for the quenching of hot steel to cool it or temper it. It should hold not less than five gallons of water, and should be deep enough so that a long, hot section of steel bar can easily be quenched.

The Oil Container. This too should be generously large, holding not less than five gallons. It must have a hinging lid that can be closed quickly to snuff out any flash fire. (Sometimes, through misjudgment, too large a piece of hot steel is quenched in too small a quantity of oil, which could bring the oil smoke up to its flash point.)

The oil-quenching container should remain either out-of-doors or in a separate area of the shop, away from wood or other combustible objects. A metal, or metal with asbestos, sheet should surround the forge and oil bucket, between them and the wall. If the floor is wood, metal sheet should cover it where hot steel or coal might fall accidentally. An earth or stone floor and walls are ideal for this area of the shop.

AUXILIARY TOOLS

Cutoff hardies, hot chisels, top and bottom swages of various sizes, top and bottom fullers, heading plates, hot punches, and forming dies all are useful and often necessary tools. These and others will be introduced in succeeding chapters as they are needed to make the things we want. The blacksmith's craft thus proves perfect for making just about any "tool to make a tool."

RECOMMENDED POWER TOOLS

The Mechanical Steel-cutting Saw

Although steel-cutting bandsaws and reciprocating hacksaws can be bought fairly cheap new, the hobbyist may be challenged to make his own.

I recommend converting a salvaged 12-inch-diameter woodworking bandsaw by first reducing the saw speed, following the scheme in the illustration. Bandsaw blades with fine teeth serve best. The hardness of the teeth is the same for the cutting of wood or mild steel; hardened steel must never be cut on such saws. (For hardened steel, use the abrasive cutoff wheel.) You will find that these worn, discarded mechanical saws, found in secondhand shops, generally have all their vital parts in good condition: the motor, wheels, bearings, pulleys, and adjusting mechanisms. The parts requiring renewal usually are the rubber wheel-linings on which the steel bands ride, the drive or pulley belts, and the two small brass guides between which the bandsaw rides.

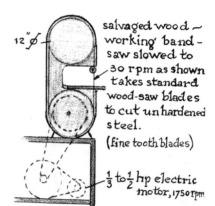

salvaged wood-working band-saw slowed to 30 rpm as shown takes standard wood-saw blades to cut unhardened steel. (fine tooth blades)

⅓ to ½ hp electric motor, 1750 rpm

The Rubber Wheel-Linings. Rummage through a tire-repair shop's scrap can and salvage a large rubber inner tube. With scissors, cut it into large rubber bands the width of the bandsaw wheel. You may need two or three for each wheel. Forcibly stretch these over the wheel rim. If tight-fitting, they do not need to be bonded together with any cement, and the wheels do not need to be dismounted for this operation.

Drive Belts. Belts that are not too worn and frayed are often found strewn around auto-wrecking yards. Keep a collection in various sizes and adapt them to your shop improvisations for driving odd transmission setups.

Adjustable Brass Band-Guides. Remove the old ones (but only if you see that hardly anything is left to warrant prolonging their use). If no brass is available to you, salvage some harder variety pot-metal parts of cast instruments and machine housings found lying around scrap yards. They can easily be hand-sawed into the size of these insets. If you find that they wear down too fast, look for scrap fine-grain cast iron from which to saw out the parts.

The Abrasive Cutoff Machine

Follow the illustration exactly to make this indispensable tool for cutting steel of great hardness. The skeleton frames of many discarded home utility machines (dishwashers, washing machines, bench-level refrigerators) can be adapted for this purpose. In combination with the converted bandsaw, the two machines will meet all of your power-cutting needs, saving a great deal of time, as well as your back and energy.

Cutoff Discs

These can be located as waste items in large steel construction plants that use discs 18 to 24 inches in diameter, ⅛- to 3/16-inch thick. The washers that clamp these large, high-speed discs securely are approximately 6 to 8 inches in diameter. In time the part of the disc *outside* the washer becomes worn down to the washer rim, therefore becoming useless in those plants. Barrels full of the remaining 6- to 8-inch discs therefore become waste. The company will either give you some, or sell them to you for much less than if you had to buy them new in that size. These industrial wheels, being of the highest quality, will not shatter easily. Use 1750 rpm wheel speed as further precaution against accidents, however.

abrasive cut-off wheel on shaft in 2 bearings below hinged 3/16" steel plate

hinge

dishwasher frame

motor hangs in belt & slides between frame posts.

Although high-speed machines may save time, they are also more *dangerous*. After all, we are not in that much of a hurry. Therefore, a one-to-one drive by a secondhand 1750 rpm, 1/4 hp motor will work fine and be *safer*.

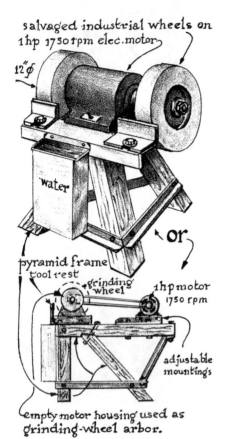

salvaged industrial wheels on
1hp 1750rpm elec.motor

12"φ

water

pyramid frame
tool rest

grinding
wheel

1hp motor
1750 rpm

adjustable
mountings

empty motor housing used as
grinding-wheel arbor.

The Large Motor Grinder

Seriously consider having one large power grinder in your shop; it will give you great satisfaction. You might choose either one of the illustrated setups, whichever fits your personal circumstances.

First Example. Your electrical wiring must be able to pull a 1 hp, 110 V, 1750 rpm motor easily. Use secondhand remnants of industrial grinding wheels, about 2 to 3 inches thick, 8 to 12 inches in diameter. The setup shown has proved to be a very great asset around my shop. Once you have it, you will wonder how you could ever have done without.

Second Example. I have also used this system very satisfactorily, although it is somewhat more complicated to set up. The center part of the old motor housing is cut in half horizontally on the abrasive cutoff machine after discarding the motor's "innards." Both bearing side-frames are kept intact. Cut an opening in the rear center, for passage of the two belts from the driver motor to the pulleys mounted on the arbor shaft within the old motor housing. Reassembled, this is the finest sturdy arbor that one could wish for.

On the metal-turning lathe, turn a shaft to fit the salvaged motor bearings. (If you have no such lathe, you may need to get help on this.) Thread both ends of the shaft to receive a fine-grain wheel on one and a very coarse-grain on the other. Two V-belt pulleys are mounted on the center of the shaft.

The Cotton Buffer and the Rotary Steel Brush

You will find that these are invaluable mechanical devices. The arrangement shown is constructed from scrap pipe, with a ball bearing used at the ends. The buffer or steel brush is mounted on one end, and the other holds the driven pulley.

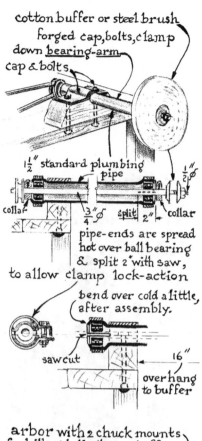

cotton buffer or steel brush
forged cap, bolts, clamp
down bearing-arm.
cap & bolts

$1\frac{1}{2}''$ standard plumbing pipe $\frac{1}{2}''\phi$

collar $\frac{3}{4}''\phi$ split $2''$ collar

pipe-ends are spread
hot over ball bearing
& split $2''$ with saw,
to allow clamp lock-action

bend over cold a little,
after assembly.

sawcut $16''$ overhang to buffer

The Double-ended Arbor

A drill chuck is mounted on one end and the other may hold whatever grinding wheel is needed. (The table drill press can be adapted for this purpose, but it is preferable to have the double-ended arbor to save the drill press from overwork.) The arbor will accommodate the endless variety of small auxiliary insert grinding points, sanding discs, and sleeves. With it, the widest range of grinding problems can be met. Tool blades or odd-shaped freeform articles that start as blanks from the forge are rough-ground, then refined, on the double-ended arbor inserts, and finally polished on the buffer.

Arbors can be bought ready made through mail-order houses. Be warned, though, to avoid the types with plastic bearing sleeves that wear down in short order when abrasive grit dust gets into them. As a rule, sleeve bearings lack a good seal against abrasive matter. Make sure to choose only arbors with well-sealed ball bearings, if you can afford them. If a bronze sleeve bearing comes your way, improvise a simple seal with oiled felt wrapping, binding it around the bearings with string. This will be effective even if it looks junky. Always be sure to keep oiling holes closed against abrasive dust.

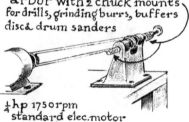

arbor with 2 chuck mounts
for drills, grinding burrs, buffers
disc & drum sanders

$\frac{1}{4}$ hp 1750 rpm
standard elec. motor

$\frac{1}{2}$ hp table model drill press

Using the Drill Press as a Wood Lathe

Temporarily converting the drill press into a wood lathe is a simple arrangement, particularly useful for those craftsmen keenly interested in expanding their projects with the making of carving tools. As shown in the illustration, the parts needed are not very difficult to make.*

With this setup, tool handles can be made without a horizontal lathe. However, well-functioning wooden handles also can be freely shaped with saws, chisels, and disc sanders, and often look more attractive than lathe-turned ones.

*In my book *The Making of Tools* (pages 7–91), greater detail is given on this subject.

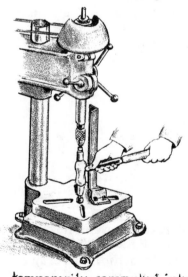

temporarily converted into
a woodworking lathe to
turn tool handles.

2. Hammer and Body Motions in Forging

hammer is
above head
at start

⊙ shoulder
joint

Right

shoulder is down & stationary
all muscles & joints are at
maximum use.

You must try to make each hammer blow as effective as possible with a minimum expenditure of body energy. The way to accomplish this is not as obvious to the beginner as it may seem, but must be learned through practice, until it becomes automatic. Any effort that concentrates action in the shoulder alone should be avoided, as the illustration clearly shows.

You stand at the anvil with legs spread enough to brace yourself firmly, one foot a little back and the other forward under the anvil overhang. Bend your head directly over the anvil, but hold it a little to the side to make certain that the hammer swings safely past it. There is a real danger that the hammer, bouncing back from an accidentally missed stroke, could hit your head. At the same time, keep your head *close* to the work, in order to have a clear view of every mark made on the hot steel by the hammer. This allows you to judge where to strike next.

Caution: During such close work it is wise to *squint* your eyes to protect them from ricocheting steel particles or oxide scales. General practice and experience will prove that, thus protected, the eyes are seldom injured. However, floating cinders and ash dust, or little oxidation scales flying off the steel during forging, sometimes do get into one's eyes. The logical question is: should you wear *goggles?* My answer is to do so if you feel apprehensive, but realize that most smiths probably do not wear goggles during forging and would rather put up with squinting and occasional eye-washings. Without goggles we have a full 180-degree view of the shop. I myself only use them while motor-grinding steel. But take warning! You must use your own judgment, and hold only yourself responsible if accidents occur which might have been avoided.

PRACTICING CORRECT HAMMERING

Use a piece of cold mild steel as a practice piece. Hold it in your hand, or tongs, or visegrip pliers, and strike it with the hammer as if it were hot. Go through the phases of the hammering movements again and again to loosen up any awkwardness.

Take special notice of the exact finger and hand positions, as these "loosely" hold the hammer at the very start of the stroke. Its weight should first be held cradled in the crotch between thumb and forefinger, while the other fingers, standing a little outward, line up along the hammer stem. The hammer is angled backward. The first motion is the contraction of the fingers, giving the hammer its initial movement, followed by the arching path of the arm until the hammer meets the steel.

With an accelerating force, all the other muscles of the body now follow that first finger-pull. (The finger-pull on the hammer stem at the

correct movements when
using light weight hammer
create a high-velocity
"snappy" blow

start of the hammer stroke is also the required technique when riveting small rivets, or hammering delicate objects such as lightweight nails into fairly hard wood.)

The drawings aim to help the student visualize a live demonstration. The *wrong way* of hammering is also shown.

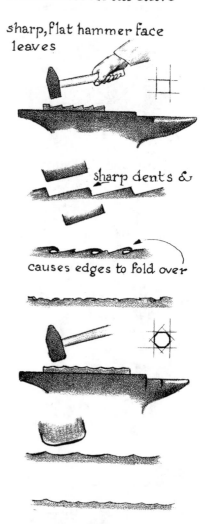

shoulder is raised high & hammer is below head at start

Wrong
shoulder does most of work, arm & wrist are less active

sharp, flat hammer face leaves

sharp dents &

causes edges to fold over

USE OF FORGING HAMMERS

No matter how skilled the smith, hand-hammering is always less accurate than machine-hammering. Therefore, the design of the hammer will affect the results.

A sharp-edged, perfectly flat hammer face will damage the surface of hot steel, making sharp dents and ridges. With continued hammering, the standing edges will begin to curl over in leaflike portions which, in turn, are pressed down into the steel below. The damage is multiplied progressively, resulting finally in a completely chewed-up surface. The only way to repair it then is by completely grinding off or filing away the damaged surface.

If the hammer face is slightly convex (crowned), with its edges somewhat rounded and its corners beveled to an approximate octagon, then the hammered surface is indented without sharp edges. It leaves only shallow concave impressions (valleys) and gentle, low ridges (summits). Continued hammer blows with this same hammer will *compress* the summits into succeeding shallow valleys. A skilled smith, working to refine that nearly flat surface, will apply gentler and gentler precise blows until finally an almost accurately smooth flat plane results.

octagon & slightly crowned hammer face leaves approx. flat final smooth surface

combination peening &

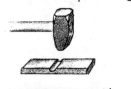

flattening

draws out steel fastest

On thin steel the flat — but not sharp-edged — hammer face can safely draw out (stretch) steel. On *thick* steel, the cross peen is more effective. It draws out an even surface with firmly pronounced ridges, but not as sharp as the ridges left by using a sharp-edged flat hammer face. Off and on, the hammer, in uninterrupted movement, is flipped into reverse position in mid-air so that the next series of blows are struck with the flat face on the ridges left by the cross peen, until gradually the steel surface becomes approximately flat once more. This alternating treatment is repeated as long as enough heat is left in the steel.

peening one side only
curves a bar

Since drawing out steel means stretching it, hammer-peening on one edge of a flat bar inevitably curves the bar, with the resulting thinner edge outward. Further stretching that side, by flattening the ridged surface with the hammer face, will curve the bar still more.

drawing out steel
(stretching thick into thin steel)

There are several methods of drawing out steel from thick to thin, narrow to wide, and various combinations, such as widening a piece in all directions while making it thinner.

To lengthen a bar without use of hammer peen, place the heated bar over the anvil horn, and pound with the flat face of the hammer. Next place the steel on the anvil face as shown, and hammer the ridges out flat. Continue drawing out the steel in this way until the desired length is reached.

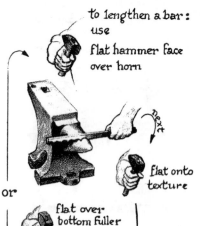

to lengthen a bar:
use
flat hammer face
over horn

next

flat onto
texture

or

flat over
bottom fuller
& next ➜

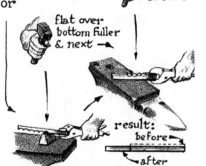

result:
before
after

Another way of drawing out is to place the bottom fuller in the hardy hole and hammer the bar out on it instead of on the horn. Flatten ridges on the anvil face, etc.

To widen a bar, use the peen end of the hammer spreading the steel *sideways* only. While there is still enough heat, flip hammer over in mid-air, as described before, using the flat face to drive down the ridges left by the peen. This stretches the workpiece sideways once more within one heat.

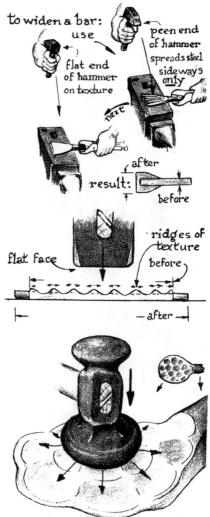

To spread the steel in all directions, first use the ball end of the ball-peen hammer, then the flat end, to spread the steel outward from the center.

It is now logical to add to the variety of hammers in your shop to allow yourself a wider choice: sharp- and obtuse-angled peens; small- and wide-diameter ball peens; shallow or more convex face, etc. Still greater variety can be obtained by adding hammers in different *weights*. The double ball-peen hammer, with two different peen sizes, is one of the most useful ones for making the blanks for woodcarving gouges (as described in Chapter 22).

flat & ball end of hammer spreads steel in all directions

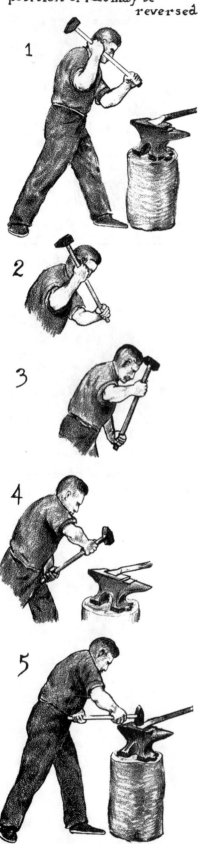

correct use of the
sledge hammer

position of feet may be
reversed

1

2

3

4

5

USE OF THE SLEDGEHAMMER

The use of a heavy (4 pounds or over) sledgehammer simply speeds up the work of forging large volumes of steel. In all hammering, body motions should combine a maximum precision of aim with the stored energy in the hammer blow.

The stroke begins with the right hand fairly close to the left hand, then sliding toward the hammer head as the stroke progresses. The drawings (steps 1 through 5) show the progression of the hammer's path and body motions during one cycle. This lets the hammer head enter into a *spiral curve* instead of a more or less circular arc. It is done by a very skilled sledger if he feels certain that by thus increasing the energy storage (the hammer head moves into a larger outer-diameter curve and therefore a longer path) he will not sacrifice the accuracy of the hammer stroke. It is this sort of combination of body movements that create a whiplash effect to increase energy storage.

Various combinations are possible, depending on the skill of the sledger. He can lift the hammer with right hand held close to the hammer head and then, letting the hammer head slide outward, bring both hands together at the end of the hammer stem. Or he can hold this final position for the maximum path of the hammer. The only danger is that the sledger might miss in his aim in this large swing.

A yet greater sledgehammer swing (not shown, since rarely needed) was practiced in olden times when the blacksmith had helpers. This was the overhead, full 360-degree swing. In my training years I often used skilled fellow student helpers, sometimes three of them simultaneously, each meshing his own cycle into the uninterrupted overhead sledging of the others during the period of one heat. It was an exhilarating experience as well as a practical way of drawing out or upsetting heavy-gauge steel parts. Circus gangs use the overhead swing with large diameter wooden (often steel-weighted) malls when driving tent stakes into the ground. You can practice heavy sledging similarly, by driving a heavy stake into the ground. Once this is well learned, practice also the 360-degree overhead swing on such a stake; you might at some time have to use this skill in helping a smith at the anvil.

The logical next step is to use the cross peen of the sledge to draw out steel. But try this only after you have learned controlled use of the sledge's flat face. *Caution:* If you should miss, and the corner of the cross peen strikes the anvil face or its edge, it may be dented or broken, a very serious setback. Such misuse can cause irreparable damage to the anvil, your work, or to *you*. You will not regret having a 6- to 8-pound sledge in the shop, but do not underestimate the necessity of knowing how to use it safely. Naturally only *heavy anvils* can take the powerful blows of heavy sledgehammers.

correct use of the sledge hammer's cross-peen when drawing out a wide bit at end of a heavy bar

$7\frac{1}{2}$ lb

4 lb
$7\frac{1}{2}$ lb
16 lb

heavier sledges to be used on heavier anvils

3. First Blacksmithing Exercises

How does the smith judge quickly each next move when he must "forge while the iron is hot"?

The student blacksmith will learn that when steel has reached forging heat, success depends on three things: *precise judgment, perfected skill,* and an instinctive *feeling* for what is right or wrong. Combined, these enable him to act decisively at the right time. When the moment has come to apply his knowledge and skill, he cannot afford to hesitate, doubt, or let the time to act pass by. He must do *then and there* what needs to be done.

A blacksmith works with hot steel that is as malleable as clay. Steel, made soft, can be pushed together (called upsetting) to make a piece *shorter and thicker*. Or it can be stretched out (called drawing out) to make a piece *thinner and longer or wider*. Therefore, both a blacksmith and a clay modeler think in somewhat similar terms when it comes to judging how to change a given volume of material from one shape into another shape. The only difference is that the potter's hands are replaced by tongs, hammer, anvil, and the other tools of the blacksmith.

It would be wishful thinking to hope that *everyone* could become a good blacksmith. My experience tells me that in this, as in other crafts, everyone has his own degree of skill. One may have great talent, another less, and still another hardly any at all. A simple and good test, which you may apply to yourself, is this: Can you drive a nail into the wall easily? Do you have a natural feel for choosing the correct weight of hammer to pound a sturdy, a slender, or a very thin nail into soft, medium, or hard wood? If you have the combination of dexterity, mental judgment, and feeling to hammer the nail well, it may indicate to what degree you may be successful as a blacksmith.

The following exercises acquaint the beginning student with the techniques and results of hammering hot steel so that in time he also will experience a feel for what can and cannot be done in the forging process.

STRAIGHTENING A ROUND BAR

From the scrap pile choose a ⅜-inch round rod about 15 inches long that is not quite straight and is bent a little.

Start a small, clean fire as described in Chapter 1. Place the bent section of the rod in the center of the fire-mound. With tight-fitting tongs or visegrip pliers, withdraw it when the bent section is *dark yellow hot*. (If the bar is long enough, and the end has remained cool, it can be hand-held.) Place it, humped up, on the anvil face.

With a 2-pound hammer, straighten the rod with gentle taps on the hot curve. Learn to check for accuracy by rolling the rod over the anvil face and watching for spots that are still unaligned. Tap with the hammer on those humped-up areas until the rod rolls on the anvil face in even contact all around.

This simple straightening exercise will give you the feel of the hot steel's malleability, and show you how to strike accurately.

SQUARING A ROUND BAR

Steps 1 through 6 in the illustration show how by a 180-degree tumbling method, a round rod progressively becomes a square one. First one side of the rod is hammered to create a flat face. At the same time the underside of the rod which contacts the anvil will also become a little flat. But because the cold anvil cools the steel on contact, that side will be less malleable. This is why the rod is tumbled over frequently in that heat period. Then the hammering can even out the little differences to maintain the symmetry.

Each time the steel cools beyond a *visible heat glow* (as seen in a semi-dark shop), the steel should be reheated.

Assume now that you have succeeded in making two evenly flat faces on the round rod in one heat period (referred to as a "heat"). The critical point is that each time the rod is heated and replaced on the anvil, it must continue to be kept *accurate*. During the next step, the already flat, parallel faces must be kept at an exact 90-degree angle to the anvil face. The hammer blows thus must also be accurate, striking the remaining round sections exactly parallel to the anvil facing. An inaccurate stance, rod position, hammer stroke, or anvil placed at a slant can easily create a parallelogram instead of a square in cross section.

It is therefore necessary, if this is your first experience, to check quickly the result of the first blows. Look at the end of the rod "head on" to see in what way it may be necessary to adjust the position on the anvil, and what corrective blows must be applied to square the rod properly.

If at first it seems that much is made of little, remember that you must learn corrective actions now in order to apply the gained skill *instantly* later on.

The pride of the good smith is to do a piece of work in a minimum of heats. Endless fussing with steel with many reheatings and ineffective motions indicate the novice's hesitant beginnings. With self-confidence, the faster 90-degree tumbling method, as shown, will soon tempt you to speed up the squaring of a rod end. Quick corrective blows during uninterrupted hammering will prevent the forming of a parallelogram cross section. You will soon improve your skill here, and end up doing perfect square ends.

Steel can undergo many reheatings before its essential properties are affected, provided it never reaches a *white heat*, or burns. It should not be worked *after* the visible heat glow has disappeared. (It can still be safely *bent*, however, after the steel has become too cool for forging.) If the steel should burn in too hot a fire, you will unmistakably see sparks fly from it looking exactly like sparklers on the Fourth of July. As nothing can be done to restore steel quality when burning has occurred, make sure to *cut off every trace of it,* as follows:

Heat the spot where it is to be cut to yellow hot. Place the cutoff hardy in the anvil hardy hole. Lay the hot steel on the hardy. With a well-aimed hammer blow, the hardy's knife edge will make its first mark on the steel. Replace it in the exact groove for the second blow. (In a quick rolling movement over the hardy you will "feel" that it is in

squaring the end
of a round bar

in approx. 2 to 4 heats

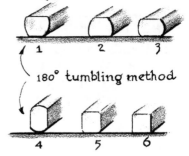

180° tumbling method

90° tumbling method

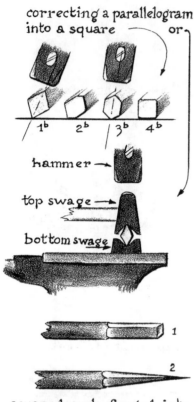

correcting a parallelogram
into a square ——— or

1ᵇ 2ᵇ 3ᵇ 4ᵇ

hammer →

top swage →

bottom swage

1

2

squared ends forged into
tapered ends become
1½ to 2 times longer

the correct position.) Just before the piece falls off with the final blow, gently strike it a little *outward* from the hardy edge, and it will be sheared off. This prevents striking the hardy and dulling it or damaging the hammer.

Another way to cut off a section of steel is to place it on the anvil's soft cutoff table using a hold-fast tool which frees both hands for cutting with a chisel head.

If you have ended up with a parallelogram instead of the intended square, correct the error as follows:

If top and bottom swages in the correct size are available, the simplest way is to "push" the hot workpiece into a square between them. Without the swages, however, simply use corrective blows using the hammer and anvil as shown in steps 1b through 4b. But this will reduce the size of the square somewhat.

TAPERING THE END OF A SQUARED ROD

Begin with a squared section of rod. Heat it and place it on the end of the anvil horn. With the flat of the hammer, deliver strong blows, first on one side, then at 90 degrees on the next side, etc., tumbling the rod 90 degrees for each succeeding blow. Deliver these blows as the rod is gradually pulled toward you until the end is reached. This action "pinches" the rod end, as one might pinch a similar section of clay to make it longer. This is *drawing out* the steel.

Using the flat side of the hammer, remove any corrugated hammer marks to make the surface texture summits flush with the valleys. This again pinches the steel, making it longer still. Continue hammering gradually toward the end so that the square cross section is finally shaped into a taper.

The blacksmith's craft is so flexible that there are, as a rule, several ways of doing such things as making a taper. For instance, instead of using the end of the anvil horn, you can use a hardy-type tool with rounded ridge called a *bottom fuller*.

Another way is to hold the rod *away* from yourself, over the anvil face, and hammer with the cross peen, which is easier than with arms positioned over each other (see also Chapter 12). Or tumble the rod in 90-degree sequence using very heavy blows of the hammer-flat without letup at the angle of the taper. The skilled smith will often use this last method for tapering rod ends in sizes up to ½-inch diameter. These very rapid, heavy blows put so much energy in the form of heat *back* into the steel that it stays hot long enough to finish the task in one heat. All that is needed besides skill, in such an operation, is ample muscle and lungs.

SHAPING THE END OF A SQUARE ROD INTO A ROUND CROSS SECTION

The procedure (not illustrated) always works best if the square is first hammered into an octagon. Next the octagon ridges are hammered with rapid gentle blows, one ridge after another, into an approximation of a smooth round surface. (Follow a similar method to make a hexagon bar into a round bar or to shape a square taper into a round taper, etc.)

While doing these first exercises in blacksmithing you will have become aware of one of the most outstanding characteristics of the craft: the *rest period* during reheatings of the steel is in fact an active time. You must plan, deduct, decide, invent, remember, conclude, and keep all "thinking-pistons" firing before the steel has become hot once more. Then the hand and eye must proceed to translate this thinking into the actual forging activity.

4. Upsetting Steel

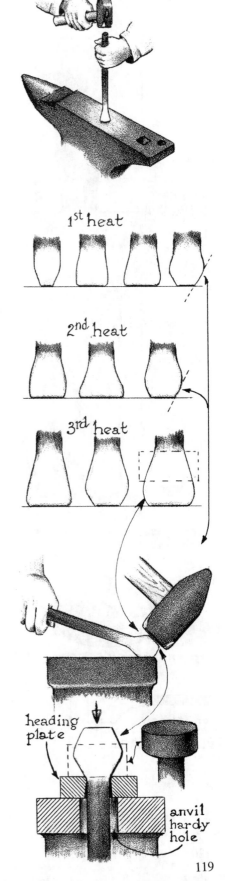

UPSETTING A ROD END TO FORM A BOLT HEAD

Use a ⅜-inch round rod, 17 inches long. Taper ¼ inch of both ends a little, which will tend to keep hammer contact at the exact center of the rod.

Heat 2 inches of one end and hold it down vertically on the anvil. With a 2-pound hammer, strike the cold end with rapid, medium-heavy blows. Aim the blows in the exact vertical direction of the bar to prevent side-glancing and consequent bending of the hot section. If it does begin to bend, *immediately* place it horizontal on the anvil face and straighten it, taking care not to strike too hard, thereby losing the increased thickness you have just gained. *Caution:* Never hit the anvil face with the hammer edge or the peen. Such misdirected blows will leave scars on the anvil that will show as texture on your workpieces.

Hammer the flared-out edge back into a slight taper before returning it to the fire for every next heat. This brings the hot end-face into closer alignment with the center of the rod, and lessens the chance of bending it again during the *next* upsetting action. Apply this corrective treatment throughout the upsetting procedure.

During the third heat, enough volume of steel should be upset to make the size of bolt head desired. If three heats have not accomplished this, however, four or five may be required.

To form the bolt head, place over the hardy hole a heading plate that fits the ⅜-inch rod (see the heading plates illustrated in Chapter 1). Take care that inaccurate blows do not tend to "pull" the malleable mass sideways as you hammer. This danger is real, because you cannot see the resulting rod and head alignment until after the damage may have been done. It is therefore a good practice to look for the slightest sign of eccentricity of the *visible* part you are shaping, making sure it stays aligned with the visible shape of the heading plate. As soon as you notice the head wandering out of line, slant your blows in the opposite direction to put it back in center.

A slight eccentricity between head and shank will not weaken the bolt head's holding strength appreciably. If the head meets the rod with a slight curved edge you can reforge the head back into line.

(high carbon) steel

1″ dia.

75 lb. cast-iron flange

straighten & align after each heat

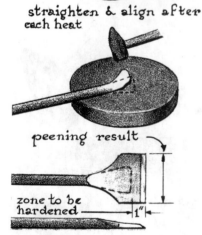

peening result

zone to be hardened

1″

USE OF THE UPSETTING BLOCK TO FORM A BAR TOOL

The ends of long and heavy bars can be upset without hammers by using the weight of the bar itself as a hammer. Dropping it vertically in a free-fall onto a heavy cast-iron flange upsetting or "stomping" block will in due time compact and thicken the end. Adding *muscle* behind the free-fall makes this upsetting technique very effective and often proves less tiring than the hammer-upsetting of much smaller workpieces.

If sideways bending of the heated end occurs, realign it by hammering it, using the upsetting block as an anvil.

When enough steel volume has been upset on the end of the bar to make a wide, sturdy blade, peen it into the needed width on the anvil with a sledgehammer (see Chapter 2, page 113).

The blade can be *hardened* by heating one inch at the end to a dark cherry red glow and quenching in oil or brine, or it can be tempered by drawing to a straw oxidation color after a brittle quench. If the blade is to be used for cutting trenches in earth without gravel or rocks, it can be tempered brittle hard (as hard as a file). In this case, the cutting angle of the bevel must be somewhat wider to keep the edge from cracking if it should accidentally hit a rock. Under constant use, tools that dig in earth wear better when they are as hard as possible. If you plan to *pry* with the tool, the quenching liquid should be oil, to toughen the blade. For a large heated volume of steel, such as this bar end, the oil container must be deep enough to prevent the oil from becoming too hot and causing a flash fire. Refer to Chapter 6 for further details on tempering steel.

FORGING A HEXAGON BOLT HEAD

To forge a hexagon bolt head freehand, use a previously forged, well-centered, round-headed bolt. Follow steps 1, 2, and 3. In the progressive heats, all angles, if kept correct, will in time result in a sharp-edged hexagon.

If incorrect angles do begin to form, *immediately* place corrective blows as shown. Postponing corrective measures results in the hexagon becoming smaller and smaller.

However, if a small, accurate, but *too-high* hexagon has resulted, it may be forged shorter and wider again by dropping the bolt into the heading plate and hammering the head down to its intended size. This manipulation widens the head as well and may save the project without weakening the bolt to speak of, provided no folds and invisible cracks in the steel have formed.

In any event, final accuracy can be realized by alternately reducing the bolt-head height in the heading plate and after that hammering the hexagon sides on the anvil, until an exact wrench size for the bolt head has been reached.

To thread the bolt shank, clamp the head in the vise and, with a threading die (using lard as cutting agent), cut the thread as shown.

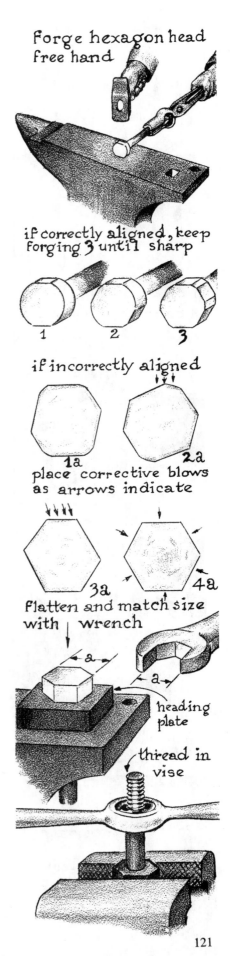

Forge hexagon head free hand

if correctly aligned, keep forging 3 until sharp

1　2　3

if incorrectly aligned

1a　2a
place corrective blows as arrows indicate

3a　4a
flatten and match size with wrench

heading plate

thread in vise

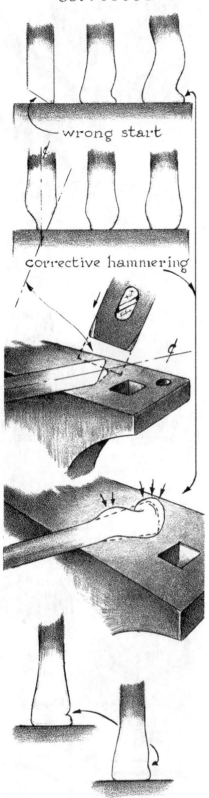

upsetting errors
corrected

wrong start

corrective hammering

grind or file every
forming fold out
before continuing

CORRECTING UPSETTING ERRORS

If the end of a bar is not at an exact right angle to its length, the heated section will unavoidably bend out of line during upsetting. Therefore, it must first be ground or forged into a 90-degree end.

You can meet the difficulty halfway by rounding the rod ends or hammering them into a slight taper. Such ends will receive hammer blows closer to the rod's true center so that even hand-held rods and slightly inaccurate hammer blows will reduce the danger of sideways bending.

Should the hot end bend, it will also start to *fold,* and this must be corrected at once. The fold must be filed or ground out after the end has been straightened and before any further upsetting is resumed. There is no other way to meet this difficulty.

It should be noted that, because of natural human error, all hand-held rods are, to some degree, placed out of line with the anvil. And, of course, hand-hammering also never is as precise as machine-hammering. Therefore, all upsetting actions that rely on hand and eye alone, without the aid of machines, jigs, dies, etc., must constantly be interrupted with corrective hammering before further upsetting can safely be continued.

5. Upsetting with the Aid of an Upsetting Matrix

MAKING AN UPSETTING MATRIX

Often the smith wishes to use tools which serve as "short cuts" when a series of identical articles is to be forged.

The upsetting of steel into bolt heads is a typical operation which uses such a special tool, called a *header-matrix* or *upsetting matrix*. It fits into the hardy hole of the anvil. Illustrations 3 and 3a on page 124 show it in place and in section.

To make a large matrix, a heavy truck axle is good. (Salvaged car axles in all sizes have proven to be excellent material for the making of hammers, swages, fullers, hot punches, as well as matrixes.) A section can be cut, to place in the anvil's 1-inch-square hardy hole for the forming of steel, with a mechanical cutting saw, provided the test with a file (see page 126) shows that the steel is not tempered too hard. High-carbon-steel car axles come in a semi-annealed state and have an inherent resilience that resists breaking. (Only the splines at the end of the axle may have been hardened somewhat to prevent wear by the gear differential movements).

If the car-axle section has a diameter equal to the diagonal of the hardy-hole square, grind the axle to fit the square using the largest grinding wheel.

If you prefer to forge the section into a square as the illustrations show, then the hot section can be driven into the hardy hole as soon as it begins to fit. Whatever does not precisely fit will yield (the steel being malleable) under precise hammer blows. It is now that a 4- to 7-pound sledgehammer should be used (if the anvil weighs 100 pounds or more) to drive the slightly oversized hot metal into the hole to make the best possible fit.

It is good planning to let the square section protrude below the anvil thickness so that the finished matrix can easily be knocked out later. During the making of the matrix, however, when the square section is yellow-hot, it may inadvertently be upset locally should you try to knock it out from below. That is, the end would be made *thicker*, as in riveting, thus locking the matrix into the hardy hole. To avoid this, bevel the end of the square section a little. To be doubly safe, place a ¾-inch bar (smaller than the hardy hole) against the locally beveled and protruding matrix blank, as a driver to knock out the blank.

Assuming all has gone well so far, reheat the piece. Drive it once more into the hardy hole, using a 3½- to 4-pound hammer to upset a little shoulder on the matrix blank where it meets the anvil face. This will prevent the matrix blank from later wedging so tightly into the hardy hole that it might chip the edges of the hole. The matrix must fit snugly without undue strain.

Remove and slowly anneal the piece by placing it in ashes. Since this takes about an hour, working on another project will overcome your impatience.

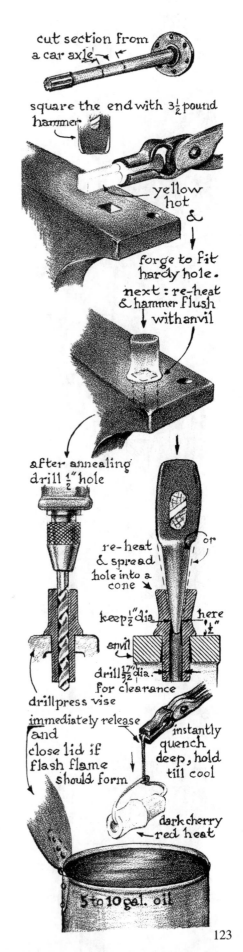

cut section from a car axle

square the end with 3½ pound hammer

yellow hot &

forge to fit hardy hole.

next : re-heat & hammer flush with anvil

after annealing drill ½" hole

re-heat & spread hole into a cone

keep ½" dia.

here ¼"

anvil

drill 17/32" dia. for clearance

drillpress vise

immediately release and close lid if flash flame should form

instantly quench deep, hold till cool

dark cherry red heat

5 to 10 gal. oil

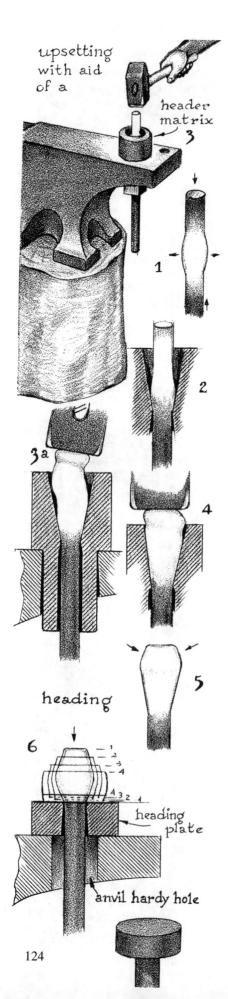

upsetting
with aid
of a

header
matrix
3

1

2

3a

4

heading

5

6

heading
plate

anvil hardy hole

Once it is annealed, test the matrix with the file for hardness to make very sure it is soft enough for drilling.

Place it in the drill vise and drill a hole through the full length of the matrix. Begin with a small drill; the size of the final drill should be the size of the rods you plan to use in the future.

Assume that ½-inch rods are to be used: Clamp the drilled blank upside down in the drill vise, and enlarge the ½-inch hole with a $^1/_{32}$-inch oversize drill to a depth which stops just ½ inch *above* the anvil face when the matrix is placed in position in the hardy hole. Now heat in the forge the section of the matrix that extends above the anvil. The shoulder where it touches the anvil should be a dark cherry red while the rest, above it, shows a yellow heat.

Quickly now, insert the taper of a hot punch. With heavy, precise blows, drive it in until it reaches the ½-inch-diameter end of the matrix section (½ inch above the anvil face). It may be necessary to use two or more heats to arrive at this exact position, checking meanwhile with a ½-inch-diameter rod whether it is about to slip through or not, and at what point it does so.

Once you are satisfied that the matrix blank is as the drawing shows, the next and final step is to *harden* it.

String a length of baling wire through the matrix and then heat the whole matrix to a dark cherry red. Holding the wire in tongs as shown, immerse the blank quickly to the deepest part of the 5-gallon bucket of oil. The oil should be at room temperature or a little cooler. The oil bucket should have a hinging lid to snuff out any possible flash fire. Agitating the blank sideways at that depth may promote the process of hardening the outside layer of the steel a little deeper. The inner core will remain a little less hard, just the same, thus preventing it from breaking in use.

UPSETTING STEEL INTO BOLT HEADS WITH AN UPSETTING MATRIX

To upset steel into bolt heads, first heat a ½-inch-diameter rod for about 2½ inches from its end. Then quickly dip the end ½ inch in water before upsetting. The resulting local bulge formed at 2 inches from the rod's end should measure ¼ inch greater than the rod diameter (see illustration 1). Place the matrix in the anvil hardy hole.

Reheat the rod end 2 inches to a light yellow heat, leaving the bulge a dark cherry red. Drop the rod, in this state, into the matrix, where the bulge will be hung up by the cone's narrow throat (see illustration 2). For once, you need not feel too rushed, because you want to let the bulge cool a little on contact with the cold matrix. The purpose is to *lodge* the bulge here, *not* drive it through the narrow throat.

While lodged firmly, the upper light-yellow-hot portion of the rod is very malleable and will respond fully to upsetting, soon filling the cone space (see illustrations 3a and 4).

Forceful hammering thickens the ''locked-up'' steel, driving it into whatever space within the cone is still not filled. It calls for keen, instant judgment as to how you must redirect the hammer, blow by blow, to avoid too much sideways bending of the rod end.

Illustration 5 shows the end result of the symmetrically upset rod-head. It is now ready for heading with the heading plate (see illustration 6).

Caution: Read carefully how to prevent eccentric wandering of the head during this operation as described in Chapter 4, page 119.

MAKING A CARRIAGE BOLT HEADING PLATE
and DECORATIVE BOLT HEADS

A carriage bolt is used to tie wooden members in structures together. These bolts have square cross sections just below the head, so that the corners of the square, pressed into the wood, will "lock" the bolt against turning when the nut is tightened on it.

To make the heading plate as illustrated, use a *square* hot punch, which leaves a square tapered hole. You may use a *round* hot punch to make a round tapered hole first, and follow with the square one. This makes it easier to obtain a more precise placement and lessens the danger of tearing the steel at the punch's exit.

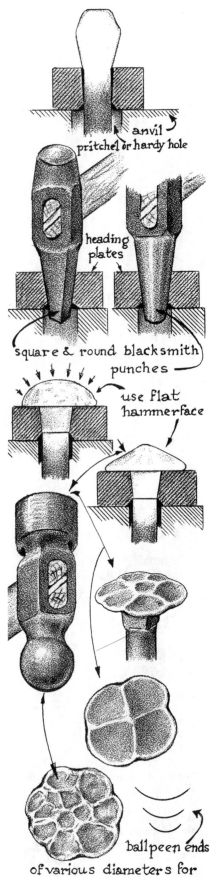

anvil

pritchel or hardy hole

heading plates

square & round blacksmith punches

use flat hammerface

ballpeen ends of various diameters for texturing steel surfaces

Both square and round-holed heading plates can be used with the small end of the hole either up or down. Used with the small end down, as illustrated, the bolt head is formed after the tapered hole is filled. If the small end is up, then the hot upset steel is driven partly through while "shearing off" some excess steel before the head begins to form. This makes a parallel square section below the bolt head instead of a tapered one. In either case the forging can be knocked out easily.

The shape of the bolt head may be forged as the illustrations show, but the head must be sturdy enough to match the holding strength of the bolt shank. Then it can be decoratively textured to suit your taste. For further details on decorative treatment, see Chapter 9.)

6. How To Temper and Harden High-Carbon Steel

A piece of high-carbon steel can be "chance"-hardened in one quick step; or it can be hardened by controlled tempering for specific hardness in a series of steps.

The blacksmith, as a rule, prefers methods of hardening and tempering which are simpler than those of the specialists who make delicate cutlery, woodcarving tools, or other small articles needing more precise techniques. You must, of course, become familiar with all the ways steel-tempering can be done, and then apply whatever method is most effective for your particular purpose.

In this chapter I present three methods: one, a simple and direct way; and two more elaborate, controlled ways. These include the basic principles applicable to all other methods of tempering.

FIRST METHOD OF TEMPERING

By this method you can harden a cold chisel without tempering.

In the forge fire heat ½ inch of the beveled cutting end of a chisel to a dark cherry red heat glow (as judged in a semi-dark shop, at the moment it is withdrawn from the fire). At this moment, quench the whole tool in water.

This first method is the quickest way of hardening high-carbon steel that can be devised, and most smiths that I have known rarely resort to any other way. In his average daily practice, the smith forges larger, heavier-caliber articles than a specialist in cutlery-making and for most of his work he will use this "one-shot" method.

The File-tip Test for Hardness and Temperability

This test is always a reliable way to test the hardness of steel.

Assume that you have quenched the cold chisel at a *dark cherry red heat glow*. With a sharpened file-tip, press down on the quenched end of the chisel. If it slides off like a needle on glass, the steel is hard enough and can cut mild steel or annealed high-carbon steel.

If the file-tip "grabs" or can "pick" at the surface, it is too soft. In that case, reheat the ½ inch of the chisel end, this time to a medium cherry red heat glow. If it is still not hard enough after quenching, repeat the same procedure once again, this time quenching at light cherry red.

When you are dealing with high-carbon steel, this last should bring correct results. However, if the steel still is not hard, you are then justified in suspecting that it is *mild steel*, lacking the needed 0.2% high-carbon content to make it temperable.

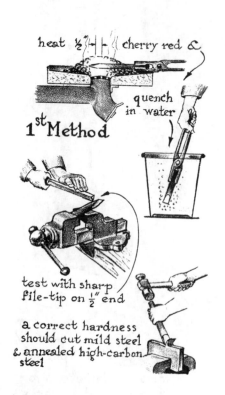

heat ½ — cherry red &

quench in water

1ˢᵗ Method

test with sharp file-tip on ½" end

a correct hardness should cut mild steel & annealed high-carbon steel

SECOND METHOD OF TEMPERING

This method introduces more elaborate controls, using the temper colors of the oxidation color spectrum as a gauge for degrees of hardness.

In the forge fire, heat 2½ inches of the bevel end of the cold chisel to a dark yellow heat glow (again as judged in a semi-dark shop, at the moment the steel is withdrawn from the fire).

Now hold only 1 inch of the end in the water or brine (in this case *not* in oil because if the hot steel is not *completely* submerged, the oil would flame up). Within a few seconds you will see the visible heat glow of the unquenched part disappear. Only then withdraw the tool.

Immediately rest the chisel, slanted downward, over the anvil's edge, and without loss of time, rub the bevel vigorously with an abrasive stone. This puts a silvery sheen on the steel. (Only on a shiny steel surface will the oxidation color spectrum become clearly visible when held to catch the light.)

The *remaining heat* in the tool now acts as a reserve, spreading gradually, through conductivity, to the bevel end of the chisel. The first color that appears is a *faint straw*, followed by *straw*, *bronze*, *peacock*, *purple,* and finally, *blue*. This color sequence, for all practical purposes, never varies in the average high-carbon steel.

Once the color you want arrives at the cutting edge, arrest the process by quenching the *whole* tool in water. At this point the reserve heat still in the tool will be cooled enough not to brittle-harden it.

In observing the color scale, note that each color must be thought of as a *hardness* indicator.

Caution: If the tool is withdrawn *too soon* after the first 1-inch quench, the reserve heat could still be too great, with the color running *too fast*. In this case, arrest it by a sudden quench. The danger is, however, that this may brittle-harden the too-hot section of reserve heat, causing the chisel to break there during use. To avoid this mistake, never deviate from this rule: Hold the hot tool end 1 inch in the quenching bath *until the visible heat glow has disappeared* before proceeding.

Another mistake is waiting *too long* before withdrawal, since then the reserve heat may not be great enough to make the full color spectrum visible, and it then would be necessary to start all over again.

If the color-run on the bevel should come *slowly to a stop* at the color of your choice, do not quench, but let it continue to cool *slowly*. This has an advantage, particularly when tempering slender tools. The slow cooling creates the least amount of tension, whereas any sudden quench always causes sudden shrinkage, with unavoidable greater tensions in the steel's structure.

Some steels are made to be hardened in *oil only* and others in *water only*. Since a scrap pile of steel will be a mixture of these, it sometimes happens — though rarely — that the faster cooling water quench "cracks" the steel across.

Assuming that all has gone well, you have now learned the basic principles of a more controlled hardening method — the process is called *tempering*. This, combined with the simpler first method, allows the smith, as a rule, all the leeway he needs in hardening and tempering steel.

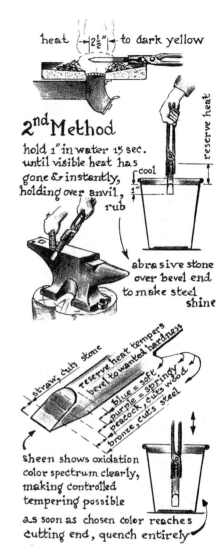

heat ⊢2½"⊣ to dark yellow

2ⁿᵈ Method

hold 1" in water 15 sec. until visible heat has gone & instantly, holding over anvil, rub

reserve heat

cool 1"

abrasive stone over bevel end, to make steel shine

straw cuts stone

reserve heat tempers bevel to wanted hardness

blue = soft
purple = springy
peacock = cuts wood
bronze = cuts steel

sheen shows oxidation color spectrum clearly, making controlled tempering possible

as soon as chosen color reaches cutting end, quench entirely

THIRD METHOD OF TEMPERING

This refined method of tempering (not illustrated) is applied to light-caliber forgings, and is the method used by makers of delicate tools and instruments.

Use here a very slender thin cold chisel.

First heat the tool end a light cherry red. With this tool, the *whole tool* is quenched. This leaves the tool brittle-hard where the visible light cherry red was, while the rest of the steel remains softer.

Carefully clean about 1 inch of the brittle tool end on a fine-grit rubber-backed, abrasive disc insert. Polish it to a mirror sheen in order to reveal the oxidation color spectrum prominently. This reduces error to a minimum when judging color.

DRAWING TEMPER COLOR

First, reheat the softer part of the tool over the blue flame of a gas burner, holding the brittle part safely *outside* the heat core of the flame. When the soft part becomes hot enough so that the heat travels outward through the steel by conductivity, you will see (as in the second method of tempering) the oxidation color spectrum appear. This time, however, you have control over the *speed* of the process, manipulating the tool by holding it *in* the flame, or *above* it. Depending on the type of steel composition, oxidation colors can be brilliant or faint. When the color for the specific required hardness reaches the bevel end, *quench the whole tool*.

QUENCHING LIQUIDS

The three most frequently used quenching liquids are water, salt brine, and oil (old motor oil will do), *always used at room temperature*.

In water the steel is cooled quickest; in salt brine, a little slower; in oil, at its slowest.

In all three quenching liquids the hot steel, meeting the identical room temperature, end up with the same *outside* hardness. However, since the boiling points of these liquids vary, the hot steel cools towards its *core* at different rates of speed. The quicker the core is cooled, the harder it becomes. The slower it is cooled, the softer it will be.

A softer core makes the tool tough. The explanation is simply that an *outside hardness* may be kept from breaking under strain when cushioned by an *inside softness*, which combination keeps the tool tough.

In addition it should be recognized that one kind of steel cools faster or slower than another during quenching. Each one thus has its own *coefficient of conductivity* which must be taken into account. It is the combination of these variables which you must be aware of during the tempering procedures.

Effect of Quenching Liquids on Hot Steel

Water boils at 100° C. (212° F.). This cools the *core* relatively fast and the heated chisel bevel will have an almost *uniform hardness;* it will be brittle all the way through after quenching.

Brine boils at a temperature a little higher than water (depending on salt concentration). Therefore the *core* cools a little slower, giving the same *outside* hardness while the core itself, somewhat *softer*, leaves the bevel end a little tougher and less brittle.

Oil boils at about three times the boiling point of water. Therefore the *core*, being cooled slowest, will be *softer still*, while the outside hardness remains the same as tools quenched in water or brine. The oil quench therefore will give the toughest bevel end.

Since it is always difficult to grasp the underlying principles of hardening steel, you should read the foregoing procedures again and again in order to understand clearly what takes place.

Once the three methods of hardening steel have been understood and practiced successfully, you can expand on them in your own way. You will then also begin to recognize how the ingenious methods of smiths in many foreign lands produce many effective results. The various swords, krisses, machetes, etc. that they make not only keep their sharp edges, but their fine blades defy breaking under severe strain.

I once witnessed a Philippine smith bring his bolo knife to an even heat in an elongated charcoal fire, then sink ¼ inch of the curved knife edge into the matching curve of a fresh squash and leave it there to cool. It gave the edge the exact hardness he was after (based upon his judgment as to the proper heat of the steel when sticking it into the squash). The softness in the remainder of the blade graduated from the knife edge to the back. I have owned that very knife all these years, using it for pruning tree branches and for clipping off nail heads in the vise instead of using a cold chisel. The soft steel of the back edge by now has cauliflowered over from hammering on it.

Another time I saw a soft needle-point hardened by heating it in a candle flame. When the very point had a soft glow, it was immediately stuck into a potato!

All over the world we can find endless examples of inventive ways of hardening and tempering steel. All of them, when correctly analyzed, are based on the principles you have now learned to understand.

use $\frac{3}{8}$" to $\frac{1}{2}$"dia. rod for this exercise

Keep small inside curve

step 1

use 1½ to 2lb hammer

step 2

small curve

or

small curve

small curve

small curve

step 3

or

rapid telling blows with 1lb hammer

small curve

keep well away from anvil's edge

corrective blows before continuing

small curve

anvil horn

is kept until final step

7. Making a Right-Angle Bend

Making a right-angle bend is like a warming-up exercise. Just as the pianist practices his scales preparatory to giving a concert, with this particular exercise, the smith sharpens his judgment, hand skill, timing, and coordination. In short, his accumulated know-how is brought into play. After these exercises he will be better prepared to undertake the next forging tasks more successfully.

When a rod is clamped in the vise and bent over with a hammer into a right angle, the *inside* will be sharp but the *outside* of the bend will have a rounded curve.

To bend the rod under a right angle and make both inside *and* outside angles sharp is the problem, and one difficult to master. At first this seems easy to do, but the major effort must be to accomplish the task in a minimum number of heats.

Try to visualize what the hammer blows must accomplish. They must *upset* the steel at the bend while not opening the curve further. Use a light hammer with high-velocity blows, pushing the malleable steel together locally while at the same time reducing the *curve* of the bend somewhat. Too heavy a hammer used with driving blows will not increase the upsetting much and tends to open the curve (which you *must* avoid).

To make a right-angle bend, use a ⅜-inch-diameter rod and a light, 1½- to 2-pound hammer and proceed as follows:

Step 1. Heat the rod locally where the right-angle bend must occur. While working on the bend, do not stray from that location. Cool ½ inch of the hot end in water, leaving hot and malleable only the bend location where the hammer blows will strike. Now bend the rod on the anvil horn as illustrated.

Step 2. With a lightweight hammer, use rapid blows in alternate series, hammering first on one end of the rod and next on the other.

Step 3. After sufficient upsetting has resulted, as shown, and the rod at the bend location has become 1½ times thicker, it is time to hammer only the bulge locally at the bend. Place the bend on the anvil, overhanging the edge but well away from it, as illustrated.

Now use a 1-pound hammer, allowing the "mass" of the rod to counter the blows. The mass of the rod can be increased when it is held in well-fitting heavy tongs. Constantly aim to keep the small inside curve of the bend from becoming sharp. At no time must it be "pushed" together into a fold. If that small inside curve becomes a fold at the bend, not much can be done to save the project. As soon as mistakes are observed, *use corrective blows* to prevent further damage.

130

Step 4. When the outside curve has become sharp, but still is slightly oversize, the whole bend can be safely refined, as the following steps in the illustrations show. Be warned, however, that the almost-finished right-angle bend flush with the anvil's edge may still accidentally be hammered *beyond* that edge. This would drive the overhanging side of the bend down, while flattening the part on the anvil face. All your previous work then is beyond repair.

Examine the common errors closely, as these are shown in the illustrations, so you will be able to avoid mistakes as you work.

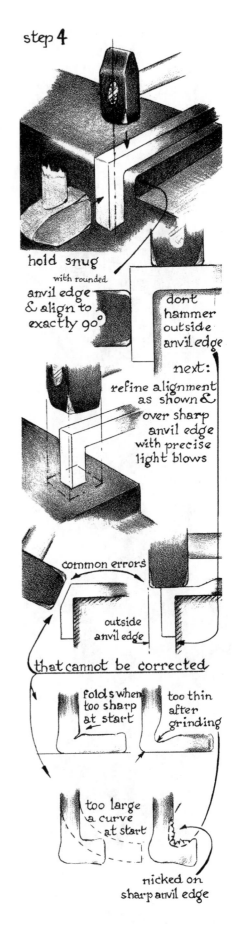

step **4**

hold snug
with rounded
anvil edge
& align to
exactly 90°

dont
hammer
outside
anvil edge

next:
refine alignment
as shown &
over sharp
anvil edge
with precise
light blows

common errors

outside
anvil edge

that cannot be corrected

folds when
too sharp
at start

too thin
after
grinding

too large
a curve
at start

nicked on
sharp anvil edge

131

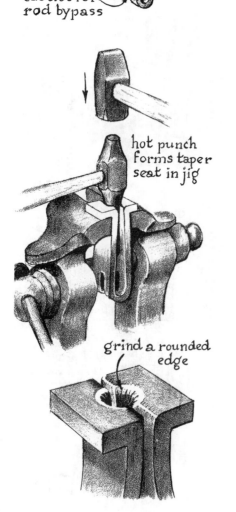

fold

3/8"

4" hot

2"

20"

hammer down
hot over vise jaws

1/2" spacer

cut slot for
rod bypass

hot punch
forms taper
seat in jig

grind a rounded
edge

MAKING A JIG TO FORGE A RIGHT-ANGLE BEND

As stock for the jig, choose a ⅜- by 2-inch leaf spring of a car. Cut off a piece 20 inches in length.

Heat 4 inches in the middle and bend it, as shown, clamped between vise jaws with a ½-inch-thick spacer.

Heat the ends of the jig to yellow heat and clamp it once more in the vise with spacer between. Next, quickly hammer the 1½-inch protruding ends outward over the vise jaws, as shown. This results in ⅜- by 2-inch lips with well-rounded edges.

Next, cut a slot at the bend of the jig for the rod bypass. This can be done on the abrasive cutoff wheel, or with drill and hacksaw.

Reheat the ends yellow hot and place the jig once more in the vise. This time, instead of the spacer, insert a ¼- by ¾-inch hot punch, clamp all firmly together and quickly hammer the hot punch down. This forces a tapered depression into each jaw. The depression can be enlarged by squeezing the vise a little tighter at each repeated heating and further hammering down the punch. Thus enlarged, the tapered depression will take a ½-inch-diameter round rod. Round off (grind) the sharp edges at the top of the hole to prevent angle bends from becoming marred during upsetting.

The jig should open just enough to insert the rod and then be tightened firmly in the vise before upsetting.

A RIGHT-ANGLE BEND USING VISE AND JIG

Upsetting using a vise alone, as in illustrations 1a and 2a, also speeds up freehand forming of a right-angle bend, but often leaves scarred surfaces when the hot steel is hammered over the sharp edges of the vise jaws. Using the special jig, as in 1b and 2b, prevents such scarring, and the upsetting is done as fast.

Once enough volume of steel has accumulated around the bend, refine and sharpen the piece as shown in illustrations 3, 4, and 5.

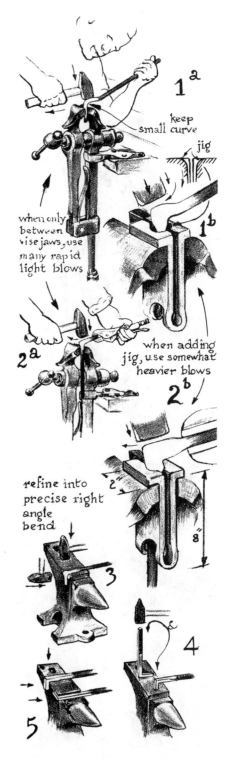

1ª

keep small curve

jig

1ᵇ

when only between vise jaws, use many rapid light blows

2ª

when adding jig, use somewhat heavier blows

2ᵇ

refine into precise right angle bend

2"

8"

3

4

5

keep small inside curve until step 4
use vise-grip pliers if tongs fail to hold firmly
3-4-5 sharpens all edges & aligns to 90° angles

heat mid section.
yellow hot

hammer top
flat under
90°, letting
side bulge

quench L tip, but
keep bulge yellow hot,

&

hold over anvil edge, hammer
bulge down, which upsets
steel locally

1st heat

90°

bend back
at each next step

2nd

3rd

4th

refine all sides & angles
on anvil with 2 lb. hammer

MAKING A RIGHT-ANGLE BEND IN A MILD-STEEL ANGLE IRON

Before you begin, make certain that the angle iron is mild steel. Test it with the file tip after a quench (see Chapter 6). Heat the center of the bar to yellow hot and clamp one flange side of the hot section between vise jaws. Hammer it into a precise 90-degree angle with the vise. During the bending the other flange will bulge outward.

Reheat the *bulge only* and quickly quench the flange with the finished right angle halfway in the water, leaving the bulge hot. Immediately place the finished angle over the anvil's 90-degree corner, as shown, with the hot bulge resting on the anvil face. Hold down firmly with visegrip pliers. Using a 1½- to 2-pound hammer, pound the bulge down with rapid, high-velocity blows wherever smaller bulges keep forming.

Even though you have cooled the finished right-angle flange, it will not be rigid enough to resist *some* bending under the strain of flattening the bulge. Therefore, reheat the whole bend and clamp once more in the vise and correct that right angle should it have opened up somewhat. Again, dip-quench that flange only and replace as before on the anvil to further compress and flatten the hot bulge as needed.

When you have arrived at step 4 you will notice a slight curve remaining where the two inside flanges meet. This can be forged into a sharp right angle over the anvil's edge if your plan requires it.

Bending the angle iron over the anvil only, *without* using the vise, will make *both* sides bulge out, but only half as much. One half-size bulge must then be flattened, however, and the other hammered freehand at a 90-degree angle. It can be done, but it is a struggle because of the constant corrective hammering necessary to realign the workpiece.

It is therefore good practice to cool one bulge first in order to keep it rigid while upsetting the hot one. Then alternate the procedure when correcting the other bulge.

MAKING A RIGHT-ANGLE BEND IN A HIGH-CARBON-STEEL ANGLE IRON

If the angle-iron bar is of high-carbon steel, and the vise is used as an aid, the cooling of the exact 90-degree flange should be restricted to a ½ second in-and-out quench in concentrated hot brine instead of water. It is safer against cracking, since in hot brine the steel cools more slowly. This very brief cooling will stiffen that side considerably without brittle-hardening it. Then finish up as in steps 1 through 4.

Note: Do not cool high-carbon steel by a prolonged quenching while making a right-angle bend. If you fear that brittleness has resulted at any given spot, anneal it before proceeding. If a brittle section is a few inches distant from hammer blows during forging, the steel will break off like glass.

8. Some Tools that are Simple to Forge and Temper

A COLD CHISEL

One of the most useful tools to have in a metal-crafts shop, the cold chisel is one of the easiest to forge.

Use as stock a ¾- to 1-inch-diameter high-carbon-steel bar, round, hexagon, octagon, or square.

Heat ¾ inch of the end to a dark yellow glow. Hold it on the anvil face at the angle you want the chisel's bevel to be (a blunt angle for heavy-duty work and the cutting of hard metals; a sharp angle for lighter, delicate work and the cutting of soft metal).

In finishing the tapered cutting end, move it to the anvil's edge (1). This allows the flat face of the hammer to bypass the anvil's face while refining the almost knife-sharp beveled end with light tapping and prevents damage to the anvil should the hammer miss.

Cut off the desired length of the chisel (see use of cutoff hardy, page 106). Next, grind the bevel to a finished state (2), and temper both ends as described in Chapter 6.

THE CAPE CHISEL

This is actually a very narrow cold chisel. The conventional design allows the cutting edge to be *wider* than the flattened bridge. Therefore, when cutting with this chisel (a key slot in a shaft, for instance), the cutting edge will not bind. The strength of the chisel is great because the flat bridge preceding the cutting edge is wide.

Cape chisels are tempered in the same way as cold chisels.

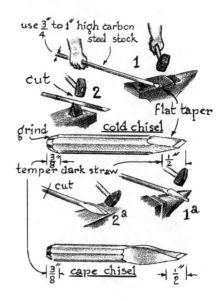

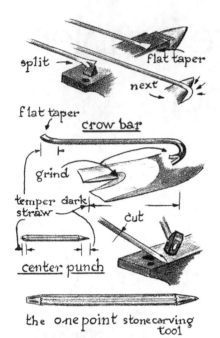

split

flat taper

next

flat taper

crow bar

grind

temper dark
straw

cut

center punch

the one point stonecarving
tool

tongs in
left hand
are to
twirl
around
&
around
during forging
the cup end.

the hardened sharp crater edge
cuts into
soft steel
stone carver
hammer

cup-end of tool

anvil edge
flush with cup-end while forging

A CROWBAR

Forge the end of a ¾-inch, high-carbon-steel bar to a slender bevel about 1½ inches long.

Reheat and split ¾ inch of this end on the cutoff hardy, and grind the forked end as shown. Reheat again for a length of about 4 inches and bend that section into a curved claw over the anvil horn. The claw end is used to pull nails out of wood, but it can also withstand the maximum strain when used to pry with as well.

The other end of the crowbar is forged in the same way as the cold chisel bit and bent only a little so that it can be used to pry with. Each end is to be drawn to a dark bronze temper color, while the remainder of the bar is kept annealed.

A CENTER PUNCH

Although this tool is easily forged, it can instead be *ground* into its final shape if your motor grinder has a large coarse stone which cuts steel rapidly. Temper as a cold chisel.

A ONE-POINT STONECARVING TOOL

This tool resembles a center punch and is made similarly. However, the end that is struck with a mild-steel hammer can be forged with a sharp-edged "cup" shape. This sharp, hardened edge bites into the hammer and prevents it from glancing off.

To make the cup-shape, stand at the anvil as shown, using the hammer and tongs as the arrows indicate. Temper both ends as a cold chisel.

136

9. Decorative Treatment: Rosettes and Wallhooks

DECORATIVE ROSETTES

Decorative rosettes can be made from all sorts of small steel-scrap items. The examples shown here are discs (slugs) of steel like thousands ejected by presses that punch holes in steel plates.

When these slugs are heated *singly* in the forge, they too easily slip down to the fire grate; to retrieve them will upset the fire. Therefore, to heat such small items, put them in a one-pint tin can and place it in the middle of the forge fire, deep enough to have hot coals come up to half the height of the can. All of the slugs will then become equally hot. Do a whole batch at a time if you plan to make several rosettes.

With ⅛-inch thin tongs, pick a hot slug out of the can and place it on the anvil. Heavy pounding with the flat of a 3½-pound hammer thins and flattens it. The slug's diameter is now considerably larger. Place it in the little mound of ashes and cinders *below* the forge to anneal. Treat all hot slugs in this way if they are of high-carbon steel (annealing is not necessary if slugs are mild steel).

After annealing, drill a hole in the center of each disc to fit the thickness of the nail to be used in it.

Next, make a heading die for the texturing of the rosette, as shown. (See also illustrations in Chapter 5).

In this case only the small hole to fit the nail and the larger clearing hole below are drilled.

Place the die in the hardy hole. The heated flat disc first receives the cold nail, which nail in turn is held by a pair of tongs. Together they are placed in the die hole. Once nail and hot disc are in position, use the special large double-ball hammer to deliver well-aimed, forceful blows to shape as well as texture the rosette.

Flip the hammer over in mid-air, using either the small or large ball to give the rosette the desired texture. In this action the nail head becomes embedded in the hot center, automatically seating it to perfection.

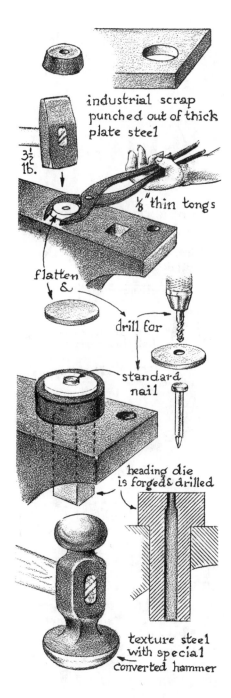

industrial scrap punched out of thick plate steel

3½ lb.

⅛" thin tongs

flatten &

drill for

standard nail

heading die is forged & drilled

texture steel with special converted hammer

Patination

Prepare the rosettes for patinating with oxidation colors as follows:

With the power sander, gently smooth all textured ridges with a small, flexible, rubber-backed fine-grit paper or cloth disc abrasive. A rubber abrasive will do also. Once the outer ridges shine like a mirror, polish them lightly on the buffer. In this process the "valleys" are hardly touched, so the natural black forging oxidation is kept intact. Finally, clean the rosette with a solvent to remove all wax residue and dry with a clean rag.

To draw the oxidation colors, hold the rosette with the thin tongs and heat it in the hot core of a gas flame, as shown. Soon the first color appears — a light straw yellow. If color moves too fast, hold the rosette higher above the flame. For variety, you can hold *only* the edge in the heat core. It will become dark bronze, then purple, and finally blue. Each rosette then can be variously colored to suit your taste.

Color the textured nail head separately, or simply polish it on the buffer to remove the forge-black. It will then look like a silver button in the darker-colored rosette. Or, if the nail head is patinated a straw yellow, it will shine like gold surrounded by rainbow colors. Such rosettes can be made into drawer pulls and other artifacts.

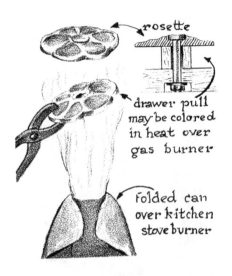

rosette

drawer pull may be colored in heat over gas burner

folded can over kitchen stove burner

Other Simple Decorative Forgings Made from Scrap Steel

Discarded automobile-engine valves are ideally shaped for making rosette spikes, as the illustrations show.

Although this steel is very hard, it is not temperable or hard enough for making cutting tools. For small decorative items, these valves must be heated to a very hot light-yellow-to-white before this special composition of steel will become malleable enough to yield to heavy hammer blows: We are dealing here with a type of steel that is made to stay *hard* while hot (the engines in which they are used create such heat). This is the reason why you must heat valves to white heat for forging.

Using your own imagination, you can make many other articles from such valves.

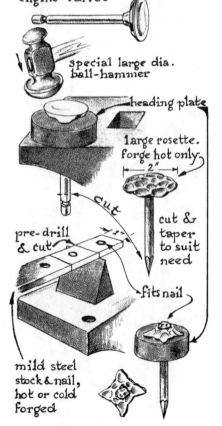

simple decorative forgings from ready scrap-steel items

discarded worn automobile engine valves

special large dia. ball-hammer

heading plate

large rosette. forge hot only. 2"

cut

pre-drill & cut

cut & taper to suit need

fits nail

mild steel stock & nail, hot or cold forged

Small rosettes are easily made in quantity from rectangular bars and nails. On a 20-inch piece mark off a series of squares and drill a hole in the middle of each one. Cut each square section *almost* through. Heat the first one to yellow heat and slip a cold nail into the hole, then quickly place the assembly flush with the heading plate with the nail in it.

Tear off the cold bar from the heated square and immediately hammer out the rosette to your taste. The nail head automatically becomes textured. Patinate the whole finished piece.

The old-fashioned wood stove lid-lifter made from a bolt is another example of how many ready-shaped items on the scrap pile can be translated into entirely different forged forms.

The same bolt, used inventively, can undergo an entirely different treatment and become a wall bracket for hanging a flower pot or a light fixture — or something else you may need or wish to create.

standard $\frac{3}{8}$" bolt for lid-lifter

forge hollow rounded & flatten

thread texture leaves good grip

flat end fits stove lid

flatten bolt head & texture underside with ball-hammer that fits.

Wall hanger for flower pot or light-fixture

decorative curving of steel
without aid of jigs

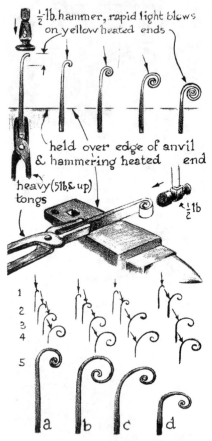

½ lb. hammer, rapid light blows on yellow heated ends

held over edge of anvil & hammering heated end

heavy (5 lbs. & up) tongs

←½ lb

1
2
3
4
5

a b c d

a-b-c-d designs need 1 to 5 heatings to complete. Arrows show direction of hammer blows

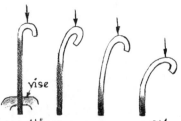

vise

resulting curves vary with heat range, steel size & shape,

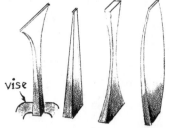

vise

direction of hammer blows & weight of hammer

FREEHAND CURVING OF STEEL

The freehand technique of curving steel is a challenge and a pleasure for the smith. By this means he shapes bars of hot or cold malleable steel into beautiful curves. In contrast, when fixed pattern jigs are used, a machine-like character is injected: the true blacksmith wants to avoid this. Such pre-arranged jigs are used to bend cold, evenly milled steel bars. In the modern gates, grills, and panels sold as "wrought-iron" work, the mechanical quality becomes evident.

No two hand-rendered curves are quite so precisely alike that they are exact duplicates. When bars are heated and tapered at the ends, as shown, they can be curved into a great variety of beautiful designs. In each step, the curving metal is affected by several factors: the weight of the hammer, the direction of the blows, the cross section of the steel where it bends, and how malleable (hot or cold) it is at a particular spot. With freehand skill, then, you can aim for a *combination* of these variables, learning as you work just which will give the most satisfying results. Examine and analyze the illustrations: they are guides to the endless possibilities in this type of workmanship.

A DECORATIVE WALL HOOK

To make a decorative wall hook, cut 6 inches from a round or square ⅜-inch-diameter rod and forge the end into a taper about 3 inches long (step 1).

Heat 1 inch next to the taper. Clamp in the vise between round-edged insets (step 2), bending and upsetting the hot section.

Or, you may use an *upsetting die* (made as illustrated in Chapter 5), which speeds up the shaping of the hook (see drawings 3 and 4) and setting a "head" on top of the tapered end. Hold this head in tongs while drawing out the rod end (step 5).

Further forging of this portion can be varied to suit the taste of the smith, who must now visualize the final product he has in mind.

If the hook end is to be curved, as shown, it calls for gradual widening while at the same time thinning it toward the end. You will start with a tight curve which progressively becomes a more open curve toward the thicker steel (7 and 8).

Texture the spike head with the ball peen (6).

The final surface finish can vary from a simple steel brushing and rubbing in of linseed oil to any combination of treatments for patination (see Chapter 9).

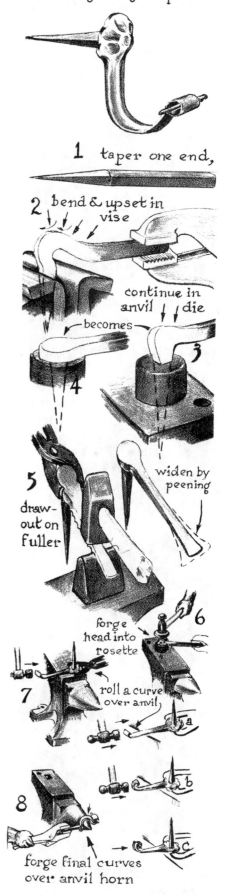

decorative wallhook
with right-angle spike

1 taper one end,

2 bend & upset in vise

continue in anvil die

becomes

4

3

5 draw-out on fuller

widen by peening

6 forge head into rosette

7 roll a curve over anvil

a

8

b

c

forge final curves over anvil horn

10. Hinges

making a hinge joint
without machining

cut hot

1

or

2

3

spread
over bottom
fuller

&

4

after on anvil
horn

5

bend together
& space over plate

6

anvil

7

curl hot,
free hand

or

8

refine with

9

forming
pin

10

other hinge half

11

use forming
pin & spacing plate for fit

The more skilled you become in blacksmithing, the more you will realize that there are several ways to forge an article. A hinge, in particular, lends itself to inventive design and forging once you have understood how it must work and how to make it.

To describe all the possible types of hinges would fill a book. Therefore, only a few of the most frequently used kinds are offered here as examples. Once you have made these, you will be prepared to meet successfully whatever hinge problem comes your way.

MAKING A HINGE WITHOUT MACHINING

Select a 16-inch-long bar 2½ inches wide by ⅛ inch thick. Heat 3 inches at the end and make a split 2½ inches long with the chisel head (1), cutting it on the soft anvil table (or on the anvil face after covering it with a protective mild-steel jacket). The bar can also be split, as shown (2), on the cutoff hardy.

Reheat yellow hot and spread the branches. Round off the sharp crotch, first on a bottom fuller (3), and further on the anvil horn (4).

After the fork has been opened wide, bend the first branch (5). Bend the second branch parallel to the first. With the aid of a spacer, adjust the two precisely (6). If the thickness, width, and length of the branches become somewhat uneven, refine them freehand on the anvil face. Once more, check for correct spacing between the two.

Heat one branch and curl it freehand (7). Then curl the second branch.

For a temporary hinge pin that will serve for forming, select a piece of round rod ⅜ inch in diameter. Curl both branches completely around it (8). Further fitting is done on the anvil face or over its edge to complete the hinge-pin seating (8) and (9).

Heat the second hinge half and forge it to fit the space between the branches of the first one. Curve it hot over the round rod also (10).

Cut the correct length from the rod for a permanent hinge pin. Rivet a small head on one end to keep it from falling out of the assembled hinge.

Heat both finished hinge halves and assemble with the permanent, headed hinge pin. If the assembly has become somewhat unaligned, the still-hot malleable parts will yield easily to many rapid blows; the self-seeking alignment over the cold pin will "set" the two hinge elements.

While the hinging area is still visibly hot, work the hinge blades back and forth to ensure easy movement.

ORNAMENTAL HINGE DESIGNS

Once the hinge halves have been assembled and work properly, the hinge blades can be forged decoratively to suit the particular areas they must fit (doors, lids, gates, etc.). These hinges are made flat and are bolted onto flat wood.

With visegrip pliers, clamp the hinge blade onto the adjustable steady-rest bolted to the anvil stump. Use a cold-chisel head to cut the hot steel. The soft anvil table allows the chisel to cut clear through the bar as shown. If *mild steel* (not more than 3/16- to ¼-inch thick) is used, it can be held between the bench-vise jaws and cut cold in a *shearing* action with a sturdy cold chisel.

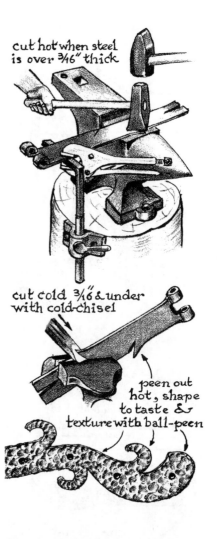

cut hot when steel is over 3/16" thick

cut cold 3/16" & under with cold-chisel

peen out hot, shape to taste & texture with ball-peen

Heat the pointed prongs locally and bend them temporarily sideways so that you can reach them easily with the hammer. Forge the desired curves and decorative pattern you have in mind, using as many heats as you need.

You may have a preconceived design in mind, but often the curves that result *naturally* during bending, peening, and flattening become unexpectedly more attractive, and you should feel free to improvise during the successive steps. The surface textures that also result automatically during the forging of the hinges are attractive in themselves. If you wish to apply added texture on the finished piece, you can deliberately do so using hammers with cross peen *or* ball, or cross peen *and* ball. It is best not to overdo this, as it may then lose its original appeal.

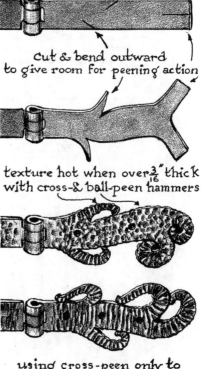

cut & bend outward to give room for peening action

texture hot when over 3/16" thick with cross-& ball-peen hammers

using cross-peen only to shape & texture.

upset for shoulder & tenon

upset to form head

forged from 5/16" x 2" stock

using 3/4" rod

rustic wood door & wall.

bend hinge arm hot over slab forms

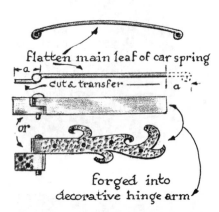

flatten main leaf of car spring

cut & transfer

or

forged into decorative hinge arm

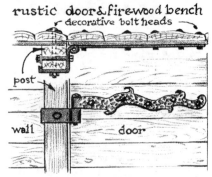

rustic door & firewood bench

decorative bolt heads

post

wall

door

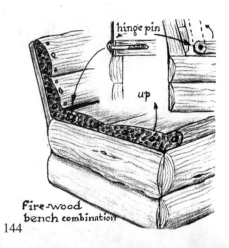

hinge pin

up

Fire-wood bench combination

A GATE HINGE

This hinge design can also be used to hang rustic doors and is both strong and practical. The bolt head acts as both bolt part and hinge-bearing socket.

Upsetting the head on a l-inch-diameter rod will give enough volume to shape the shoulder, the hinge-bearing socket hole, and the decorative head end as well.

After this part has been forged, as illustrated, draw out the remaining section of the rod into a ¾- to ½-inch-diameter bolt shank. Thread it at the end so that, with a nut and washer, it can be used to tie wall and post together also.

Drill the hinge-pin hole, either partly or all the way through the bolt head.

Several greased washers placed underneath the hinge shoulders make smooth bearing surfaces. Adding, or removing, a washer makes it easy to adjust the hanging of the door accurately.

To locate the exact hinge-bolt positions, hang a plumb line along the door post and scribe off the correct heights, one above the other, for drilling the bolt holes.

Adjust the hinge bolts inward or outward to hang the door accurately, relative to the true vertical alignment.

Next, assemble top and bottom hinges, place door in its allotted space, and scribe off on it the exact hinge-bolt locations for the fastening of the door hinges.

I have installed several such hinge arrangements, fitting irregular wood-slab doors to slab posts and walls, and have found them to be about as easy to place as conventional door hinges. If it becomes necessary to remove such a hung door, it is easily accomplished by lifting it in an upward movement out of the bolt head sockets when the door is in *open* position.

A HINGE MADE FROM A LEAFSPRING

This hinge makes use of the existing calibrated and curled-over hinge ends of the main leafspring of a car.

Heat the leaf and flatten it. Anneal and cut it into two sections, one short and one long, as illustrated. Make a headed hinge pin that slides easily in the leafspring hinge sockets.

After the two arms of the hinges have been forged decoratively, assemble the two parts with the hinge pin. Place in position over door and post and mark off the location of the bolt holes. The fastening bolts, used for such special hinges, can have decorative hammering on the heads as well.

In a variation of the foregoing gate-hinge design, one element is forged to fit *around* the post (see the cross section). Used with a tie bolt, it clamps wall and post together. This is an example of the opportunity you have to design hinges to suit special situations.

HINGE FOR A WOODBOX BENCH

This practical hinge design is used on a woodbox bench, as illustrated. A long hinge pin, driven into the wood of the bench, secures the hinge arm at the back of the bench to the end of the other hinge arm that binds the box lid together. The opposite hinge of the lid is placed in approximate alignment with the first hinge pin. This box lid acts as a bench seat as well.

11. Hold-Down Tools

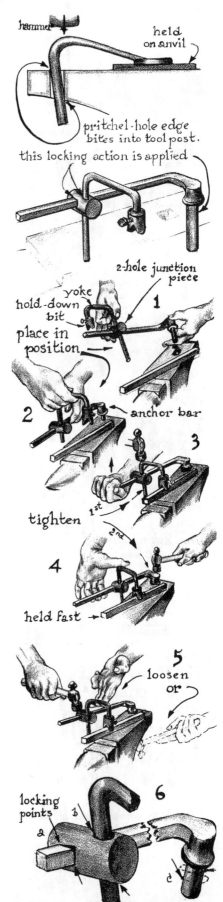

Using a hold-down (or hold-fast) tool to steady a workpiece on the anvil leaves both of the smith's hands free to work. This tool replaces the apprentice assistant who was the standby helper in former times.

Hold-down tools are based on a combination of bending (leverage), friction (locking), and twisting (torsion). A careful analysis of the illustrations will reveal that the fit of the three parts in the hold-fast shown is purposely made *loose*. Once their contact points bite in during the twisting action, the tool and workpiece will be locked together. Therefore, a few light taps of a 1½-pound hammer will hold the workpiece firmly down on the anvil.

HOW TO USE THE HOLD-DOWN TOOL

1. Slip the anchor-bar footing *halfway* into the pritchel hole.
2. Slide the junction piece out of the way from the hot part of the workpiece.
3. Hand-lower the yoke and bit onto the workpiece while driving it down further through the junction piece with a 1½-pound hammer, as much as the tension in the assembly will allow.
4. In this position, the anchor footing in the pritchel hole is rammed down flush with the anvil in a final cinching.

With a little practice these adjustments will take only a few seconds. And now all forging on the workpiece can be carried out during one heat. The smith, with both hands free, can swing a sledge or manipulate a hot punch, a hot chisel, a flatter, a set hammer, and so on.

To loosen the assembly, tap *downward* on the junction piece, or tap *upward* on the anchor footing below the pritchel hole.

The *locking points* in illustration 6 show clearly the principles in the foregoing.

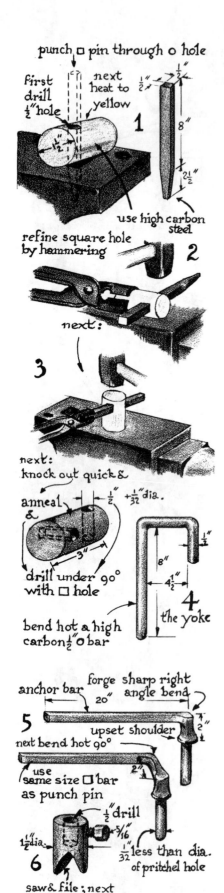

punch □ pin through o hole

first drill ½" hole

next heat to yellow

use high carbon steel

1

refine square hole by hammering

2

next:

3

next: knock out quick &

anneal &

½" + ½₃₂" dia.

drill under 90° with □ hole

bend hot a high carbon ½" □ bar

4 the yoke

anchor bar

forge sharp right angle bend

20"

upset shoulder

2"

5

next bend hot 90°

use same size □ bar as punch pin

½" drill

⁵⁄₁₆

½₃₂" less than dia. of pritchel hole

1½" dia.

6

saw & file; next harden by quenching in oil at light cherry red heat

HOW TO MAKE AN ADJUSTABLE HOLD-DOWN TOOL

First prepare a square punch-pin tapered at one end from a piece of high-carbon steel, ½ x ½ x 10½ inches.

Cut off the 3-inch-long junction piece of the hold-down from a salvaged car axle 1½ inches in diameter. Anneal it and make a square hole in it as follows:

Drill a ½-inch-diameter hole through it as shown (1).

Now heat the junction piece yellow hot around the hole (2). Place it over the hardy hole or the pritchel hole of the anvil, and quickly drive the square punch-pin through with a 4-pound hammer. Without a moment's loss, use a 2-pound hammer to pound *around* the inserted pin to bring the hot steel in closer contact with the pin's square sides.

During this action, to prevent the pin from heating up to the point of malleability, hammer it down progressively further through the hot junction piece and, without stopping, refine the fit of the pin into the square hole (3). After the junction piece has lost its forging heat, knock the punch-pin out with one of a smaller size kept handy for this purpose.

At right angles with the square hole, drill a 17/32-inch-diameter round hole, as shown. The ½-inch-diameter yoke bar (4) must fit into it loosely when assembled. The yoke bar is bent hot and left annealed to remain "springy."

To make the anchor bar (5), cut a ½-inch-square cross section high-carbon steel bar 28 inches long. Upset it 5 inches from the end to form a ¾-inch-diameter shoulder. This shoulder functions as a "stop" when the hold-down assembly is hammered down flush with the anvil surface to lock the workpiece in place.

Offset the rest of the square anchor bar 2 inches in order to overhang the anvil edge, well out of the way of most workpieces held fast by the tool.

The hold-down bit can be made with a flat serrated surface *or* with a V slot (as illustrated in drawing 6) to straddle a workpiece. Use a ⁵⁄₁₆-inch setscrew to secure it to the yoke end.

All dimensions for the hold-fast tool are approximate, to fit an average 100-pound anvil. You may have to adjust them to the anvil in your shop.

Note that the fit of the square anchor bar into the square hole of the junction piece is purposely a *loose* one.

All three loose fits of the locking points permit easily adjusted tool positions before the assembly is hammered tightly onto the workpiece. It is locked in position by the twisting action of the tools' parts which bite into each other during such twisting. If the hold-down's parts were to fit *precisely*, the locking action could not occur and all would become undone instead of holding the workpiece fast onto the anvil.

12. A Fireplace Poker

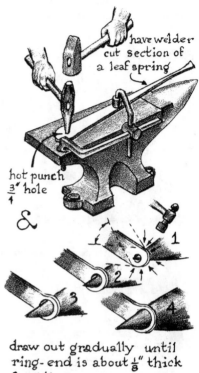

have welder cut section of a leaf spring

hot punch $\frac{3}{4}$" hole

&

1

2

3

4

draw out gradually until ring-end is about $\frac{1}{8}$" thick & 2"dia.or more

The stock used for this lightweight poker comes from a leafspring of a car. Such a poker can be used easily by those who find many fireplace tools too heavy and unwieldy. Made from this tough steel, it is strong enough to move and pry heavy logs in the fire. The tempered and knife-edged lip is useful for scaling charred wood off burning logs.

Start by making the hole for hanging the poker. With a section of leafspring cut by a welder and held with the hold-fast tool, drive a tapered hot punch through the hot end of the steel, enlarging the hole as much as the taper allows. Enlarge it further by drawing it out over the anvil horn until a thin ring about 2 inches in diameter has been formed (illustrations 1–4). The round ring may be reheated and bent into free-form pattern if you prefer.

Caution: Always stop forging high-carbon steel as soon as its *visible* heat glow disappears. Coming in contact with the cold anvil, the thin, hot, high-carbon steel quickly cools down. This process resembles quenching, leaving the steel brittle. Therefore, you must heat the steel often and for as many heats as you may need to finish the blank. At the same time you must be careful never to overheat the steel because if brought to white heat, the metal will burn.

Next, draw out the handgrip section. Later, two hardwood hand pieces will be fitted to it.

Bend the end lip as shown. If instead you prefer a point and prong at the end, proceed as in the next illustration.

The lip or prong of the poker is tempered a bronze color. The two wood handle sections are riveted to the steel (see procedure for the fireplace shovel, page 172).

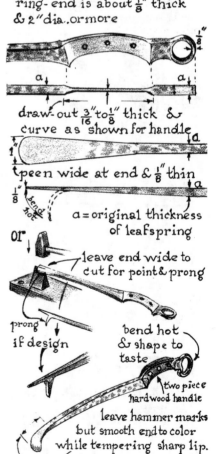

draw out $\frac{3}{16}$" to $\frac{1}{8}$" thick & curve as shown for handle

peen wide at end & $\frac{1}{8}$" thin

a = original thickness of leafspring

or

leave end wide to cut for point & prong

prong if design

bend hot & shape to taste

two piece hardwood handle

leave hammer marks but smooth end to color while tempering sharp lip.

rivet wood sections & rasp, file, sand-down·polish

13. Fire Place Tongs

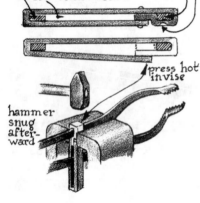

To make a pair of fireplace tongs without hinges, cut a 60-inch length of ⅜- to ½-inch-diameter high-carbon steel stock. At yellow heat, forge the middle 8-inch section, peening it out to a width of 1½ inches. This will leave you about 1/16-inch thickness there.

Smooth and *hollow* the full 8 inches lengthwise at a radius of about 1 inch. This will stiffen its spring-action when in use. This spring-action allows the tongs to open and close, thus replacing the conventional hinge.

Flatten the ends of the rod and forge them spoon-shaped. File teeth along the rims so they will grip charcoal or wood in the fireplace or stove. Flatten the remaining parts of the rod to a 90-degree angle with the middle and end sections. The ⅜-inch-diameter rod should now become approximately ¾ x $^5/_{16}$ inches. (Or, if you are using ½-inch rod, it should become 1 x $^5/_{16}$ inches.) This stiffens the tong arms to prevent their bending.

Next, at dark yellow heat, forge the curve of the tong arms as suggested by the illustration. Bend the hot middle section into a 1¼-inch radius hollow swage, using a ball-peen hammer. Then curve the same section progressively until an evenly smooth bend results.

Without a guide strap, the ends of the tongs would spread open too wide. With the slotted guide strap, the jaws are restricted to a certain maximum opening. It also prevents them from bypassing each other when squeezed together. Examine the illustration carefully to make and assemble this strap.

To make the hardwood handles, cut the wood pieces ¼ inch wider than the steel it will attach to. Make in it a slot only as deep as the width of the steel over which it is slipped. Fasten the wood piece in place with three ⅛-inch rivets and countersunk brass washers. Or, if you prefer, make the wooden handle in two pieces as with the shovel, page 174, or the poker handles, page 147.

Lightweight brazier tongs (without hinges) can be made from the wide band steel of the type used to package stacks of lumber or other extra-heavy merchandise for shipping. Literally, miles of this steel are discarded daily at destination points. These tongs should not measure more than 18 inches in length (unless they are to be used to reach for hot coals in a fireplace). The longer they are, the deeper the stiffening grooves must be forged into the arms to prevent them from bending in use.

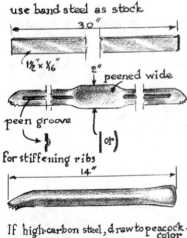

14. A Spatula Made From a Section of Coil Spring

A thin, flexible spatula is very useful for work in plaster casting, paint mixing, spreading glue, and many other jobs.

Spatulas and similar light-gauge tools can be made from stock of approximately ¼ to $5/16$ inch diameter. Good temperable steel of this size can be salvaged from discarded coil springs used in garage doors, garden swings, and many other items. The springs can be handled best when cut into 5-inch sections on the cutoff wheel.

I have invented a practical gadget to unwind these coil sections into straight stock. (See illustration, and also see photograph of this device on page 183). Bring the 5-inch section cut from the spring to a yellow heat in not-too-hot a fire. It is *very important* that during the heating you keep turning this short section over in the fire to make sure the heat is *evenly* spread. Lock a visegrip pliers firmly on ½ inch of the end of the coil, and slip that hot coil over the reel. With one quick and forceful pull, the limp steel will unwind into a straight piece. (If the coil has been heated unevenly, it will unwind as a wavy rod and will have to be straightened later.)

A simpler coil-straightening device can be made by using a ¾- x 10-inch-long well-greased bolt clamped in the vise. Over the bolt, place a section of 1-inch plumbing pipe. The easy turning pipe will function as a reel when a section of hot coil spring is slipped over it and then pulled off.

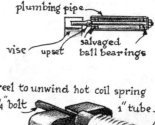

plumbing pipe

vise   upset   salvaged
ball bearings

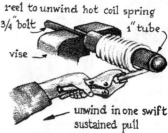

reel to unwind hot coil spring
3/4" bolt   1" tube
vise

unwind in one swift
sustained pull

slip evenly
heated coil over
reel and pull out
in one quick draw

1

heat 2½" yellow hot
& flatten in 1 heat
with heavy hammer

2 **3**

trim contour
on grinder

&

graduate
blade thickness
evenly

4

heat to dark cherry red
& quench vertically
in oil

OIL

5

cool constantly
while

next:
polish &
draw color first
here

7

6

grinding to
final thinness
without losing
hardness

next

finally draw blade
to center of heat, held
still higher till color

becomes
even
purple

8

farther away
from flame
to slow color run

9

giving blade spring
flexibility

To make the spatula use ¼-inch-diameter stock from an uncoiled spring.

1. Saw off a 12-inch length and heat 2½ inches at the end. Flatten it to $\frac{1}{16}$-inch thickness with a 4-pound hammer, using rapid, forceful, and accurate blows. Do it in *one heat* if possible.

2 and 3. Grind the forged blade to the shape you want. Clean-grind each side but keep the blade to that $\frac{1}{16}$-inch *even thickness*.

4. Reheat the blade to dark cherry red, then, holding it vertically, quench it in oil. If the blade is an even $\frac{1}{16}$-inch thick, it should emerge without warping. If it is irregular in thickness, or is paper thin, warping may result, requiring reheating and realigning. This time grind very carefully to maintain an even thickness. Measure for accuracy with the calipers if there is any doubt, before brittle-quenching once more. Now the blade should emerge from the quench perfectly aligned and brittle-hard.

Illustrations 5 and 6 show the blade being ground thinner. Try not to lose the brittleness through overheating. Hold a finger lightly on the blade while grinding; if it becomes too hot to touch, interrupt grinding momentarily, holding the blade $\frac{1}{16}$ inch from the stone to allow it to cool in the air thrown off by the stone as it revolves. Repeat this finger-testing and cooling method as often as necessary until blade is as thin as you want it. (This is one time you could wish to have an old-fashioned watered grindstone moving at slow speed. Such a stone will never heat up the steel.)

Once the blade is thin enough (experience will tell you later how flexible you can make it), it can be tempered as shown in illustrations 7, 8, and 9.

15. A Door Latch

This door latch calls for the forging of three keepers that are riveted onto the base plates, and a latch bolt which slides through the keepers.

To make a keeper without a jig is possible, but cumbersome. With a jig, several of them can be made quickly. Two types of jigs are described here. The third method described is a compromise.

Keepers, in one shape or another, are required for many different projects. Therefore, the time taken to make these jigs for them is not wasted, since they are welcome accessories for the shop.

FIRST METHOD: A SLOTTED JIG

Use a section of steel salvaged from a leafspring of a heavy truck, approximately ⅜ inch thick, 3 inches wide, and 18 inches long. Straighten and anneal the spring.

With the abrasive cutoff wheel, make two lengthwise slots in the bar, leaving a middle strip as shown. (The thickness of the keeper stock must be equal to the thickness of the cutoff wheel that cuts the slots.) This section is bent out (hot or cold) to make space for the heated keeper to be inserted under it.

The keepers are partially forged into a decorative shape. Hold the jig down on the anvil and insert the yellow-hot keeper. Immediately press down the set hammer on the center of the jig above the keeper and deliver a hard blow with a 4-pound hammer. This forces the keeper's center uprights and sides into its final shape in one stroke.

The jig's inherent spring-steel resilience makes it unnecessary to harden or temper its "working" end. The only possible trouble it might give is if the stock for the keeper is too thick, which could force the jig's sides outward.

SECOND METHOD: A SPRING-ACTION DIE SET

In this method a separate small die *(a)* is cut out with a saw from a solid, high-carbon steel bar. Or you may forge it into the desired shape. File it to the exact size of the latch bolt and rivet it onto the end of a leafspring of a car. As a rule these springs measure about 2½ to 3 inches wide and ¼ to ⁵/₁₆ inch thick. Cut the jig about 30 inches long.

At the other end of the spring make a slot wide enough to leave room for the thickness of the keeper on each side of the die's boss. The 6-inch middle section of the leafspring, prepared in this way, is heated about 4 inches in its center and bent over, as shown.

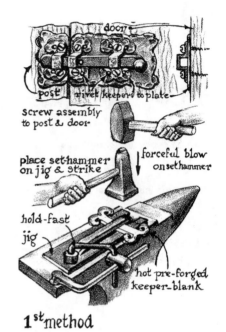

door
post — rivet keepers to plate
screw assembly to post & door
place set-hammer on jig & strike
forceful blow on set hammer
hold-fast jig
hot pre-forged keeper-blank

1ˢᵗ method

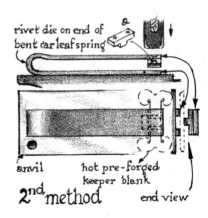

rivet die on end of bent car leafspring
anvil
hot pre-forged keeper blank

2ⁿᵈ method   end view

This die set can be placed free on the anvil face when you are using a set hammer. Press the set hammer down on the end of the jig, forcing the die down onto the hot keeper blank, before striking it with a 4-pound hammer.

If you do not want to use the set hammer, clamp the hot keeper between the jaws of the jig as if the jig were a pair of tongs, and place it squarely over the middle of the anvil. Hammer the die ends together with the flat face of a 4-pound hammer.

THIRD METHOD: A VISE AND A BAR

Open the vise jaws to the width of the bolt plus *twice* the thickness of the keeper stock. Use a bar as thick as the bolt to force the keeper blank down onto a prop below. Once the hot keeper is forced down in this way, remove and reheat it. Clamp the now folded keeper firmly on the bar in the vise to the depth of the keeper. Both your hands are now free to spread the two ends of the keeper outward with a chisel wedge. Hammer the ends down flush with the top of the vise jaws.

An alternate method is to replace the bar used in the above description with the same stock as that used for making the latch bolt. Place that bar over the hot keeper which straddles the open vise jaws, and simply hammer it down. This action automatically results in the correct final shape of the finished keeper.

Many other methods are possible, depending on your available equipment and your ingenuity. The guidelines to improvisation should be to consider *many* methods instead of only the suggested ones.

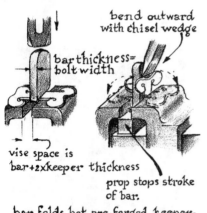

bend outward with chisel wedge

bar thickness= bolt width

vise space is bar+2×keeper thickness

prop stops stroke of bar.

bar folds hot pre-forged keeper-blanks between vise-jaws

3rd method

MAKING THE BUTTON ON THE DOOR-LATCH BOLT

Using a bar that fits the size of the keeper, upset it at one end to get as much volume as possible. Several heats will be necessary to gain enough material to commence the shaping of the spherical button over the jig and the horn. Use a fairly light hammer (½ to 2 pounds), alternating the strokes between the flat face and ball-peen ends.

The knob end of the bolt may be curved outward to suit your taste. Keep in mind that it should be easy to operate.

clamp hot bar end in vise to upset

use 2 hammers to keep aligned

hold heavy hammer on one side & straighten with light hammer

clamp between jig with rounded edges & upset more

vise

form over rounded edge of anvil next

form button & throat over anvil horn

bend throat

leave hammer-mark texture

or

vary design to suit taste

16. Making an Offset Bend in a Bar

standard
one piece
hold-down
tool

3 to 4 lb

set-hammer
or flatter

yellow hot

3/8"   3/4"

high-carbon steel jig

clamp in vise

offset

or
hammer

anvil

twist

forge the
bending forks to fit
154 hardy hole in anvil

Often offsets must be forged in special hinges, door-bolt receptacles, wall hangers, straps that join boards of different thickness, and so on. To forge an offset bend, proceed as follows. Heat the portion of the bar to be offset to yellow hot and place the bar on the cutoff table so that portion overhangs the table. Fasten the cold part to the anvil with a hold-down tool.

Place a set hammer or a flatter on the hot overhanging part and, using a 4-pound hammer, strike it with one or two heavy blows. This offsets that portion accurately and simply. To increase or decrease the amount of offset or jog, build up the anvil face or the cutoff table with plates of various thickness.

Another method of offsetting is to use a jig, as illustrated, which allows you to shape the piece without using the anvil. Clamp the jig in the vise. Hold the cold section of the bar with tongs and place the hot part in the jig slot. Hammer successively each side of the offset 90 degrees, flush against the jig's side, to complete the offset.

Next, reheat and true it up by squeezing the whole assembly between the jaws of the vise (assuming your vise is heavy and strong enough to exert the needed pressure).

Still another way of offsetting is simply to place the combined jig and heated bar, held together with visegrip pliers, on the anvil face, and hammer all down into proper alignment.

Bending an offset in a rod or bar can also be quickly and easily done in specially made bending forks placed in the anvil's hardy hole (see illustration). These forks are designed in a variety of shapes to solve a wide range of bending problems. Heated sections of rods and bars held in the appropriate fork can easily be twisted in this system. But your anvil must be firmly bolted down on a well-anchored wood stump or strong base and the fork must fit snugly in the hardy hole. Additional trueing-up is usually needed with hammer and anvil or in vise jaws.

17. Blacksmiths' Tongs

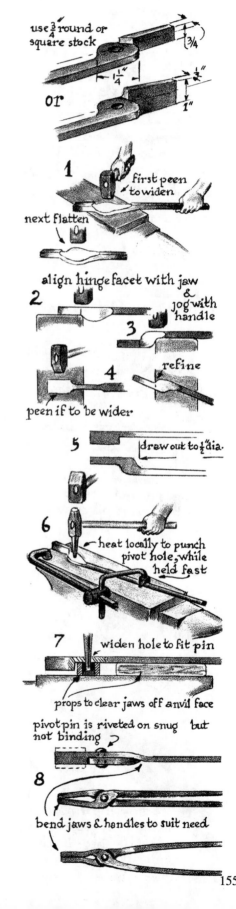

The modern smith faces a permanent puzzle when he makes a pair of tongs without resorting to welding. In tongs the thick, strong jaws and hinge must be combined with handles that are relatively thin to keep the weight of the tongs down without losing the tool's strength. In selecting stock for this tool, therefore, you will have to compromise and use an in-between size to avoid having to weld thin handles on too heavy jaws (as old-timers often did).

To make a pair of blacksmiths' tongs, use ¾-inch square or round stock. It can be either mild or high-carbon steel. For these symmetrical tongs, each half is to be forged *identical* in shape to the other. Illustrations 1 through 5 show how to ready each half.

The hinge pin should preferably be ⅜ inch in diameter for the average (medium-sized) tongs. The hinge-pin hole is hot-punched. Since the area around the hinge-pin hole is subjected to great strain, and must be sturdy, try to leave about ½ inch of steel around the pin. The total hinge area for such tongs then would be approximately 1¼ inches in diameter.

Heat the hinge area and place it over the hardy hole. Secure it with a hold-fast. Punch the pivot hole through with the ¼-inch hot punch (illustration 6).

Reheat and, with the same punch, widen the hole from the other side over a section of plumbing pipe (illustration 7). This does two things. The bulge created when the hole was driven through is bent back in place, while at the same time the hot punch tapers this end of the hole and widens it to ⅜ inch. When assembled, the riveted hinge-pin then fills the cone portion, with a holding action. This, in addition to the rivet heads on the pin, tends to prevent a wobble in the tongs at the least strain. (I venture to say that wobbly, loose hinges are the rule in most blacksmiths' shops, caused by undersized, ill-fitting, and weak hinges.)

When the parts are permanently assembled, heat only the jaws in the fire, without taking the tongs apart. Each jaw can then be bent or reshaped as needed to fit any workpiece for which you may not have special tongs.

blacksmith tongs
most useful jaw types

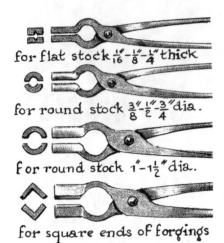

for flat stock $\frac{1}{16}" - \frac{1}{8}" - \frac{1}{4}"$ thick

for round stock $\frac{3}{8}" \frac{1}{2}" \frac{3}{4}"$ dia.

for round stock $1" - 1\frac{1}{2}"$ dia.

for square ends of forgings

to hold thin flat discs

to hold parts with hole in end

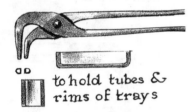

to hold tubes & rims of trays

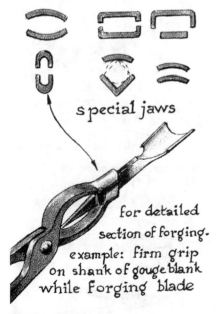

special jaws

for detailed section of forging. example: firm grip on shank of gouge blank while forging blade

Specific workpieces demand different jaw shapes in the tongs used. The shapes shown here simply aim to give an idea of the kinds of stock different tongs are suitable to hold. For instance, the woodcarving-gouge blank shown required the narrow-slotted tongs to match the gouge's upright shank. These tongs can, of course, be used for other things as well.

It is always possible to reforge an old pair of tongs whenever a new project demands a new shape. But whether tongs are new or reforged, it is greatly to your advantage to increase the range of tong shapes and sizes in your collection.

18. Making Milling Cutters, Augers, and Drills

Small hand-forged items in need of some machining can be speedily and accurately refined and finished without benefit of a true metal-turning lathe if you have a good power drill press. The makeshift methods offered here will be completely sufficient for most "precision" work around the modern blacksmith's shop.

By resorting to positioning by sighting, you can, as a rule, get accurate results without the often laborious instrument-measurings in a "make-ready" procedure. Once understood, you can gain a new independence simply by making use of your body's built-in measuring instruments: the eyes (sighting) and the fingers (calipers, etc.), combined with judgment, know-how, and inventiveness.

A MILLING CUTTER

First, accurately forge a 3-inch-long round-headed bolt from ½-inch-diameter round high-carbon steel rod, as described in Chapter 4.

Clamp it in the drill-press chuck and adjust the spindle speed to about 3000 rpm.

Next, tightly clamp, upside down and exactly vertical, a ¼-inch high-speed steel drill in the drill-press vise. The drill must be perfectly sharpened for this function, and should only protrude above the vise jaws ½ inch. Align the drill point by sighting it with the drill-spindle center, after which, with a C clamp, secure the drill vise onto the locked drill-press table.

Switch on the motor and lower the bolt head so that it barely touches the sharp drill point. This is done to clean the contact spot between the bolt head and drill. It "shaves" it evenly, ready to begin the hole to be drilled.

At this point, loosen the C clamp a little to let the vise follow the central pull on the drill point. Then, as you lower the bolt head, the drill will begin to "bite" into the steel. Automatically it will be drawn to dead center of the bolt's rotation.

Your previous sighting of the drill and chuck spindle most likely is now off very slightly (but, as a rule, not more than $1/32$ of an inch). It is this slight inaccuracy that allows the drill to seek a center position automatically, pulling the following vise with it. As you hold the height of the bolt-head position steady for 30 seconds at about ⅛-inch drill penetration, you will see that initial slight movement of the drill point come to a stop. Lock the vise with the C clamp on the drill-press table, and switch off the motor.

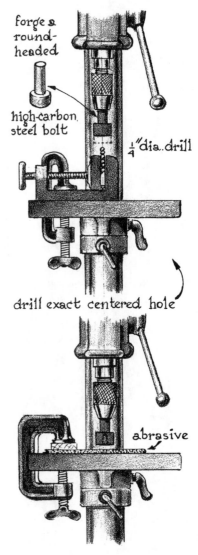

forge a round-headed

high-carbon steel bolt

¼" dia. drill

drill exact centered hole

abrasive

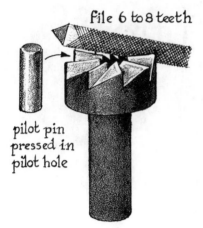

file 6 to 8 teeth

pilot pin
pressed in
pilot hole

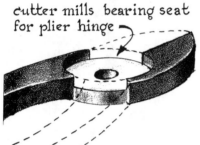

cutter mills bearing seat
for plier hinge

Now reduce speed to about 1000 rpm or less, and drill the pilot hole at that speed until the bolt head comes flush with the vise jaws. Raise the bolt head, stop the motor, and remove the drill, but keep the vise tightly clamped on the drill-press table.

Again lower the bolt head until it rests flush with the vise jaws. This time clamp a flat file tight in the vise, so its sharpened, squared-edge end is vertical and touching the rim of bolt head. This file end will now act as a cutter to reduce the bolt head to proper size and to accurate alignment with the pilot hole and bolt shank.

With drill spindle raised, switch on motor and adjust to 1000 rpm or less. Pump the bolt head up and down along the file cutting edge. When you are satisfied that all excess steel has been removed, with the file end at this position you may still judge that more has to be trimmed off. By simply tapping with a 1-pound hammer on the side of the vise, it will yield just enough for you to see a bit more steel being cut off the bolt head. As soon as it is clean, measure its diameter for size.

Next, replace the vise with a flat abrasive stone clamped on the drill-press table with a little wooden cleat as a cushioning block. This time adjust spindle speed to its highest and switch on the motor. Gently lower the bolt head onto the abrasive. This will grind its plane at 90 degrees to its length exactly and smoothly.

This finishes the machining of the cutter blank. Now remove and clamp it, head up, between the jaws of the workbench vise, in order to file the 6 to 8 teeth. Temper to a dark bronze color.

The illustration shows the end result, with the ¼-inch-diameter pilot pin ready to be pressed into the cutter's central hole. Should the pin fit too loosely, dent the portion to be inserted with the hammer peen so that it can be forced in with a wooden mallet.

You now have a milling cutter, a very useful tool. It will prove to be one of your most valuable tools for seating hinge-joint surfaces for pliers, shears, tinsnips, etc.

With the pilot pin removed, such a cutter can become a *router* (provided each cutting tooth is also sharpened on its upright side). Steel sections then can be surfaced and deepened that must be exact and smooth.

AUGERS

Forge the auger blank (illustration 1) and anneal it. Grind it smooth and of an even thickness, with sharp 90-degree edges and sides.

If it is mild steel, twist it cold for an even screw pitch. If it is high-carbon steel, twist it when evenly hot. If it is not heated *evenly,* an irregular screw pitch will result and you will have to start over again. If the alignment should be incorrect, place the still-hot blank on a wood stump and align it with a wooden mallet. If done well, it will not distort the screw shape.

Bend and file the end of the auger as in illustrations 2 and 3. The leading cutting points must be razor-sharp and on the auger's exact diameter, plus $1/64$ inch over. *Only the cutting end of the auger is to be tempered and drawn a peacock to bronze color.* (If you make the auger of mild steel, the cutting end must be case-hardened.)

WOOD DRILLS

The simplest, but most effective wood-drill bit can be forged and tempered in a short time (illustration 4). For extra-wide cutting blades, upset the stock first and peen it wide.

Temper wood-cutting edges between peacock and bronze color when the bit is made of high-carbon steel. If made of mild steel, it must be case-hardened.

DRILLS FOR CUTTING STEEL

The difference between this steel drilling bit (illustration 5) and the above (4) is that it must have a thicker blade and a cylindrical pilot lead. This pilot must be filed precisely to fit into a predrilled hole in a steel workpiece to create an accurate seating (as in the making of the hinge-bearing surface of pliers, page 170).

If the stock used is high-carbon steel, temper the last ¼ inch of the blade at the cutting edge a bronze to dark straw. The rest of the blade may be drawn purple to blue.

If the stock used is mild steel, case-harden the bit locally.

All kinds of makeshift wood drills can be made from various sizes of nails (illustration 6). (Remember that nails are mild steel. Such drills, if dull, will heat up and "burn" themselves through wood.) It is perhaps a practice which is frowned upon, but if you are in a hurry and if no great accuracy is required, I see nothing against it.

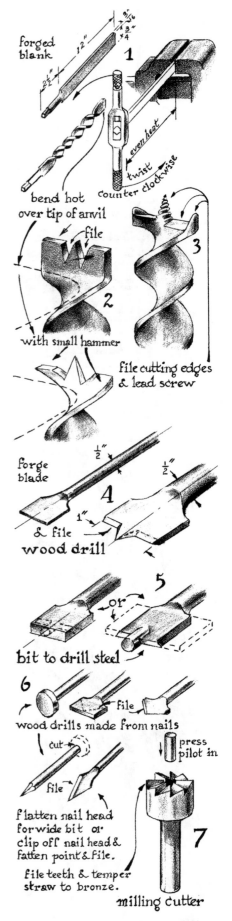

1 forged blank
even heat
twist clockwise
counter clockwise
bend hot over tip of anvil
file
2
with small hammer
3
file cutting edges & lead screw

forge blade
4
& file
wood drill

5
bit to drill steel
6
file
wood drills made from nails
press pilot in
cut
file
flatten nail head for wide bit or clip off nail head & fatten point & file.
7
file teeth & temper straw to bronze.
milling cutter

19. Stonecarving Tools

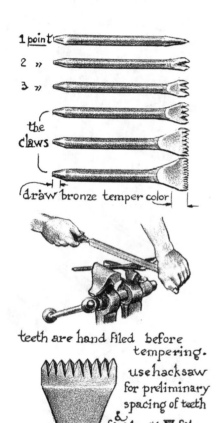

1 point

2 "

3 "

the claws

draw bronze temper color

teeth are hand filed before tempering. use hacksaw for preliminary spacing of teeth & finish with ▽ files

the bush tool, used mainly on granite

The most commonly used stonecarving tools are simple tools, little changed in design from earliest times. As a rule, stonecarving tools measure about 8 to 10 inches long, and ⅜ to ¾ inch thick. All are made from high-carbon steel.

ONE-POINT STONECARVING TOOL

The one-point tool is made and tempered in the same way as a center punch (Chapter 8) except that the end that is struck by the hammer has a small cup-shaped crater with a hardened knife-sharp rim. A mild-steel hammer, striking this end, will never glance off because the cup "bites" into its face.

TWO-POINT, THREE-POINT, AND THE CLAWS

The teeth are filed into the tapered blade. To temper, heat ¾ inch of the blade end and, after brittle-quenching, put a sheen on the blade with an abrasive. Draw to bronze color over gas flame.

BUSH TOOLS

These tools generally have nine points only, when used for shaping the first forms in stone too hard for the one-point. Others can be made with many very fine teeth for creating smoother textures on the stone's final surface. Held and struck at a 90-degree angle to the stone, it crushes the hard surface. The end of the bush tool that is struck with the hammer is slightly crowned and then hardened. These tools are especially effective when used with air hammers. A bush hammer has a face with nine points and functions as bush tool and hammer combined.

DRIFTS

These tapered round rods are used for splitting stone. A series of holes is drilled in the stone with star-drills or carbide-tipped drills and the tapered drifts are wedged into them. One after the other they are driven gradually a little deeper until they exert the total force necessary to split the stone. They are made of high-carbon steel left in its annealed state.

CLEAVING CHISEL

This stone-splitting tool is made by upsetting a ¾-inch round rod, then using the cross-peen hammer to widen it. It can also be made by beginning with a much thicker (1¼-inch-diameter) bar, upsetting and widening one end and drawing out the other.

The last ¾ inch of the wide blade is tempered to a bronze color. The other end may be hardened and somewhat crowned to keep hammer contact at the tool's center.

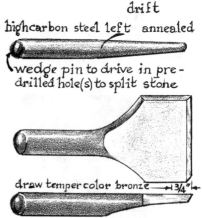

drift
highcarbon steel left annealed

wedge pin to drive in pre-drilled hole(s) to split stone

draw temper color bronze ⟶ 3/4" ⟵

cleaving chisel to crack off segments along precut groove.

20. Wrenches

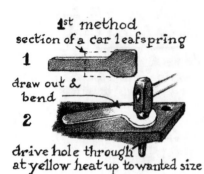

1st method
section of a car leafspring

1

draw out & bend

2

drive hole through at yellow heat up to wanted size

3

cut, file to size & temper

4

2 step-wrench

2nd method

5 cut

& spread over anvil ridge where angle is 120°

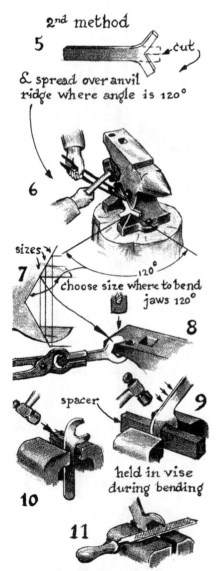

6

sizes

7 120°
choose size where to bend jaws 120°

8

spacer **9**

held in vise during bending

10

11

anneal & file to exact size
temper color: bronze to purple

You will sometimes find yourself tempted to whip out a needed, easy-to-make tool that you have lost or mislaid. I have often made a duplicate wrench, screwdriver, hammer, garden tool, rather than spend the time searching for it. The happy result is that wherever I happen to be working — in the shop, studio, house, or garden — there is always a spare one nearby.

OPEN WRENCHES

First Method

An open wrench is easily made. The blank is cut from a salvaged leafspring of a car.

With a hot punch, make a hole and enlarge it to the size you need (illustrations 1 and 2).

Cut away excess and file to exact size (3). Temper to purple color.

To make a two-step open wrench (4), simply forge a duplicate at the other end, and bend them in the opposite direction to each other. The ends are cut open hot with chisel head and hammer, *or* they can be cut by sawing them cold after annealing.

Second Method

As shown in illustrations 5 and 6, cut the end of the hot piece and spread it open over the anvil base where the angle measures approximately 120 degrees. (It is practical to use any part of the anvil which lends itself to the forming and bending of steel. I frequently use the hollows between the horn and base, or between the two anvil footings, to shape articles matching the needed curves.)

Study the *diagram* (7) and choose the size at which to bend the jaws to a 120-degree angle. Shown are the three locations that correspond with three wrench sizes (8). Bend one jaw while holding the wrench horizontally on the anvil's edge. Repeat with the opposite jaw. Hammer it down flush with the anvil facing.

If the wrench head must have a jog with its handle, bend the handle hot just *below* the jaws (9 and 10.) A *spacer* is handy to arrive at an exact size, with further refining by filing the annealed jaws (11).

After brittle-quenching in oil, draw temper color to bronze or purple.

This second method works well to make an *extra-wide* wrench; for example, one that fits the hexagon pipe-locking ring on a plumber's gooseneck trap.

BOX WRENCHES

Draw out the center section cut from a piece of spring steel over the anvil horn or a snub-nosed hardy (illustration 1). When the desired length for the wrench has been reached, punch a hole in each end large enough to receive a hexagon-sided hot punch made especially to convert round holes to hexagonal ones. (You can make this punch by grinding it from a high-carbon steel bar.) Use it with or without a handle. This hot punch is tapered and graduated for size so that it can be driven through prepared smaller holes up to the marked size you need (illustration 2).

Knock out the punch and reheat the wrench head.

Next, use a case-hardened bolt head (4) instead of the punch. It too is somewhat tapered at its end to act as a starter, which can seek a proper seating with the prepared, but slightly undersized hexagon in the box-wrench or head.

If the bolt is long enough, it can be hand-held. Place the hot end of the wrench on a section of pipe which has an opening large enough for the bolt head to be driven through. This "calibrates" the wrench to a perfect fit for that bolt size.

Temper the wrench heads peacock to purple, leaving the center bar annealed.

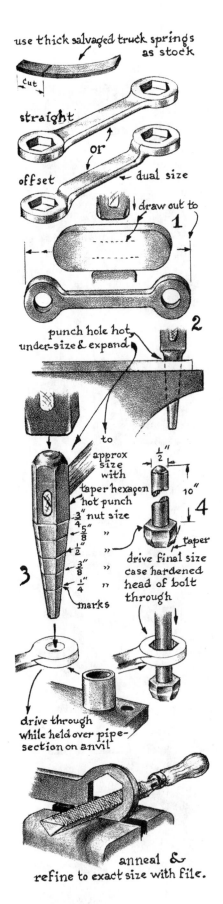

use thick salvaged truck springs as stock

cut

straight

or

offset

dual size

draw out to

1

punch hole hot, under-size & expand

2

to approx size with taper hexagon hot punch

$\frac{3}{4}$" nut size
$\frac{5}{8}$" "
$\frac{1}{2}$" "
$\frac{3}{8}$" "
$\frac{1}{4}$" "
marks

3

$\frac{1}{2}$"

10"

4

taper

drive final size case hardened head of bolt through

drive through while held over pipe-section on anvil

anneal & refine to exact size with file.

21. Accessory Forging Tools

temporary fill plug

1

chipped face
hammered
smooth into
forms
to

2

match
swage curves

3

clamp
in vise &
on prop below

4

5

6

7

hammer forcefully

8

Swage to stiffen
gouge blades

hardy-hole
holds all special
swages, bicks,
matrix parts,
etc., that
must be anchored solidly

9

Various bicks

temper bicks purple

Worn hammers made of good steel may be saved and made usable. Such a hammer, if you find no deep cracks anywhere, can be forged into any specially shaped hammer you may need.

Most of the double-ball hammers that I have made to form the curves of gouge blades are converted from secondhand hammer heads.

Hammer a temporary steel plug into the hammer-stem hole (illustration 1). This allows you to squeeze the hammer (heated yellow hot) between the vise jaws without collapsing the thin-walled middle section. A steel prop, placed below it and resting on the vise bar, takes up the blows as you reshape the hot hammer head and relieves the stress on the vise jaws. At the same time, the vise bar below also is protected by the prop (2).

Once the ball end is shaped, heat the other end, and convert it also to the shape you need. These hammer shapes should match the curves of swages (3).

Swages can be modeled from secondhand ones (if you are lucky and can find them). But you can also forge new ones from sections of very-large-diameter salvaged truck axles, in shapes as shown on page 106).

Having a choice of special hammer shapes, as shown in illustrations 4 through 7, is a great advantage when curving steel, or when decorative texturing of surfaces needs to be done.

Illustration 8 shows a forming swage used to forge a reinforcing rib on thin blades. This rib down the center of the blade reinforces a light tool to make it strong.

Bicks (9) resemble anvil horns. They are indispensable when projects call for bending tubular parts, cone-shaped sheet metal, and many other jobs that cannot be done on the anvil horn if this part of the anvil is too large. Bicks can be of several sizes, shapes, and lengths, and are quite easy to make. From square stock that fits the hardy hole, forge them into cone-shapes and then bend. Temper them peacock to purple color.

22. Woodcarving Gouges

A CONE-SHAPED GOUGE

The advantage of the cone-shaped gouge blade shown here is that this tool will not bind when cutting deep curves in wood, whereas the conventional cylindrical blade will.*

The steel that I have found to be quite satisfactory for the forging of gouges is the high-carbon round or square stock salvaged from coil springs that have been cut and straightened (see Chapter 20).

Follow the steps in illustrations A, B, C, D, and E to make the cone-shaped blades. You will find that the main difficulty in the forging of tool blanks is to keep blade, shank, and tang in precise alignment.

After the blank is formed, refine it by hand-filing and motor-grinding with coarse- and fine-grit wheels, grinding points, and the rubber honing wheel. Finally, polish it with the motor buffer, temper peacock to bronze, and attach the wooden handle.

Instead of using coil springs as stock, you will find the work somewhat simplified if your scrap pile has some high-carbon steel bars, ¼ x ½ inch or ⅜ x ¾ inch. Illustrations 1, 2, and 3 show clearly the advantages of using stock with dimensions already close to those of the planned workpiece: this makes the forging easier and less time-consuming.

Proceed as follows:

1. Forge the shank to $^3/_{16}$ x ½ x 4 inches.

2. Peen the blade to 1 inch in *width* and at exactly 90 degrees to the shank. On the anvil's face, "set" the blade flush with the shank, so that the bottoms of blade and shank form a *straight line*.

3. Hold the blade in tongs or visegrip pliers and forge the tang, leaving a shoulder between shank and tang. This shoulder is made to rest on the ferrule and washer when the tool is assembled with its wooden handle.

In the making of conical blades for gouges of various sizes, you will find that a wide variety of bottom swages and top fullers will make your work easier, as will a choice of specially shaped hammers that fit the swages. However, standard swages and fullers can be used during preliminary steps, the blade being finished freehand after that.

*I have discussed the cone design in my book *The Making of Tools.*

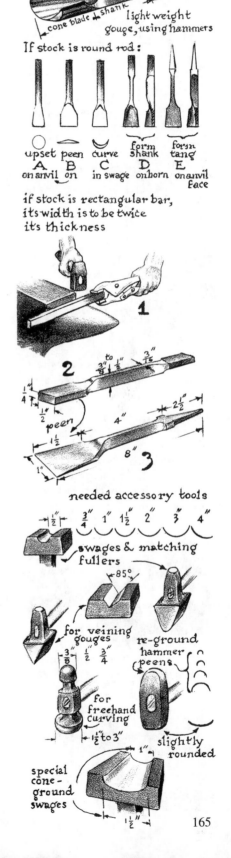

basic design
hardwood handle
tang
cone blade · shank
light weight gouge, using hammers

If stock is round rod :

upset peen on anvil / A
peen on / B
curve / C
form shank in swage / D
form tang on horn / E on anvil face

if stock is rectangular bar, its width is to be twice its thickness

needed accessory tools

swages & matching fullers

for veining gouges

85°

re-ground hammer peens

for freehand curving
1½″ to 3″

slightly rounded

special cone-ground swages

165

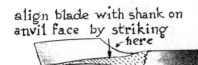

align blade with shank on
anvil face by striking
here

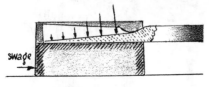

heavier blows where blade
is thickest, until bottom of
shank & blade are straight

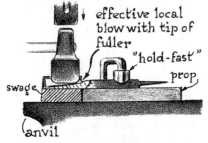

swage

effective local
blow with tip of
fuller

"hold-fast"

prop

swage

(anvil

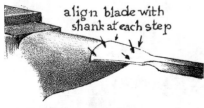

align blade with
shank at each step

close or open blade curve when
needed during refining over horn

important steps in forging
conical blades of gouges
in cylindrically ground
swage grooves.

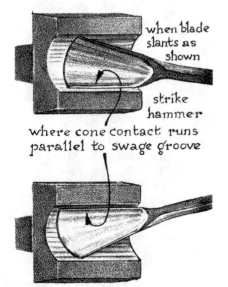

when blade
slants as
shown

strike
hammer

where cone contact runs
parallel to swage groove

Assume that you have forged the blank with a flat blade. Place this blade in the nearest-size swage groove and gently peen it down without marring its smooth surface. You will notice that during this forming of the blade the thicker part becomes humped up relative to the shank, because it resists the force of the hammer blows. It is this thicker part which now must be hammered down to restore the alignment with the shank.

If you have a hold-fast tool, your alignment task is simplified, as the illustration shows.

Next, form the cone shape of the blade by opening or closing the curves over the horn and in the swage, until, as you progress, the exact curve and alignment have been reached.

The conical-blade gouge blank is now ready for refining.

Shaping the Blade with a Standard Swage

If you do not have a special cone-grooved swage, you can use a standard cylindrically shaped one by employing the special techniques illustrated here.

Note particularly how first one side of the tool is placed parallel to, and hammered against, the swage's upright side; then the *other side* of the tool against the *opposite* upright. The hammer peen must be ground dull to peen out the cone shape in this way, to prevent sharp textures on the steel.

However, no matter how logical the illustrations and text may seem, in practice this will prove to be a persistently confusing procedure. *It is advisable, therefore, to first make a pattern* that resembles the gouge blade. Cut it out of sheet metal with tinsnips.

Practice hammering this flat mock-up, cold, into a conically shaped blade using a cylindrical swage. Once the technique is mastered, you will understand the correct procedure to follow in order to avoid making typical errors and having to correct them.

CORRECTING COMMON ERRORS IN THE FORGING OF GOUGES

Unfortunately it happens again and again to the beginner that halfway through the project, a good start is ruined by a mistake. A little more thoughtful planning before each move can save the day.

The illustrations show what happens if you fail to aim your hammer blows with precision at the very outset, or if you hold the tool blade slightly slanted in the swage so that a precise central blow then bends it askew.

Another serious error is to use too large a curved hammer with a swage, resulting in the blade being marred by the sharp edges of the swage.

An incorrect alignment between blade and tool shank (an often repeated error) can be corrected by one of the following methods:

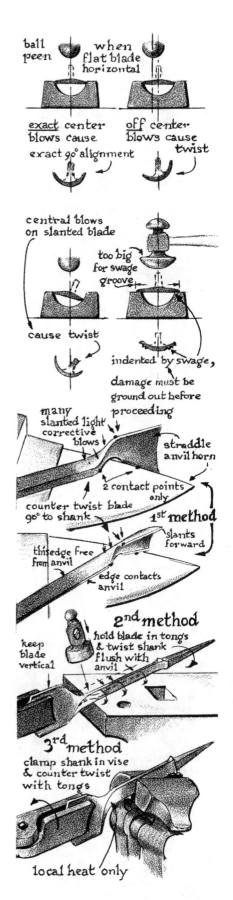

ball peen — **when flat blade horizontal**

exact center blows cause exact 90° alignment

off center blows cause twist

central blows on slanted blade

too big for swage groove

cause twist

indented by swage, damage must be ground out before proceeding

many slanted light corrective blows

straddle anvil horn

counter twist blade 90° to shank

2 contact points only

1st method

slants forward

this edge free from anvil

edge contacts anvil

2nd method — hold blade in tongs & twist shank flush with anvil

keep blade vertical

hold blade in tongs & twist shank flush with anvil ×

3rd method — clamp shank in vise & counter twist with tongs

local heat only

First Method

The more experienced smith can hold the tool diagonally over the horn of the anvil in such a way that a few light but telling hammer blows will twist the blade back into alignment.

Second Method

If the tool is very much out of line, use heavy tongs to hold the cooler part of the blade so their mass acts like a heavy vise. Locally heat the area between the blade and the shank a dark yellow. Hold the blade vertically and hammer the incorrectly slanted shank flush with the anvil facing while holding the blade and tong steady. This then untwists the twist.

Third Method

A similar counter-twisting can be done without the hammer. Clamp the cool shank between the vise jaws, while holding the cool blade firmly with the tongs. The yellow-hot part in between permits you to twist and bring the blade to exactly 90 degrees with the shank upright. In that corrected alignment, quickly "set," and re-shape somewhat (using a light ball-peen hammer) the still hot part, which will have become a little deformed in this twisting action.

Whenever corrections or refinements in forging are called for, you will realize how important it is to apply *very exact* hammer blows, delivered on well-planned locations. It is here that you should make use of every second of the period between heats to judge and plan your next moves. This, combined with skill, makes the good smith. Bad judgment will set you back, no matter how strenuous your hammering may be.

VEINING GOUGE

The forging of a veining gouge can be done *freehand,* but is not as easy as it might seem. Therefore, I have devised a fairly simple way of making this tool with the aid of *dies* held in a spring clamp.

Making a Die Assembly for the Veining Gouge

Use a salvaged car leafspring measuring about 3 inches by ¼ inch in cross section and 40 inches long.

Forge a hole at one end with a square hot punch to match the shank of a V-grooved bottom swage. This end of the steel must be well annealed.

With the abrasive cutoff wheel, cut the other end of the spring clamp lengthwise in half for a distance of 5 inches. Heat it and spread it to form a fork, as shown. This fork must be precisely made, so that its arms are filed *exactly parallel and evenly thick.* Leave it annealed so that the slotted male die can be forced into the fork and held firmly by spring tension.

Install both dies in the prepared ends of the leafspring. Next, heat 6 inches of the exact middle of the spring and, since the die ends remain cool enough to hold by hand, simply bend the spring over, placing one die into the other.

In this position, clamp the die ends firmly between the vise jaws, and rapidly hammer the heated bend into the curve, as shown. The hammering "sets" the spring-clamp position, removing all tensions.

Immediately — before the visible heat glow of the bend has disappeared — transfer the assembly to the anvil, and quickly fasten it with its capscrew and its slanted pipe-spacer. (If it has cooled too much, reheat the bend for this step.)

Now pull the upper arm of the clamp up, just enough so that the hot steel at the bend "gives" a little. When released, the male die should then be spaced about ¼ inch from the female die. If you have pulled it up too much, bear down on it to reseat the male die properly. With a 1-pound hammer, tap with many very light blows all around the bend to remove all tensions.

Release, and check if the ¼-inch spacing between the two die parts has been correctly established. Let all cool slowly.

The ¼-inch spacing allows you to insert the hot tool blank easily and quickly. Also, because of the inherent springiness of the annealed bend, you can pull it up high enough to extract the female die (the swage) in order to replace it with another if needed.

It is important to secure the die assembly *firmly* onto the anvil; any looseness might interfere with the actual forging. To anchor the assembly, drill a hole into the swage shank, as shown, and thread it to fit a ⅜-inch capscrew. The capscrew holds the swage on the anvil with a washer and slanted pipe section. This V-grooved swage acts as the *female* part of the die set. The *male* die is made to match it as follows:

Use a section of a car axle (round or square) measuring about 1¼ inches in cross section. Its working end can be ground to fit the groove of the swage exactly. Slot the die as shown to fit snugly into the forked end of the spring clamp. This arrangement makes it easy to change dies, should it be necessary in making other shapes of gouge blades.

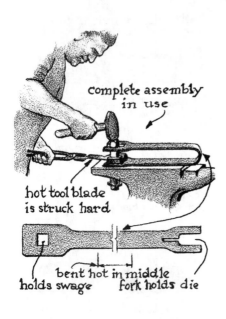

complete assembly
in use

hot tool blade
is struck hard

bent hot in middle
holds swage   fork holds die

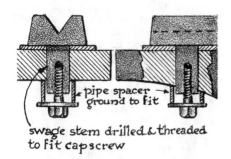

pipe spacer
ground to fit

swage stem drilled & threaded
to fit capscrew

Forging the Gouge Blade

1. Firmly clamp the die assembly to the anvil.

2. Fold the hot blade of the tool blank, as shown, to make sure that the fold and the shank are aligned. This preliminary fold will automatically seek an aligned seating in the bottom die, thus preventing a lopsided positioning between it and the top one.

3. Reheat and insert the hot folded blade while sighting the blank for alignment with the die assembly.

4. With a 3- to 4-pound hammer, deliver a few exactly vertical, heavy blows, *or* many lighter rapid blows, depending on how thick the steel of the blank is.

If all has been constructed correctly, and the forming action done well, the V-gouge blank is ready to be filed, ground, and prepared for tempering.

Note: As an advanced student, you will have noticed by now that if the die surfaces are rough, pitted, or inaccurately matched, every such mark is transferred onto the blank. It means that much work in filing and grinding will have to be done to make up for a poor set of dies. Therefore, if you do decide to spend time making a die assembly, it will pay to make a good one. The die surfaces must be accurate and smooth, exactly aligned, and well tempered.

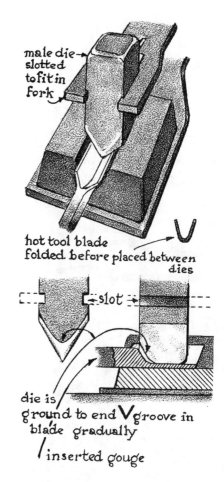

male die slotted to fit in fork

hot tool blade folded before placed between dies

slot

die is ground to end V groove in blade gradually

inserted gouge

23. Forging a Pair of Pliers

left-over core

use 3/16" drill

holes leave sharp ridges after break

grind these off, preventing during hammering a mushrooming or a folding effect

cut & forge into

2 identical elements

File or grind accurately flat

drill 3/16" pilot hole to lead seating cutter

resulting seat

use same dia. bar as jig

file sharp teeth

clamped on seating in vise & file off excess around seat-rim.

temporary hinge pin

apply oil & abrasive compound

& work by hand up & down till smooth

enlarge hinge hole to 5/16" dia., disassemble & temper each element separately: jaws a dark straw, hingeseats purple, handles pale blue.

For this pair of ½-inch-thick pliers you can use a 4-inch-wide, heavy-caliber leaf of a truck spring.

Cut off a 12-inch section, heat it and flatten it, and then anneal it slowly, buried in ashes.

Scribe off on the steel the curved pattern of the pliers. Along these lines, make evenly spaced center-punch marks for holes to be drilled with a ³/₁₆-inch high-speed drill at slowest speed. The holes must be close enough to leave only *paper-thin* divisions between them. Clamp the core section in the vise and knock off the outer pieces. Use the core section of these pliers. Tie the two outer pieces together with baling wire, and hang them on a nail for future use.

The sharp ridges resulting from the drilling must now be filed off. If they are not removed, they will fold down into the surface during forging. Such flattened ridges would later be revealed again during filing and grinding.

Draw out the stock to form the two plier handles. Then cut the piece in half as shown, and forge the two identical blanks.

Next forge the hinge sections locally into their approximate final dimensions, leaving room for the full diameter of a hinge-seating cutter to mill the bearing sockets.

When the blank for each plier half is finished, refine the flat surfaces on the motor-driven side-grinder. (However, if your skill in flat-filing is sufficient, the pliers' surfaces can be filed perfectly flat by hand.)

Clamp one blank on the drill table. Place a ³/₁₆-inch high-speed drill in the exact punch-marked center of the hinge area, and drill the hole to guide the leading pilot of the seating cutter. Do the same with the other blank.

Each plier half is now ready to be milled with the seating cutter to the depth of exactly one-half the thickness of the plier blank.

In order to fit the two halves together, the excess steel at the *rim* of the hinge-bearing diameter must first be filed off smooth with the aid of a file-jig. The jig is a section of a round bar having the same dimension as the seating cutter. This bar (cut at an exact right angle to its length) is clamped in the vise onto the plier half. File back and forth flush with the bar-jig, removing precisely all excess steel from the rim around the hinge-bearing surface.

Trim the other plier half accurately in the same way. The two halves should now fit together in a temporary assembly. At first they will bind somewhat. To remedy this, smear a coarse-grit valve-grinding compound and some 3-in-One oil over the bearing surfaces. Assemble the plier halves with a temporary hinge pin (use a ³/₁₆-inch-diameter capscrew and nut) holding the two halves lightly together.

Clamp the plier in the vise by one of its handles and work the other half up and down in a "lapping" action, progressively tightening the capscrew a little. The abrasive wears down all minute inaccuracies. Use some kerosene to flush out the metal pulp and abrasive residue. Replenish with fresh abrasive and oil from time to time.

Finish with a finer-grit abrasive and 3-in-One oil. When a smooth, snug fit has been established without binding, take the plier halves apart and clean thoroughly.

Next clamp together with two visegrips the assembled plier, but without the capscrew. Place it on an accurate hardwood prop on the drill-press table. In this position you can enlarge the temporary hole through the hinge area with a $^5/_{16}$-inch drill to take a permanent $^5/_{16}$-inch-diameter hinge pin.

Each hole, on the outer side of the blanks, is now *countersunk* to a depth of ⅛ inch, as accurately and smoothly as possible. One of the countersunk depressions acts as a bearing surface for the hinge pin, while the other, as a rule, does not move.

Separate the plier halves and temper each one as follows:

1. Heat the jaw and hinge section to a medium cherry red in a slow, clean forge fire and quench in *oil*.

2. Clean off all carbon scale left by the quench using fine-grit carborundum paper. Polish the whole tool, then draw color to dark straw and quench again, this time in *water*.

3. To adjust the temper of the hinge area, use a propane torch or a Bunsen burner gas flame to locally draw the *hinge area only,* very slowly, to peacock color, leaving the jaws a dark straw. Withdraw from flame and let cool slowly.

Assemble the pliers with a temporary hinge pin just long enough to hold the two plier halves snugly together. All remaining inaccuracies can now be ground off on the side grinder, as shown. When all other grinding, refining, and polishing is done, knock out the temporary pin.

Now permanently assemble the pliers, using an accurately fitting $^5/_{16}$-inch hinge pin long enough to cold-rivet both heads. (Now you will realize that the better the countersink has been made, the better it will function as a bearing surface.)

Place the assembly flush on the anvil face. The cold-riveting of the hinge pin is done with a ½- to 1-pound ball-peen hammer, using the ball and flat alternately. First the *ball* strikes the whole surface of the pin evenly. Then the *flat* forces down the ridge texture left by the ball. Repeat alternate hammering with ball and flat until complete countersink depression has been filled. Treat the other side the same way.

Any slightly raised surplus material can now be gently ground off without marring the finished surface of the surrounding steel. If you like the appearance of raised, hand-riveted heads, don't grind them off flat, unless they are in the way during use.

If you want to refine the pliers still further, they can be repolished on the buffer to mirror smoothness to prepare them for an application of oxidation color patina to suit your taste (see Chapter 9 on applying color patina). Keep in mind that the jaw and hinge area must not be drawn darker than straw color to ensure that this previously tempered area will remain unchanged in hardness. The handles may vary in color all the way from dark yellow to peacock to blue.

Note that two kinds of pliers are illustrated here. The perfectly symmetrical pair may *look* good but it does not necessarily *work* better than the asymmetrical pair. You will recognize, though, that symmetrical pliers are simpler to make, having identical halves; while the pliers with offset jaws calls for two *different* blanks.

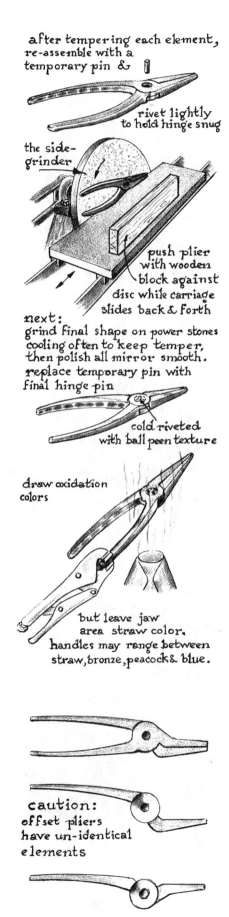

after tempering each element, re-assemble with a temporary pin &

rivet lightly to hold hinge snug

the side-grinder

push plier with wooden block against disc while carriage slides back & forth

next:
grind final shape on power stones cooling often to keep temper, then polish all mirror smooth.
replace temporary pin with final hinge-pin

cold riveted with ball peen texture

draw oxidation colors

but leave jaw area straw color. handles may range between straw, bronze, peacock & blue.

caution: offset pliers have un-identical elements

24. Making a Fireplace Shovel

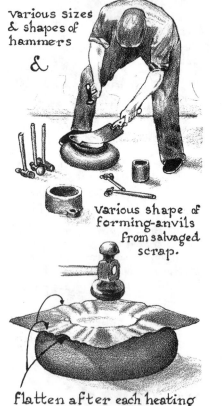

various sizes & shapes of hammers &

various shape of forming-anvils from salvaged scrap.

flatten after each heating

sand or earth

salvaged car bumper part clamped in vise acts as a forming anvil

BASIC PRINCIPLES OF FORMING HEAVY-GAUGE SHEET METAL

Hammering spherical shapes out of mild sheet steel is easiest to do when the metal is yellow hot.

When it is heated locally and hammered over forming pieces improvised from openings in heavy machine parts, such as steel rings, pulleys, cast-iron pipe sections, or any other piece of massive scrap steel, the heated metal can be *stretched* without difficulty.

As the steel yields more and more during stretching, you must reheat and peen out progressively each next area so that hammering will reach every part of it uniformly.

The final shape of the project is reached by hammering it over a mound of sand or earth and scrap parts such as an automobile bumper. (There seems to be an endless variety of shapes in bumper parts, since manufacturers annually change the "sculptural" designs of car models.)

Trim off all superfluous surrounding sheet with a slender cold chisel on the endgrain of a wood stump — a good setup for cutting steel that is too thick for metal shears.

Note: Modern mild steel, used for cold "die-stamping" of auto-body parts, can stand a great deal of cold hammering before breaking. Therefore, much of the mild steel from your scrap pile can be shaped cold. Nevertheless, from time to time you should heat and anneal the cold-hammered parts to avoid breaking the steel.

FORMING A SHOVEL BLADE

The forming of heated sheet metal into a shovel blade, as in the illustrated steps 1 through 5, is a combination of *stretching* and *folding* the steel. Sand or earth, as well as various shapes of heavy scrap steel parts, are used in shaping the blade, as described above. The anvil horn and a choice of bicks placed in the hardy hole are used to hammer out the curves of the handle socket of the shovel.

Both blade and handle socket are formed together, as shown, in a step-by-step process of heating and hammering, reheating and hammering. Final touches can often be applied by cold-hammering and occasional annealing with intermediate reheating.

The handle of this shovel is made of wood from a natural curved section of a tree branch which you may find in piles of orchard prunings, or select and cut in your own garden. First make it fit the handle socket approximately; cut and shape with saw, wood chisel, rasps, plane, and sanding discs.

Then fit it more exactly by burning the hot socket over the handle end. Heat the socket just hot enough to scorch the wood, but not so hot that the smoke will turn into flame. This would char the wood severely. Once it is fitted in, gently tap the end with a hammer to seat the wood evenly. Clamp it between the vise jaws to further pinch it together. While it is in the vise, drill a hole through both socket and wood and countersink it to receive a steel pin. The pin should be long enough to rivet shallow heads flush with the steel of the socket on each end.

Make a ferrule and ring, as shown, to hang the shovel on a wall hook.

Apply a surface finish on the metal by steel-brushing and then waxing or oiling it. Sand the wood handle down as fine as you can, then wax and polish it.

forming shovel blades over scrap steel parts, or sand, earth ---- with ball hammers

1 heated area

2

3

4

5 anvil   form over

rivet after assembly

dry natural curved fruit wood branch

heat to burn-in branch

cut handle ferrule from scrap tubing & crimp onto wood

cut ring from scrap coil spring

heat locally & bend, next:

heat locally to open & while hot

hole to fit ring

use tongs to press in position.

173

½" square mild steel 15"

sharp right-angle bend

2"

upset to form tenon

½"

1½"

½"
⅛"

¾"

draw out ¼" x ¾"

clamp in vise and twist

yellow hot

cut hot on anvil hardy

draw out hot

hammer curves hot

washers

brass countersunk rivets are nail sections

rivet handle onto blade

rasp, file, sand-down & polish wood.

rub steel with oil, but leave forge-oxide finish natural

A DECORATIVE STEEL SHOVEL HANDLE

First, make a shovel blade in the manner described on page 173.

To make the steel handle as suggested in the accompanying illustration, first upset the steel bar at one end to provide the extra thickness needed for a short shoulder and tenon. The square tenon is forged and filed into shape to fit snugly into a *square hole*. This hole is drilled and filed, or punched, in the shovel blade. In the final assembly, the handle is riveted cold onto the blade.

The decorative design of the handle begins with a right-angle bend. The portion that is to receive the wooden handle-grip is flattened and widened somewhat, as in the illustration. Between the handle and shovel blade, the steel is forged into a rectangular bar ½ inch by ¼ inch in cross section. If the metal is high-carbon steel, this section is most easily twisted *hot*. If it is of mild steel, it may be twisted cold.

Note that the advantage in twisting steel cold is that the perfectly even temperature makes the screw lead evenly spaced. If the metal is heated, the slightest unevenness of heat will make the screw lead unavoidably uneven. However, sometimes such unevenness is desirable, as it gives a hand-rendered character.

The right-angle bend is cut on the anvil hardy as shown. Its two arms are drawn out into slender tapers and curved freehand.

Drill the handle-grip section for the rivets that must hold the two wooden handle sections. Scribe off the hole locations on two identical pieces of hardwood of your choice. With a wood rasp, make these pieces fit approximately on the steel handle. Further perfect the fit by heating the steel, then quickly assembling wood and steel with short pins and clamping these three elements between the vise jaws. The hot steel should scorch the wood, not burn it. Left in the vise jaws to cool, these wood sections will fit to perfection without harming the wood.

To permanently assemble the parts, follow the illustrations. The color of the wood will be brought out when rubbed with beeswax and polished. The steel surfaces can be finished to suit your taste. They can be lightly steel-brushed and oiled to preserve the natural forge black. Or they can be slightly buffed and color-patinated, then sprayed with acrylic as an added rust-resistant.

A LARGE FIREPLACE SHOVEL

Since these shovels never undergo the strains of a garden shovel, the steel stock can be of a modern, malleable type that can be cold-formed easily, once the major deep shapes are hammered out hot.

Anyone well acquainted with the folding characteristics of cloth and of paper is well prepared to plan how to fold and bend sheet metal. Softened by heat, sheet metal yields to shaping in much the same way. The one advantage sheet metal has over paper and cloth is that it can be *stretched* and will maintain the form given it. Therefore, the smith may

174

combine the two possibilities, folding and stretching, in his forming techniques and can make beautifully shaped articles of fairly thick sheet metal. If you have made special hammers of various shapes to curve the blades of woodcarving gouges, these will now come in handy.

First cut out a pattern that is larger than the final blade is to be.

Form the handle socket over a sand or earth mound at first, and further finish it over a bick.

Next, turn the piece over, heat one side of the blade and hammer it upward. Do the same with the other upright of the blade. Form these two uprights to meet the socket portion. Several more local heatings will be needed to blend all curves harmoniously, as in the illustration of the final result.

The separate bottom section of the shovel is also hammered over sand, forming the matching part of the handle socket as well. It must fit the contours of the top half of the blade. Use as many heats as required to make the final shape, as shown.

The wooden handle is made from a curved branch of fine-grained fruit wood. Shape one end of it to fit the finished shovel socket.

If you find that the two shovel sections, placed over the wooden handle end, do not meet precisely, you will now find the malleability of the cold steel to your advantage. Clamp the top and bottom handle sections in the vise with the wooden handle inserted in place.

With the slender cross peen of a lightweight hammer, tap along the creases of the folds of both upper and lower parts. Follow up with shallow rounded ball hammers and smaller flat-faced (slightly crowned) ones. Use these to bend the top and bottom sections together in a tight fit around the wood.

Drill the holes for riveting the two shovel parts together. With a rotary file, trim off the burrs left by the drill, and countersink the holes as deep as the thickness will allow without enlarging the holes. The rivets will then hold firmly, without protruding too much on the outside or inside of the blade.

Short sections of annealed nails that fit the diameter of the holes can serve as rivets. Cold-rivet the shovel parts together while they rest on the anvil face. No grinding or filing will be needed if the length of the rivets is just enough to fill the countersunk holes.

Wherever the bottom piece does not touch the top closely, it can be hammered *between* the spaced rivets, bringing these sections flush together. This forced bending can be neatly done over a large-diameter rounded end of a branch of wood clamped in the vise.

The end of the handle can now be inserted for the permanent fit. Heat the steel blade socket just hot enough to scorch the wood lightly, and hammer the handle into the socket with rapid light blows, using a 1- to 1½-pound hammer. The heated steel will scorch down every interfering unevenness of the wood.

Next, holding the whole assembly together by its handle socket, clamp the uprights of the lower blade section onto the handle between the vise jaws and drill the rivet hole through metal and wood for the holding rivet, which is now installed and headed.

Provide the handle end with a tight-fitting ferrule and crimped-on ring to hang the shovel on a wall hook. Cut the ring from a coil spring and shape hot. Bend the ends out somewhat to reach the ferrule holes. Heat the middle section of the ring, open it, and, after inserting the ends in the ferrule holes, close it between tong jaws.

All surfaces can now be smoothed and finished as in previous similar projects.

paper folds approximately as sheet metal bends

heat locally & form sheet metal on sand pile

sand

final result

use any suitable size & shape hammer to draw-out various forms

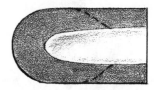

bottom half of handle-socket riveted onto shovel blade

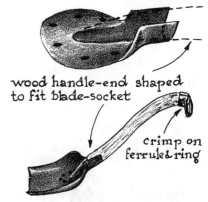

wood handle-end shaped to fit blade-socket

crimp on ferrule & ring

use natural curved fruit tree branches for the shovel handle

25. Making a Small Anvil from a Railroad Rail

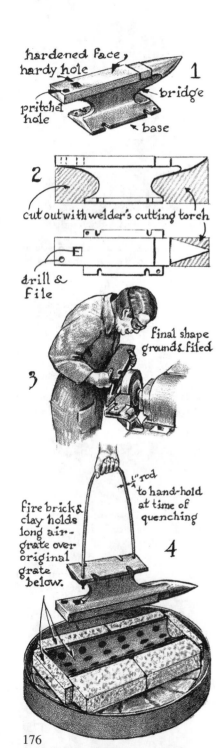

hardened face,
hardy hole
pritchel hole
bridge
base
1

2
cut out with welder's cutting torch

drill &
File

3
final shape
grounds & filed

fire brick &
clay holds
long air-
grate over
original
grate
below.

1" rod
to hand-hold
at time of
quenching
4

It is of first importance to determine if the salvaged section of a heavy-gauge railroad rail is of temperable quality. Illustration 2 shows where the welder must cut out the sections not needed. (Save these for your scrap pile, however, so they can be used for future projects.) To test the temperability of the steel, heat one of the waste sections to a cherry red and quench in water. File-test it, and if it has become brittle-hard, you will know that you can make of the blank a fine, well-tempered anvil large enough to hammer out most forgings made in the average hobby shop.

The design of the anvil in the illustration should be followed closely. The large, 1 hp motor-grinder with 12-inch-diameter hard, coarse-grit wheel will enable you to grind the proper shape of the horn and flat surfaces (3), instead of hand-filing them. The anvil face should be smooth and polished.

Since the steel comes more or less annealed, you can drill the pritchel hole. Next, the larger hole for the hardies must first be drilled and then filed into a square. The four holes, or notches, in the base are to bolt the anvil down on a wood stump. Two small holes in the base are drilled for a temporary handle with which it can be lifted when being tempered.

Since this anvil is too long to heat *evenly* for tempering in a small, centrally heated forge fire, it will be necessary to convert the forge temporarily (see illustration 4.)

To make a *long* air grate, I have found it practical to use a salvaged broken 4-inch-diameter cast-iron plumbing pipe. Cut a 14- to 16-inch section lengthwise on the abrasive cutoff wheel or mechanical saw (if you have one). Drill a dozen or more ⅜-inch holes, evenly distributed, to form an elongated air grate.

Surround this grate with firebrick, as shown. Dry, porous building bricks could also be used for a temporary project such as this. But make sure they are *dry,* as they might explode during heating if moist and not porous enough.

Plug all remaining air passages around the bricks with fire clay so that the air can reach only the long mound of coals over the grate.

Prepare to handle the anvil *upside down* by fastening a ¼-inch-diameter bent rod to the base as a yoke. Thread the rod ends for ¼-inch nuts and slip them through the holes in the base of the anvil. Make the handle long enough so that your hand is far from the fire (about 2 feet above the anvil).

Before you start the fire for tempering the anvil, be sure to have a 50-gallon drum full of water ready to quench it. Also have ready a shallow tray with an inch of water to cool the anvil face.

Make a *clean* fire with at least a 4-inch-thick layer of hot coals. Carefully place the anvil upside down on it and, with the poker, bank the coals over its "head." Cover the fire with sections of dry asbestos cement sheet (see illustrations 5 and 6). This is a practical and efficient way to enclose the fire and these sheets are often available as waste material at building-supply yards. Sheet metal, of course, would do, but its heat radiation would be uncomfortable.

It may take an hour with an even, slow, and low heat to bring this bulky piece of steel to medium cherry red heat. From time to time lift and peer underneath the asbestos cement sheets to check how the heating of the steel has progressed. Rake a little added coal to keep up the fire, when necessary, and continue the heating by cranking the air steadily, but *very slowly*. If the blower is driven by an electric fan, control the air flow to maintain this very slow, steady heat.

As soon as the moment arrives to quench, cool the top of the yoke handle with a wet rag to be able to hand-hold it. Remove the asbestos sheets and rake the coals away. Lift the hot anvil from the fire. (The exposed fire now is radiating much heat and should be quickly covered again with the asbestos sheets to make certain that nothing around it will catch fire.

Plunge the anvil into the 50-gallon drum filled with water. Sink it as deep as you can reach, while pumping the anvil up and down. This cools the steel fastest and makes the hardness penetrate the anvil face as deep as it can get (7).

Clean the anvil face of its scales and carbon, and restore the polish until it shines like a mirror. Replace the yoke handle by another one, attached as shown in illustration 8. The anvil now stands right side up.

Clean the slag out of the fire and replenish it with fresh coal evenly spread over its full length. You need a lesser fire this time; it must give just enough heat for the temper colors of light bronze or dark straw to appear on the face of the anvil.

Wait for the fire to become clean and smokeless once more. Place the anvil, suspended right side up, on the bed of hot coals. Cover it with the asbestos shields, leaving a ¼-inch space between the sheets and bridge of the anvil (see illustration 8). This allows the fire's heat to flow *evenly under and around the anvil head,* which has remained free from direct contact with the fire. The anvil horn may be ignored for the time being because it will be annealed separately later on.

Since the heat flow should be even and very slow, fan the fire only very gently. Watch for the slightest faint straw color to appear on the anvil face. It is now important to shift the hot coals with the poker to any spot below the anvil where the polished anvil face has not yet begun to show the faint straw color. This is to try to even up the distribution of the heat flow. As long as the heating of the anvil body is very, very slow, such corrective measures will be effective.

hot coals 3" layer below anvil & around

horn to stay relatively cool

5

plug with clay all open spaces

blocked-in fire shown with sheet asbestos or asb. cement shingles

6

leave ¼" space around anvil bridge

a very slow, even heat is kept. at moment that anvil body has become cherry red,

lift out & quench

30 gallons water

7

next polish anvil face mirror-smooth &

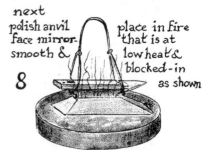

place in fire that is at low heat & blocked-in as shown

8

the steady low heat gradually reaches anvil facing, showing temper colors. When straw yellow, place upside down in in shallow water & let cool

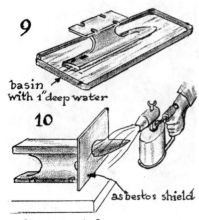

9

basin
with 1" deep water

10

as bestos shield

after torch flame anneals
horn to pale blue, let all cool
slowly.

When, at last, the color is drawn evenly over the whole anvil face to a dark straw or light bronze, lift the anvil by its strap handle, and with tongs, carefully tumble it upside down. Quickly, in this position, lift it, with a tong in each hand, and place it in a shallow tray of water (9). The anvil face is thus prevented from being further heated through conductivity by the remaining stored heat in the anvil base and bridge.

Now leave the whole to cool slowly. Once cooled, the horn, which always is to remain soft, can be annealed separately, as shown in illustration 10.

I have made four small anvils in this way. Each one was of high quality with a hard, tough anvil face and softer annealed horn. They have proved to be as good as any larger commercial anvils I have ever had. I bolt them firmly on a heavy wood stump and ignore their light caliber, using them as if they were 100-pounders.

26. The Power Hammer

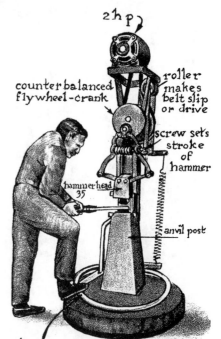

The value gained in using a power hammer is mainly in drawing out heavy-gauge steel from thick into thin, short into long, and narrow into wide, without much physical effort.

Sometimes the power hammer can also help in upsetting thin steel into thick, but this is quite tricky to do because a heavy blow will buckle thin upright parts, and a lighter blow will not penetrate to the center. As a result, a cauliflowering of the edges of the thin upright takes place, and this must be flattened immediately again if folding is to be avoided. If a fold is detected, it must be ground or filed out before proceeding.

Of course, repeated cauliflowering and flattening at long last does make the total dimension a little thicker. But again, the trickiness of this operation will generally cause you to abandon the attempt and to start all over again with heavier stock. In time you will learn that there is a limit to the upsetting of thin stock.

You can extend the uses of a power hammer by making special hammer and anvil inserts. With these you can form, with a few blows, a gouge blade for instance, which otherwise would have taken you half an hour or more to forge by hand.

A whole book could be written for the hobbyist on the use of the small power hammer. But it is my belief that this elaboration is not called for here. The reason is that by the time the student has diligently practiced the making of the projects offered in the foregoing pages, he will be fully prepared to judge whether he will benefit from adding this machine to his equipment.

My personal experience has been that the power hammer has been useful when I needed a sledger-helper. So my admonition remains: Learn first all that a blacksmith must know about freehand forging; only after that will you be able to make the greatest use of the machine as a time-and-muscle-saver *while remaining in full control of it.*

Do not underestimate the danger of machine-hypnosis. It is a trap which you must try to avoid falling into. Often the less-talented, the commercially oriented, the non-artist, and the vocational machine operator yield to this hypnosis.

I recognize, however, that for the making of mass-produced items that lend themselves to power-hammer treatment, you can increase your chances of earning a livelihood with it. Should this come about, all that you have learned to do in blacksmithing *without* the machine will enhance your work. At the same time, you will have the satisfaction of knowing you could do as good a job with the simple use of a fire, a hammer, and an anvil.

Logic, skill, and common sense, which are a good blacksmith's attributes, will guide you from now on when untried steps must be taken. You will discover, with pleasure and satisfaction, that you have become your own teacher.

down pressure controls degree of speed of hammer movement as well as force of blows, based on belt slippage to full run

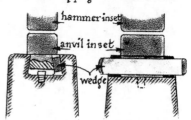

anvil & hammer-sets have identical footing

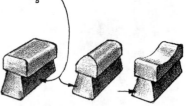

flattening   peening   forming

infinite variety insets possible to suit one's need

curving   cutting   tapering

gouge blanks easily forged on a power hammer with special insets

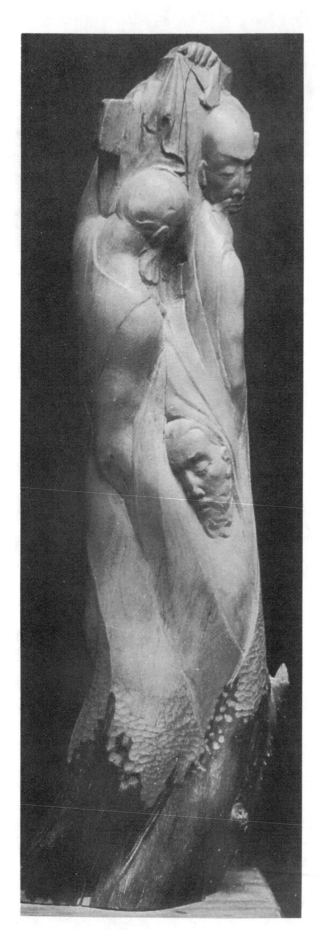

Left: **Maylaya,** lifesize, carved by the author in Java with the tools he forged in the native blacksmith shop described on page 8. *Right:* Endgrain wood engraving, cut with traditional engraver's burins which serve as a basis for the design of small woodcarving gouges, as shown on opposite page.

Bali Mother and Child, carved in sandstone with one-point and claw stonecarving tools described on page 136. Weyger's sculptures are all direct carvings, done without benefit of prestudies, drawings, models, or measuring devices. Photo: Jim Ziegler.

Descent from the Cross, madrone wood, 12 inches high, Alexander G. Weygers. This piece was carved with small hand gouges like those shown on page 181.

Sister Joanne, O.P., art teacher at Dominican College, San Rafael, Calif., a pupil learning to handforge stonecarving tools at Weyger's blacksmith shop. Photo: Jim Ziegler.

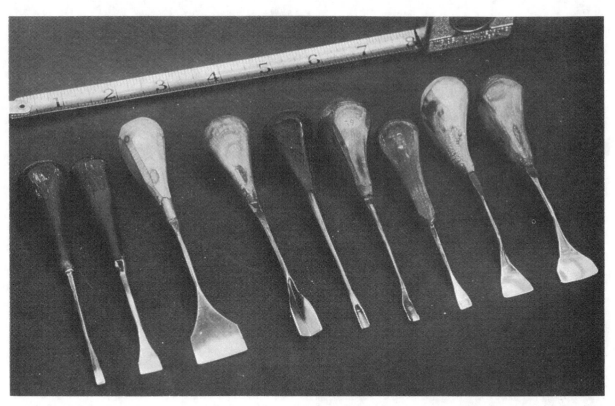

These small woodcarving gouges are designed to be manipulated in the manner of engravers' burins.

Above left: Peter Partch, a student, carves a low relief in marble using his own forged carving chisels. Photo: Jim Ziegler.

Above right: The inner bowl cavity was carved with the special tool shown. Standard sculpture gouges were used for carving the outside.

Right: Collection of tools forged by the author's students during a three-week sculpture workshop. Photo: Jim Ziegler.

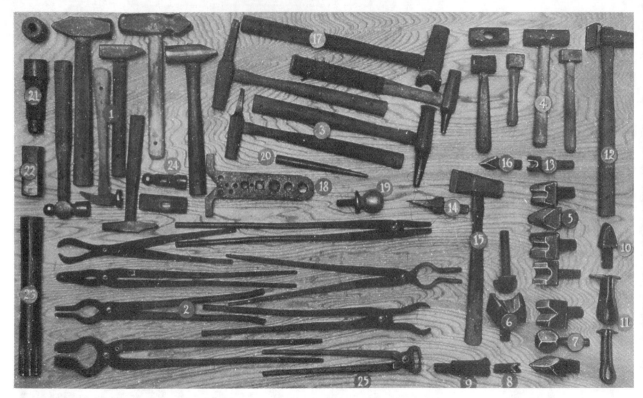

Typical blacksmith tools: 1 hammers, various weights and shapes / **2** tongs, various sizes and shapes / **3** hot punches, square- and round-ended / **4** stonecarving hammers, various weights (mild steel) / **6** cone-shaped bottom swage and matching fuller, to form blades of woodcarving gouges / **7, 8** angle-shaped swage and matching fuller, to form blades of large and small V-shaped carving gouges / **9** die to form a reinforcement rib in gouge or garden-tool blade / **11** jigs which may be clamped between vise jaws / **12** set-hammer or flatter / **13** hardy / **15** hot chisel / **17** top swage / **18** heading plate with holes of various sizes to form bolt heads (from 3/4-inch-thick scrap steel) / **19** car-hitch ball, used as forming head to shape curved blades of implements / **20** drift pin to enlarge holes started by hot punches / **21** upsetting matrix made from a section of a heavy truck axle; fits hardy hole of a large anvil / **22** squared section of a heavy truck axle from which a blacksmith hammer is to be made / **23** remnant of a heavy truck axle.

Samples of scrap steel for the modern blacksmith: 1 section of a car bumper / **2, 21** coil spring of a car / **3** section of a plow disc cut with welding torch / **4** half a 4-inch-diameter cast-iron plumbing pipe / **5** angle iron from bed frame / **6** section of a broken pick axe / **7** parts of a steering-gear rod / **8** set of car leaf springs / **9** half of a large ball-bearing race / **10** old wood-splitting wedge / **11** bent 1- x 5/16-inch bar / **12** old flat file / **13** section of a car bumper / **14, 22** old phonograph spring / **15** farm machinery spikes / **16** large bolt nut / **17** hub socket wrench / **18** motor valves / **19** industrial hacksaw blade / **20** cast-iron valve-box lid / **23** ½-inch round bar / **24** large tie-bolt washer / **25, 39** die parts / **26** half of compound shear / **27** old carpenter's flat chisel / **28** hub socket wrench / **29** double magnet / **30** lever linkage / **31** cast-iron bridge shoe / **32** ball-bearing thrust washer / **33** broken crowbar / **34** farm machine cutter blades / **35** engine pushrods / **37** old Ford car magnet / **38** thrust block / **40, 44** old wrench / **42** collar / **43** screwdriver blank / **45** hay-rake tine / **46** scrap-steel-plate remnant / **47** 1- x 1-inch high-carbon-steel bar from a harvester's drive shaft / **48** part of a carpenter's handsaw / **49** shaft end and bevel gear / **50** part of a car stick shift, and lawnmower blade / **51** ball bearings / **52** antique car springs / **53** car linkage bar.

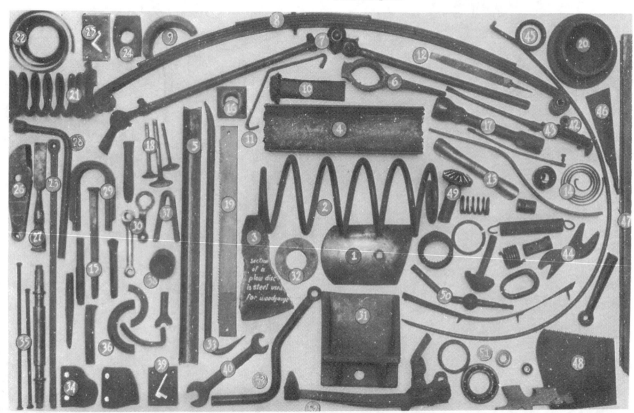

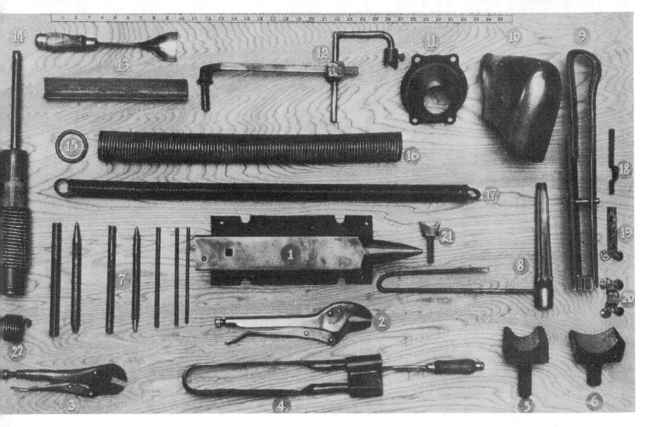

Modern and traditional blacksmith tools and scrap-steel stock: 1 small anvil (hardened face) made from a section of heavy-gauge railroad rail / 2, 3 self-locking pliers (modern vise-grip) / 4 hand-held die-set to form small, round woodcarving gouge blade / 5 standard large cylindrically ground bottom swage, used as stock to make # 6 / 6 special wide-curved cone-shaped blade for a large woodcarving gouge / 7 various sizes of high-carbon-steel rods to make one-point stonecarving tools / 8 hexagon die punch to shape hexagon holes in making box wrenches / 9 hand-held die-set to form keepers for door latches / 10 part of car bumper, which, when clamped in vise, acts as a forming block over which curves are shaped for blades of gouges, shovels, etc. / 11 salvaged spherical section of a ball-joint housing to act as forming block to shape curves / 12 universal hold-down tool / 13 section of a high-carbon bed frame that was used as stock for wood gouge (# 6) / 14 jig, clamped in vise, serves to uncoil heated coil springs into straight rods / 15 single winding of a coil spring / 16, 17 salvaged coil springs from which to make high-carbon, straight steel rod, used as stock for forging small artifacts / 18, 19, 20 keeper blank, forged into final shape / 21 bottom swage made to fit anvil (# 1).

Useful items forged from scrap steel and special hammers used to form shovel blades and gouges: 1 one-piece fireplace tong / 2, 10 fireplace pokers / 3, 8, 9 fireplace shovels / 6 stove-lid lifter / 11, 12, 4 hammers converted from standard hammer heads, to shape special curves into specially made forming swages.

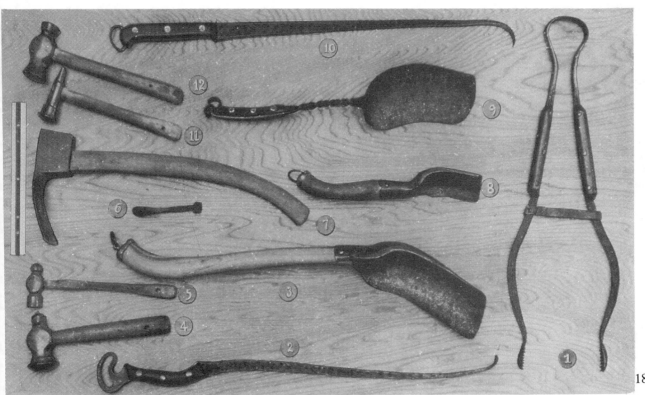

THE
RECYCLING,
USE, AND REPAIR
OF TOOLS

CONTENTS

Introduction

The scrap steel yards across the country are full of every conceivable metal object discarded for reasons of wear, obsolescence, or damage. Much of this material can become useful stock for the beginner, as well as the skilled metal craftsman, who intends to "make do" with what can be gleaned from this so-called junk.

Those who must fabricate items that can be made only from new bar and plate steel if they are to compete in the commercial market cannot afford to spend the time it takes to work with salvaged odds and ends. They will help to feed the scrap pile instead of being fed by it.

I have for many years practiced and enjoyed making something out of nothing in constructing a tool or reconditioning a junked piece of machinery. My aim in writing this book has been to compile information on how to do just that for the person with a combination machine shop and smithy.

It is through actual demonstration, seeing how to manipulate tools to make tools, that I believe the student benefits most. But short of that one can learn from books in which the illustrations come as near as possible to live demonstrations. I have tried to present the information in such a way that the reader can imagine he is watching me making things in the shop.

That approach has been put to the test in my previous books, *The Making of Tools* and *The Modern Blacksmith*. Since their publication a few years ago, readers have reported results that underscore my conviction that the same guidelines to self-teaching should be continued here. I have done this in the hope that it will be possible for those who diligently follow these instructions to advance and round out their know-how with this extended information.

If a self-employed and independent smith can be a good machinist as well, there is great promise that in time his abilities will prove to be a better security than any money in the bank.

1. How to Repair Broken Garden Tools

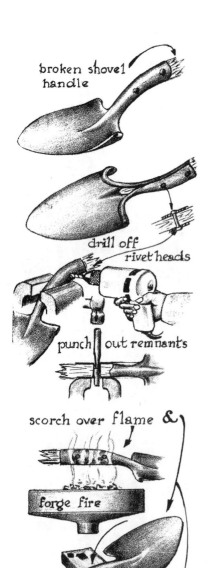

broken shovel handle

drill off rivet heads

punch out remnants

scorch over flame &

forge fire

knock out handle stump

Most long-handled garden tools break where the wooden handle enters the steel part of the tool. Not knowing how to replace a handle, most people throw the whole tool away and buy a new one. Their loss is our gain if we like to recycle waste, have a modest shop to make such repairs, and have the time to do it.

Removing the Handle Remnant from the Shovel Blade Socket

The wooden stub is held in the socket by two rivets. Clamp the socket in the vise as illustrated, showing the countersunk depression in the socket. With a 3/8-inch drill, or similar sized tool appropriate to the task, clear the depression of the rivet head. Next, straddle the rivet location over the partially opened vise and punch out the rivet remnant. The handle stub can now be knocked out. If this proves too difficult to do, heat the shovel socket over a flame until the wood begins to scorch. The stub can then easily be released and the new handle installed.

Installing the New Handle

If you live near an orchard, the farmer may let you have a branch from his fruitwood prunings. Choose a fairly straight one, thick enough to make two handles. Saw the branch in half lengthwise. After trimming, each handle cross section will be oblong, which will do just as well as handles that have round cross sections. In selecting the branch, try also to pick one with a natural curve ending that approximates the curve in the shovel blade socket. Shape the end to fit the curve closely.

Heat the handle socket to the point of scorching the wood, then quickly clamp the shovel in the vise and hammer the handle home. If you do not succeed on the first try because the handle does not match the socket curve well enough, you may have to preheat the end before hammering it into place. You can easily accomplish this heating (not shown) by dipping the curved end in water and holding it in the flame until the water has evaporated and wood begins to scorch. Repeat this operation maybe four or five times. Between the steam of the evaporating water and the heat of the fire, the entire curved end becomes heated to the core, momentarily making the wood limp. At the same time the handle socket should be heated to the point of scorching wood. Then quickly clamp the blade in the vise and hammer the limp handle into the socket. It will easily follow the curvature of the socket under the hammerblows on the other end of the handle. Following the curvature of the socket, the limp end flexes as if it were made of rubber.

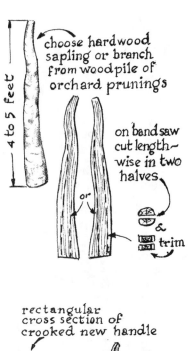

choose hardwood sapling or branch from woodpile of orchard prunings

4 to 5 feet

on bandsaw cut lengthwise in two halves.

or

& trim

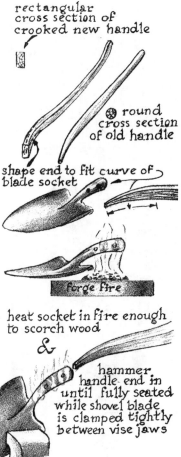

rectangular cross section of crooked new handle

round cross section of old handle

shape end to fit curve of blade socket

forge fire

heat socket in fire enough to scorch wood

&

hammer handle end in until fully seated while shovel blade is clamped tightly between vise jaws

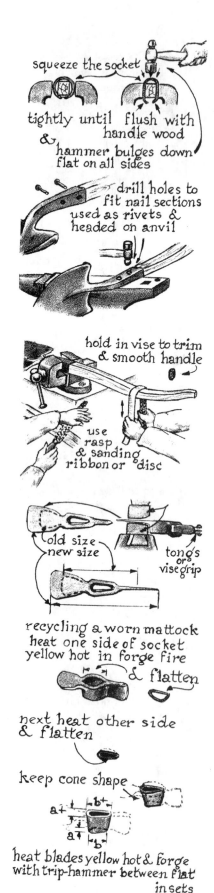

squeeze the socket

tightly until flush with handle wood

& hammer bulges down flat on all sides

drill holes to fit nail sections used as rivets & headed on anvil

hold in vise to trim & smooth handle

use rasp & sanding ribbon or disc

old size
new size

tongs or visegrip

recycling a worn mattock heat one side of socket yellow hot in forge fire & flatten

next heat other side & flatten

keep cone shape

a+ +b+

a+ +b+

heat blades yellow hot & forge with trip-hammer between flat insets

Let all cool off before taking the next step.

With the handle firmly seated, clamp the socket between the vise jaws until the steel sides of the socket press firmly onto the wood.

Since the shovel socket is round and the new handle cross section is oblong, the socket must be reshaped to hold the handle. Squeeze the socket onto the handle between the vise jaws as forcefully as possible. Hammer the bulging upper part of the socket flush with the flat part of the handle.

After the bulges on all four sides of the socket have been hammered down flat in this way, drill holes through the wood from both sides, meeting halfway to secure alignment. Choose a nail to fit the hole and clip off pieces to be used as rivets. Hammer them through the holes and, with the heads placed flush on the anvil facing, rivet them with a lightweight ball peen hammer.

Round off all sharp edges on the handle with rasps, abrasive ribbons, or small, rubber-backed disc sanders rotating in a hand-held electric drill.

I have used for years several shovels that have been recycled in this way. I enjoy their extra-long, slightly curved handles, which seem to lighten the work when I need to twist the shovel in levering action.

RECYCLING AN OLD WORN MATTOCK

Continued resharpening of the blades will in due time shorten them to the point that the mattock will no longer do what it was designed to do. All the same, the steel closest to the handle socket will remain thick enough to be peened out longer in the forge. This restores a blade long enough to make the tool useful once more.

Lengthening and Widening Both Blades

Although hand-peening the blades is not too big a job, you can do it in a fraction of the time with the regular flat-faced insets of a trip-hammer. This not only saves your back, but improves the steel's quality as well, because trip-hammer blows can be forceful enough to *pack* the steel most effectively.

Since this process makes the tool a lighter gauge than the original one, the old handle will be too large and clumsy. To narrow the opening for a lighter handle, follow the illustrations as guidelines. They show that the length of the socket hole itself will not change if one side is forged at a time and the other side remains cold (but not brittle). Hammering the hot side compacts the steel in an upsetting action. In this way, the handle socket becomes narrower, but not longer, than it was originally. The new, more slender handle will be strong enough for the more slender head.

Dressing the Blank and Hardening the Blade Edges

As always with tools of this size, the hard, coarse, and large motor-driven grinding wheel will do the best job of dressing.

Hardening the tool blade edges is somewhat simpler if (as with a mattock) both edges are far enough apart that heating one edge will not anneal the other.

Place one blade in the fire over a slow heat and bring the edge to a *cherry red*. Allow that heat to dissipate up to the handle socket. At this point, quench the whole blank in oil. Treat the other edge in the same way.

You will end up with a new, though recycled, mattock that will stand up under normal use for several years before it has worn to the point where rehardening is necessary.

grind blade ends sharp, next

heat cherry red one at a time
to harden ends only

& quench in oil one at a time

Installing the New Handle

Make a mattock handle from a salvaged fruit tree branch (as described in the making of shovel handles, page 9), fashioned into a shape to your own liking. The top of the handle socket, being somewhat larger than the bottom, allows the end of the split handle to be wedged open and seated firmly by filling the entire socket cavity.

This slender, lightweight mattock is a welcome addition to the bulkier, heavyweight ones, and you will find it useful in many different ways.

cut hardwood handle to fit
new socket & cut slot with saw

to take wedge which is
forged from flat stock $\frac{1}{8}''$

after wedge is driven in
drill hole through socket & wedge
for rivet

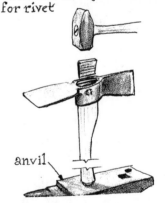

anvil

2. Making a Charcoal Brazier and Screening Scoop

charcoal brazier made from a salvaged truck head light housing & an ash basin from a large hubcap

Using a charcoal brazier to cook food on a picnic table can be very convenient. I have used the brazier described here for many years, on camping trips as well as at home.

In campgrounds where fires have been doused there is usually a quantity of unburned charcoal to be found. The screening scoop is simply used to separate the charcoal from the ashes, and in no time the brazier is full and ready to use. Good charcoal can be scooped from your home fireplace as well.

The brazier bowl is made from a car headlight, the ash basin from a hubcap. To light the fire, place a wad of crumpled paper below the perforated grate in the basin. If the charcoal is dry and the grate holes not too small, the fire should start easily. Let the paper burn completely before fanning the glowing coals with a piece of cardboard to speed up the spread of the flames. There will soon be a steady, smokeless, safe fire giving fifteen to twenty-five minutes of cooking time.

TO MAKE THE BRAZIER

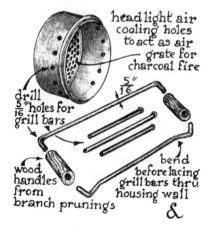

head light air cooling holes to act as air grate for charcoal fire

drill 5/16" holes for grill bars

wood handles from branch prunings

bend before lacing grill bars thru housing wall

5/16"

&

Try to find as large a truck headlight housing as you can in auto wrecking yards. The hubcap to catch the ashes should be larger in diameter than the headlight housing.

If the headlight cooling holes are too small, enlarge them to at least 3/8-inch in diameter. Place them close together to make it easier for the flames from the paper below to ignite the charcoal.

To make the grill over the charcoal fire, drill 5/16-inch holes evenly spaced along the top of the housing to take the 5/16-inch grill rods on both sides. Bend them as shown and lace them through the holes.

Assemble the two outer grill rods with wooden handles so that the brazier can be picked up safely while the charcoal is burning.

To install the four legs, follow exactly the instruction in the illustration. Their arrangement creates a self-locking hold between brazier and ash basin, spacing them in a permanent tight fit.

I find that a brazier of this type works as well as, if not better than, those that are made from heavy iron. Constructed from car parts, it weighs only a fraction of most cast-iron braziers.

install wood handles at bar ends pressed between vise jaws, thus straightening bends in grill bar

space leg holes in hub cap farther apart than in housing bottom to lock the two together

punch

file leg ends to fit snug in washer holes & rivet legs onto housing bottom with hollow end punch while leg foot-end is clamped between vise jaws

housing bottom

long riveting punch extends well beyond top rim of housing for easy hammering

hub cap basin

scoop to gather & screen charcoal from campfires or home fire places

retaining fence cut from a scrapped roofing sheet steel panel

fold along lines & flatten edges around end wires & cinch with rubber mallet on anvil

make the tines from scrap guy-wire cable

untwist into &

$\frac{1}{8}''$

cold hammer straight on anvil

THE CHARCOAL SCOOP

A piece of salvaged galvanized sheet metal that can be cut with tinsnips is easily shaped and folded as illustrated. It acts as a retaining fence while the charcoal is shoveled up. Small remnants of guy wire cable, often thrown away by electrical companies, can be used for the tines of the scoop. Untwist the cable section and hammer each wire straight on the anvil. Curve a mild steel bar slightly, as shown, for the assembly of the tines. About sixteen tines is a suitable number. Drill the holes into which they will fit snugly, then hammer-lock them tight. The hammering actually squeezes the steel bar sections onto the tine endings.

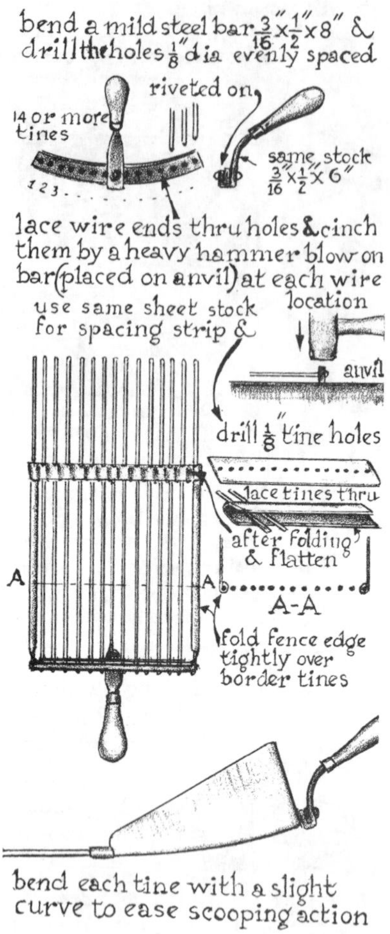

bend a mild steel bar $\frac{3}{16}'' \times \frac{1}{2}'' \times 8''$ & drill the holes $\frac{1}{8}''$ dia evenly spaced

14 or more tines

riveted on

same stock $\frac{3}{16}'' \times \frac{1}{2}'' \times 6''$

1 2 3 . . . .

lace wire ends thru holes & cinch them by a heavy hammer blow on bar (placed on anvil) at each wire location

use same sheet stock for spacing strip &

anvil

drill $\frac{1}{8}''$ tine holes

lace tines thru

after folding & flatten

A   A

A-A

fold fence edge tightly over border tines

bend each tine with a slight curve to ease scooping action

Use the same mild steel stock to bend a small handle holder, which serves as a tool tang. Rivet it onto the curved bar. The wooden handle can be burned onto the tang.

Cut a tine spacer strip of the same sheet metal used for the scoop wall. Punch through it evenly spaced small holes like the ones in the steel bar. Fold the metal strip along the row of holes and lace the tines through them.

Once the strip is in the correct position, flatten it over the tines with a rubber hammer, using the anvil facing as a base. The sheet metal strip easily yields under the hammerblows so that it is pressed out between the tines. This locks them in position so they will not obstruct the scooping up of the charcoal, yet the whole assembly is held firmly together.

3. A Candlestick

The round disc cutouts used here were cut from heavy plate steel on a factory milling machine. The plate out of which the discs were cut was used to reinforce a cylinder that was subjected to enormous pressure from within. I was fortunate to come by a great quantity of these waste discs. It proved very much worth my while to make special trip-hammer insets to aid me in making various articles from this excellent scrap material. The base and drip cup of a candlestick were made of these, as decribed in Chapter 26. The candlestick shown here makes use of these discs.

Drill a hole in the center of the large rosette that acts as the base. Tap into it a 1/2-inch thread to take the threaded candlestick column. Thread a similar hole in the drip cup to fit the threaded top of the candlestick.

Make the twist in the column of the candlestick by heating one half of the rod, twisting it, and cooling it. Then heat the other half and twist it counter to the first, giving variety to the design.

Cold forge the candle socket from a section of pipe and thread it to fit the little extra protruding thread in the center of the drip cup.

Mild steel conduit pipe can be cold forged to a remarkable extent, but because these pipes are bonded with a welded seam it is only by compacting that they can be made into cones as shown. Expanding the diameter of the pipe, however, cannot be done by cold stretching. Hammering the pipe ending over a cone would only break the weld. A very small amount of stretching can be done by hammering the end *very carefully* over a bick or an anvil horn in a peening action.

If you are lucky enough to find on a scrap pile some seamless tubing, it can be stretched outward by force to a certain extent without peening. The illustrations clarify this further.

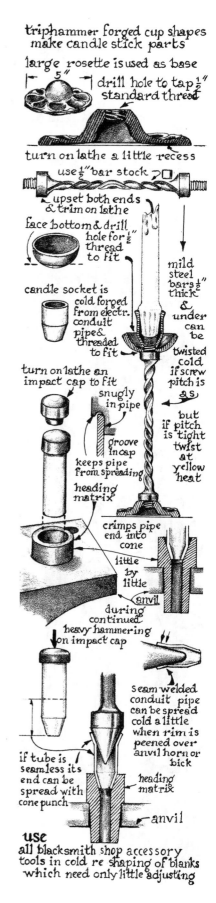

triphammer forged cup shapes make candle stick parts

large rosette is used as base

5" drill hole to tap ½" standard thread

turn on lathe a little recess

use ½" bar stock

upset both ends & trim on lathe

face bottom & drill hole for ½" thread to fit

candle socket is cold forged from electr. conduit pipe & threaded to fit

mild steel bars ½" thick & under can be twisted cold if screw pitch is as

turn on lathe an impact cap to fit

snugly in pipe

groove in cap keeps pipe from spreading

heading matrix

but if pitch is tight twist at yellow heat

crimps pipe end into cone

little by little

anvil

during continued heavy hammering on impact cap

seam welded conduit pipe can be spread cold a little when rim is peened over anvil horn or bick

if tube is seamless its end can be spread with cone punch

heading matrix

anvil

use all blacksmith shop accessory tools in cold re shaping of blanks which need only little adjusting

4. Making Tool Handle Ferrules and Shoulders

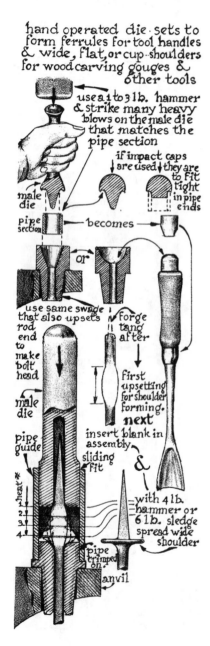

hand operated die·sets to form ferrules for tool handles & wide, flat, or cup·shoulders for woodcarving gouges & other tools

use a 1 to 3 lb. hammer & strike many heavy blows on the male die that matches the pipe section

male die

if impact caps are used they are to fit tight in pipe ends

pipe section

— becomes —

or

use same swage that also upsets rod end to make bolt head

forge tang after

first upsetting for shoulder forming.
next
insert blank in assembly

male die

pipe guide

sliding fit

&

with 4 lb. hammer or 6 lb. sledge spread wide shoulder

heat *
1.
2.
3.
4.

pipe trimped on·

anvil

Tool handle ferrules are made in the same way as the candle sockets described in Chapter 3. The only difference is that many tool handles do need to have the small diameter cone endings if tool shoulders are fairly small.

MAKING A JIG FOR STRAIGHT FERRULES

Straight ferrules without cone endings are required for tools with large diameter shoulders. Attempts to make such wide shoulders without a jig will prove to be not only very frustrating, but time-consuming as well. Thus it will be worthwhile to construct a jig, as shown in the illustrations.

Error during the upsetting of the rod to make the shoulder will require constant straightening up of the bends that are formed. The jig prevents this by acting as a guide during hammering and makes it possible to correctly distribute the upset rod section. Once any danger of bending has been eliminated, the male part of the die can be hammered down with a heavy sledgehammer to spread the shoulder enough to take a wide diameter ferrule.

TOOL SHOULDERS

The combination of ferrule and shoulder in one forging gives the handle a streamlined look but does not necessarily make it any more functional. The tool serves just as well if it has only a very small shoulder pressed onto a tight-fitting heavy washer, which in turn reaches the rim of a ferrule that has been pressed on the wood of the handle. So it is up to the craftsman who makes his own tools if he wants to spend extra time and effort simply to make his tool more attractive with such a combination of shoulder and cup washer. Shoulders can also be formed into cup shapes in a set of hand dies made for that purpose.

Once the jig has been made for the wide shoulders and ferrule cups, making the whole tool would not take much longer. It is making the jig, not making the tool itself, that takes the time.

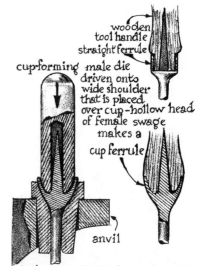

wooden tool handle
straight ferrule

cup-forming male die driven onto wide shoulder that is placed over cup-hollow head of female swage makes a cup ferrule

anvil

stock must fit die hole accurately without binding. During forging cup shoulder is formed yellow hot while hammered in place.

5. A Pump to Recycle Waste Water

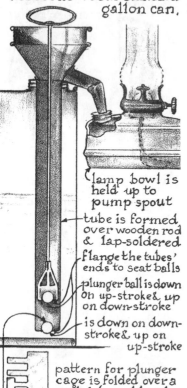

pump to fill lamps with kerosene from standard gallon can.

lamp bowl is held up to pump spout

tube is formed over wooden rod & lap-soldered

flange the tubes' ends to seat balls

plunger ball is down on up-stroke & up on down-stroke

is down on down-stroke & up on up-stroke

pattern for plunger cage is folded over a rod & lap-soldered

toy marbles act as valves

pattern for funnel part of pump top is joined with soldered lap joint

hole over which pump spout is soldered

after strip is bent in circle & soldered, it is flanged top & bottom, & folded over the funnel flange or soldered on

spout pattern is folded & bent, soldered at elbow & flanged at base to fit onto funnel & soldered to it

To make such a pump I followed a very simple design used throughout the Orient, where people pump kerosene from a five-gallon can into the little lamps they use every night.

It was in Java, Indonesia, that I saw such a little pump made before my own eyes. In the marketplace the tinsmith soldered together the various parts he had cut out with tinsnips from flattened five-gallon cans. None of the moving parts had been machined or put together in a machine-precisioned way.

The pumping is done by quick, short up-and-down movements of the plunger. There is some backflow leaking between the plunger and the cylinder, but it somehow does not reduce the pump's effectiveness. The liquid rises quickly to fill the open top part of the pump body so that the kerosene gently flows by gravity through the small spout into the lamp bowl.

It is this design, as I adapted it, that we are now following in the making of a larger pump to recycle waste water during a drought year.

200

It must be remembered that each person wishing to make such a pump must do it from the odds and ends he has accumulated and that his supply is bound to differ from mine. The final appearance of his product, therefore, will probably be quite unlike the one I am showing here.

First note that the waste water from the bathroom tub and washbowl is collected in a fifty-gallon drum. The idea is to pump the water high up into some kind of a holding tank (38) and to make a bottom valve arrangement (6–7) to hold the column of water above it—without leaking, if possible. In this way the water can flow by gravity out of the holding tank for as long as it takes the hose to distribute the water.

Place the top holding tank a few feet *above* the garden level so the water will flow easily to it by gravity. Hand-pumping from time to time will distribute the water, saving on the use of electric energy at the same time.

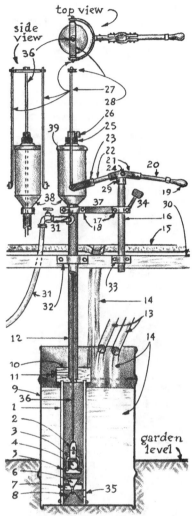

pump is to distribute rain & waste water to garden

1 pump body is made of waste plastic drainpipe material
2 copper link~3 copper wire cage~4 rubber ball valve
5 copper valve seat~6 rubber valve~7 copper valve seat
8 rubber band~9 water drum
10 wooden coupling~11 set screw
12 electrical conduit pipe
13 plastic drainpipe~14 water
15 outdoor porch~16,17,18 clamps & brace~19 pump handle
20 handle bar~21 yoke ~22-23 fork arm of handle bar
24 yoke hinge pin
25 rubber buffer~26 collar ~
27 two linkage rods~28 yoke
29 hinge body ~ 30 porch timber
31 garden hose take-off & bib
32 installation clamps~33 rain water from porch ~ 34 rubber stop for pump handle
35 intake holes~ plunger rod
37 brace   36
38 holding tank & lid 39

SELECTING THE MATERIAL FOR THE PUMP

Sections of ABS plastic drainpipes can often be found on house construction site scrap piles. This salvaged section was 4 inches in diameter.

The storage can is a fifty-gallon oil drum, often discarded or available at minimum cost in scrap yards. Coat it on the inside with hot roofing tar to make it rustproof.

Large diameter electrical conduit pipe is available from wrecking yards at very little cost or from heaps of waste thrown out during reconstruction of old buildings.

The three-gallon tank at the top happened to be on my own scrap pile. It was originally used in a milk separator machine. Anything else—a galvanized drum or small garbage can, for instance—would do.

The main pump rod (36) and linkage rods (27) are made from an unreeled coilspring heated at yellow heat and straightened (as described in my book *The Modern Blacksmith*). Hammer it out straight and remember that spring steel, even in an annealed state, remains resilient.

Large strips of steel are forged or bent into brackets or braces that can be anchored against a part of a building to brace the pump assembly spaced from the handle column in a clamping action.

All in all, the various small details shown in the illustration are likely to be accumulated, in one form or another, by most of us when gathering scrap steel, and especially by those who have geared their activities to such shop practices.

Once you have assembled the elements needed for this type of pump arrangement, begin with the major element.

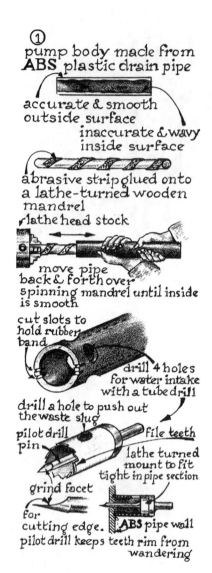

①

pump body made from
ABS plastic drain pipe

accurate & smooth
outside surface

inaccurate & wavy
inside surface

abrasive strip glued onto
a lathe-turned wooden
mandrel

lathe head stock

move pipe
back & forth over
spinning mandrel until inside
is smooth

cut slots to
hold rubber
band

drill 4 holes
for water intake
with a tube drill

drill a hole to push out
the waste slug

pilot drill
pin

file teeth

lathe turned
mount to fit
tight in pipe section

grind facet

for
cutting edge. ABS pipe wall

pilot drill keeps teeth rim from
wandering

THE PUMP BODY

The plastic pipe section I used proved to be smooth on the outside but not on the inside. Should this happen to be so in your case and should you have no other choice, you can make the inside accurate as shown in the illustrations.

First turn a wooden mandrel on the wood lathe. Glue onto it a cloth strip of coarse abrasive. It should barely fit the inside diameter of the pipe section. With the mandrel clamped in the lathe chuck, rotate it at normal speed. Slip the pipe section back and forth over the full length of this abrasive core. The pipe interior will thus be smoothed out.

THE INTAKE VALVE

It is here that you must apply some ingenuity to make do with the material that you have. My own decision was to let the total pump assembly rest upon the bottom of the fifty-gallon drum. This called for drilling intake holes 1 inch from the rim of the pipe bottom to allow the water to enter the pump chamber and placing the bottom valve just above them. I used a rubber toilet ball (6) for the bottom valve. To it I attached a strip of copper ending in a hook. I attached a rubber band to the hook to hold the ball down for a good seating. Each end of the rubber band was anchored in slots cut in the pipe rim.

NOTE: The holes in the plastic pipe body that let the water enter must be drilled with a drill that cannot wander sideways. A wandering drill in soft material might ruin the job.

Make a special drill with a 2-inch length of 3/4-inch conduit pipe. File teeth in its rim and crimp the other end on a little mandrel, which is turned on a lathe. It should have a protruding pilot pin as shown, which can act as a guide to the drill to prevent wandering. This mandrel can then be clamped into the drill chuck. Naturally, such special tools can serve for similar jobs in the future.

THE VALVE-SEATING DIAPHRAGMS

Over the years I have salvaged things made of copper, and for this job I chose an obsolete photoengraver's copper plate from which to make the valve-seating diaphragm.

Find a similar plate and cut out a disc with a cold chisel. Next, turn from a 3/8-inch thick scrap steel disc a base on which to form the ball-seating diaphragm. The illustration shows how to hammer gradually the edge of the copper disc around the 3/8-inch steel base. Once flanged over that base, the diaphragm will fit perfectly on the three-jaw chuck of the metal-turning lathe. Clamp it on and turn all hammered parts to an accurate fit after drilling a 5/8-inch-diameter hole in its center.

Next, rest the diaphragm on the anvil and place a 3/4-inch-diameter, large ball-bearing ball on the smaller valve hole in the diaphragm. Several gentle hammerblows on the ball will form a spherical seating in the copper, which yields readily to the blows.

It is for an operation like this one that I never pass up the chance to collect the large steel balls from enormous ball bearings when I come across them. Most electric motor repair shops will have some to spare. Being of extreme hardness, they take a heavy hammerblow without denting, but will themselves dent softer material on which they are hammered, thus creating perfectly curved seatings.

Making the plunger valve is a little more complicated, but it is somewhat simplified if we keep in mind that none of its parts needs to be machine-accurate. If there is some leakage through backflow of the water it will prove to be insignificant. Up-and-down pumping will open the bottom valve at each stroke and add more water to the column above it, filling the holding tank on top in no time while the bottom valve easily holds it there without leaking.

Carry out the step-by-step procedure as illustrated and connect the plunger diaphragm to the pump plunger rod with a rivet. Notice from the illustrations how the copper strips are first bent over the pin with light hammering and then clamped in the vise to tighten them together. Next solder the pin and bent-over copper strip. Tin soldering is done best if the contact surfaces of the parts to be soldered are *previously* tinned so that, in assembly, the melting heat causes all tinned surfaces to fuse together easily.

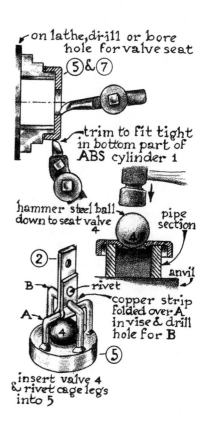

First cut corners off copper plate on vise with cold-chisel & finish edge with file

dia = inside dia ABS pipe − 2 × thickness of copper plate

vise jaw

3/8" steel disc jig

step by step hammer the collar evenly all around disc until it is flush with the disc rim

prevent folds from forming & from time to time heat copper to anneal for maximum malleability

⑤ ⑦

on lathe, drill or bore hole for valve seat

⑤ & ⑦

trim to fit tight in bottom part of ABS cylinder 1

hammer steel ball down to seat valve 4

pipe section

anvil

②

B —

A —

rivet

copper strip folded over A in vise & drill hole for B

⑤

insert valve 4 & rivet cage legs into 5

203

THE PLUNGER ROD

Widen by peening the hot end of the high-carbon steel plunger rod in a forging action. Drill a hole in the end to receive the rivet, which fastens it to the plunger valve unit (5).

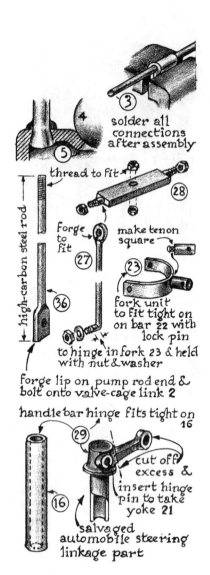

solder all connections after assembly

thread to fit

Forge to fit

make tenon square

fork unit to fit tight on on bar 22 with lock pin

to hinge in fork 23 & held with nut & washer

high-carbon steel rod

forge lip on pump rod end & bolt onto valve-cage link 2

handle bar hinge fits tight on 16

cut off excess & insert hinge pin to take yoke 21

salvaged automobile steering linkage part

The top end of the plunger rod is threaded to fit the yoke (28). This yoke has at each end a threaded stub to receive the two linkage rods (27), which in turn allow the forked unit (23) to fit the hooked-over endings of these linkage rods with little nuts and washers. The fork itself is riveted onto a bar (22). This bar is clamped with small bolts between the split yoke parts.

The *yoke* is made with two bars of suitable size and could be angled somewhat to suit the position of the operator in relation to the pumping action.

Hot forge the bar ends into a swage on the anvil to fit the roundness of the fork and handlebars. Depending on where you think the hinging point should be, drill holes in the bars to receive the hinge pins of the hinge body as shown in 29. Now all moving parts can be assembled.

It seems unnecessary to elaborate how the remaining parts of the pump assembly are made and connected, since you will never find yourself in the identical circumstances as I was when gathering material to make those parts. The main thing is to understand the working of the pump, after which inventiveness is in order. You should not hesitate to make the pump as shown; it is very much within your reach to do so, if you use the illustrations simply as guidelines.

This chapter should have made clear that the craftsman with knowledge of machine-shop and blacksmith work never needs to shy away from making everything that he needs.

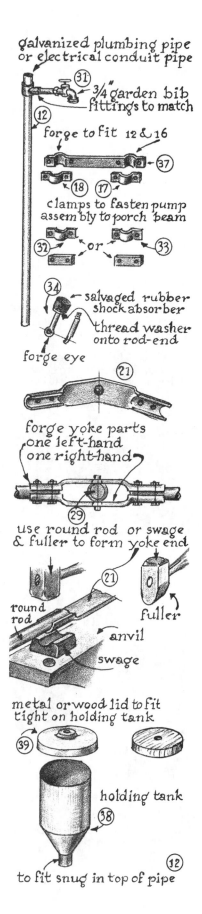

galvanized plumbing pipe or electrical conduit pipe

31 ¾" garden bib fittings to match

12

forge to fit 12 & 16

37

18 17

clamps to fasten pump assembly to porch beam

32 or 33

34 salvaged rubber shock absorber

thread washer onto rod-end

forge eye

21

forge yoke parts one left-hand one right-hand

29

use round rod or swage & fuller to form yoke end

21

round rod

fuller

anvil

swage

metal or wood lid to fit tight on holding tank

39

holding tank

38

12

to fit snug in top of pipe

6. How to Make a Wood-Turning Lathe and Lathe Tools

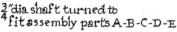

a wood-turning lathe made from salvaged materials & inexpensive surplus items

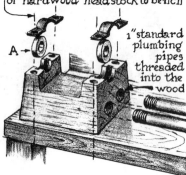

4 bolts & 2 forged straps clamp ball bearings in seat grooves of hardwood headstock to bench

A

1" standard plumbing pipes threaded into the wood

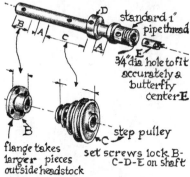

¾" dia shaft turned to fit assembly parts A-B-C-D-E

standard 1" pipe thread

¾" dia hole to fit accurately a butterfly center **E**

B

flange takes larger pieces outside headstock

step pulley

set screws lock B-C-D-E on shaft

tail stock to have sliding fit on the dual 1¼" lathe bed pipes

Although a simpler wood-turning lathe than the one here presented could be made, readers who already have a well-equipped shop may want to make one that can be used for more than the simplest wood-turning projects.

The main difference between a wood-turning lathe and a metal-turning lathe is in the rigidity of their structures. Turning the larger and heavier pieces will call for a very sturdy lathe.

The four main parts of a lathe are:

1) the *headstock*, which drives the workpiece
2) the *tailstock*, which holds the workpiece aligned and secured between headstock and tailstock
3) the *tool rest*, clasped in an adjustable socket that supports the cutting tool during wood cutting

The tool rest assembly is made so that it can slide over the lathe bed and reach the workpiece over its full length from a fixed position on that bed.

4) the *lathe bed* and its tail end *anchor block*

MAKING THE HEADSTOCK

Use a piece of close-grained hardwood (pear or maple would do) to hold securely the two headstock ball bearings that seat the headstock shaft. The illustration shows how to accomplish it.

Between the bearings the headstock shaft mounts a step pulley and a collar. This pulley is driven by a belt from a counter-step pulley mounted onto an electric motor shaft. This arrangement makes it possible to mount a faceplate outside the lathe headstock shaft extension. Large diameter pieces, which would not fit between headstock and tailstock, can then be turned outside the headstock.

I prefer ball bearings over sleeve bearings, and I use recycled ball bearings since their availability seems unlimited. In the trade, a noisy electrical motor is, as a rule, silenced simply by replacing the noisy ball bearing with a new one, the old one being thrown in the scrap bin. Most of these old bearings are completely adequate for our purpose and are as good as new after they have been greased and/or oiled. Locked up securely in the bearing seatings, they seem to last forever. I have never bought a new ball bearing for any equipment I have made that required bearings. Auto wrecking yards also have ball bearings salvaged from cars damaged beyond repair in highway accidents. We have our choice of the best if we need them, at hardly any cost.

THE HEADSTOCK SPINDLE

Since the threaded end of the headstock is to receive many lathe accessories, I use pipe thread instead of the standard thread in order to utilize much of my salvaged plumbing pipe parts.

You can cut pipe thread on the wood lathe headstock with a standard hand-operated pipe-threading unit. To do this accurately, clamp the headstock spindle in the metal-turning lathe chuck and slip over the free end the thread-cutting die unit. Hold the tailstock sleeve against the die housing to keep all in perfect alignment during cutting of the thread. The tailstock feeds the die during the threading action.

Using the metal lathe headstock in its slowest back-gear drive and steadying the die handle under 90 degrees with axis of rotation, you will be able to cut the thread in alignment to the spindle. If you do not have a metal-turning lathe, clamp the headstock spindle in a vise between brass insets in a perfectly vertical position and cut the thread bit by bit while keeping the path of rotating die handles in an absolutely horizontal path at all times.

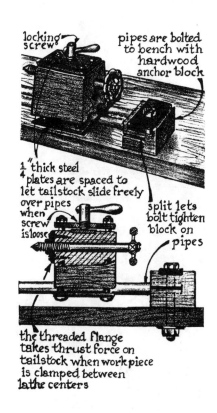

locking screw

pipes are bolted to bench with hardwood anchor block

$\frac{1}{4}$" thick steel plates are spaced to let tailstock slide freely over pipes when screw is loose

split lets bolt tighten block on pipes

Once the dies have reached their end position against the spindle shoulder, remove them and turn them upside down so that the smallest diameter thread is now placed over the cone beginning of the headstock thread. Remember that pipe threading dies leave cone threaded pipe endings. Seated and guided by the thread already cut, the same dies will then cut parallel to the axis of rotation leaving a uniform diameter thread.

Remember that cone-threaded pipes are necessary for seating accessories tightly to prevent leaking. The plumbing parts that are to fit the headstock must be rethreaded in a similar way. Do this with the tapering pipe tap by turning the accessory element around and widening its narrow part so that it can be screwed onto the headstock without binding.

the threaded flange takes thrust force on tailstock when work piece is clamped between lathe centers

CONNECTING THE WOODEN HEADSTOCK BASE TO THE LATHE BED AND TAIL ANCHOR BLOCK

The headstock base and tail anchor block are connected with plumbing pipes, which act as a lathe bed. This bed in turn permits the tool post assembly to slide along its length, to be clamped on at any spot you choose. One end of each pipe is threaded to fit corresponding holes in the headstock, which are threaded with the pipe tap. If you do not have such a tap, you can fashion a makeshift one (for threading wood only) out of the threaded end of scrap plumbing pipe. That end can be fluted with a file or an abrasive cutoff wheel to resemble a professional tap. Made of mild steel, the sharp edges of the thread at the grooved parts will cut through wood very easily.

If you should be tempted to use such a makeshift tap many times it may be useful to harden the thread in the forge fire with case-hardening compound, sold in machinists' and blacksmiths' supply houses. Follow directions on the can.

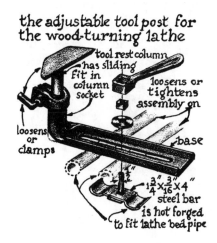

the adjustable tool post for the wood-turning lathe

tool rest column has sliding fit in column socket

loosens or tightens assembly on

loosens or clamps

base

$1\frac{1}{4}$"x$\frac{3}{16}$"x4" steel bar is hot forged to fit lathe bed pipe

THE HARDWOOD LATHE BED ANCHOR BLOCK

This also is a block of close-grained hardwood with pipe socket holes to match the ones in the headstock block. Drill the pipe sockets only halfway the length of the block and then split them, as shown,

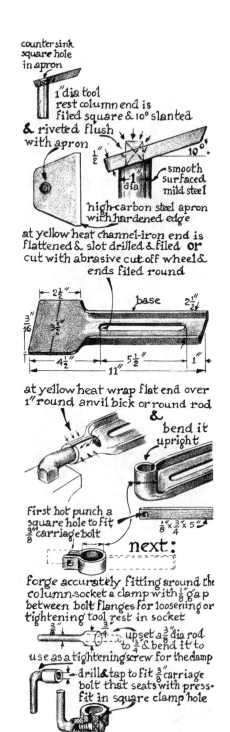

counter sink
square hole
in apron

1" dia tool
rest column end is
filed square & 10° slanted
& riveted flush
with apron

smooth
surfaced
mild steel

high-carbon steel apron with hardened edge

at yellow heat channel-iron end is
flattened & slot drilled & filed **or**
cut with abrasive cut-off wheel &
ends filed round

base

at yellow heat wrap flat end over
1" round anvil bick or round rod

&

bend it
upright

First hot punch a
square hole to fit
$\frac{3}{8}$" carriage bolt

$\frac{1}{8}$" x $\frac{3}{4}$ x 5"

next:

forge accurately fitting around the
column socket a clamp with $\frac{1}{8}$" gap
between bolt flanges for loosening or
tightening tool rest in socket

upset a $\frac{3}{8}$" dia rod
to $\frac{3}{4}$ & bend it to
use as a tightening screw for the clamp

drill & tap to fit $\frac{3}{8}$" carriage
bolt that seats with press-
fit in square clamp hole

so that a bolt placed between pipe holes will clamp the block onto the pipes as well as onto the workbench. Once the lathe bed pipes are thus anchored securely, slide the tailstock center body back and forth over the pipes, which have abrasive paper stuck to them so that the grooves of the tailstock body will become accurately seated.

Oil the wooden seating surface. The oil will penetrate the wood so that all future adjusting of the tailstock over the bed is eased.

MAKING AND FITTING THE TAILSTOCK ONTO THE LATHE BED

First drill the two holes in the tailstock block to fit the dimension of the lathe bed pipe. Bisect these holes with a saw over the full width of the block, separating the two parts. The top part is made to slide along the pipes freely, as shown.

The tailstock center spindle should be pipe-threaded over a considerable length so that maximum adjustments can be made. Fasten a salvaged wheel-shaped water faucet handle to the end of this center to hand turn it. Fasten a threaded flange with pipe thread to the block and install the pipe-threaded center spindle.

If a scrap part is available to turn a little steel flange on a metal-turning lathe, you can use standard thread taps and dies. This makes it possible to use standard bolts and nuts when making the tailstock center screw and spindle.

MAKING AN ADJUSTABLE TOOL POST AND TOOL POST BASE

A short section of channel iron serves as the tool post base. To clamp this base on the lathe bed with 1/2-inch locking bolt, cut a slot so that the tool rest apron itself can be positioned forward, just free from the rotating workpiece.

To make the slot, drill a series of 9/16-inch holes so close together that only a paper-thin thickness of steel is left between each of them. Next place the channel iron between the vise jaws flush with the hole edges and, with a cold chisel, hack in a shearing action through the hole edges. The resulting slot may then be filed clean and smooth.

Or, if you have an abrasive cutoff wheel, you can cut the slot with this abrasive disc. First cut along one side of the slot and next the other, finally freeing the steel remnant at each end with a drill. Clean any irregularities with a rotary file, cold chisel, or grinding points.

To make the tool post socket, heat the slotted channel iron at one end until it is *yellow hot.* Flatten it and, should you lack width to wrap that section around the tool rest collar or a bick of that size, peen it wider.

Next forge the tool rest socket clamp. It should fit around the newly forged column socket as shown. The clamp is to tighten the socket onto the tool apron column so that the desired position of this tool post apron may be locked into place.

One of the aligned holes in the clamp lips is made into a square to fit the square neck of a small carriage bolt. This keeps the bolt from turning when the threaded clamp handle is loosened or tightened. *Or* keep the hole round and use a square-headed bolt that is kept from turning when its square head is held in place by the clamp wall.

208

CUTTING TOOLS FOR A WOOD-TURNING LATHE

A lengthy vocabulary of names for lathe tools exists, but simpler words for special-purpose tools may better clarify their use. For instance, the first tool shown in the illustration (a) is used to cut most silhouette shapes of a workpiece. When this tool is moved straight and parallel to the lathe bed, the silhouette line will be straight, but if the same tool is moved in a combination of parallel and forward and backward movements, a variation of silhouette shapes can be turned. This is not so with the second and third tools shown (b & c), which have right-angle endings. This type of tool cuts a slot of that width or a wider slot if it is moved to right or left during cutting.

The next tool (d) is simply a narrow cutoff tool to cut off a piece entirely. Often it calls for stopping the lathe, cutting the piece off with a saw just before it drops off, then sanding the cut smooth.

The shallow-curved contouring tool (e) is designed to be held in such a way that only a portion of its width will shave off a thin sliver of the workpiece surface when the refining of a rough surface is called for.

In using all lathe tools you may combine forward motions with side motions, using a pivot point along the tool post apron close to the cutting edge when making circular paths inward or outward to create hollows or spheres.

The tools in the second series of illustrations immediately suggest their uses when we visualize, for instance, what a sharp-pointed tool can do and what round and flat tools cannot do. It is here that we begin to realize that reading long, elaborate explanations is superfluous soon after the beginner finds out for himself the best use of these tools. A little practice with a few basic tools prepares him in a short time to do justice to all the other wood-turning tools designed for special jobs.

Once skilled in working with wood-turning tools, the inventor-craftsman will want to extend his tool collection with a still wider variety with special contours. These can, in one stroke, accomplish a task that would otherwise take several cuts with narrower tools.

To make wider tools, flat leafsprings of cars are ideal. Simply grind them into the needed profiles, taking care during the grinding that the spring steel temper is not lost through overheating. The tools illustrated can in one continued stroke shape a spool, a sphere, a little burin handle.

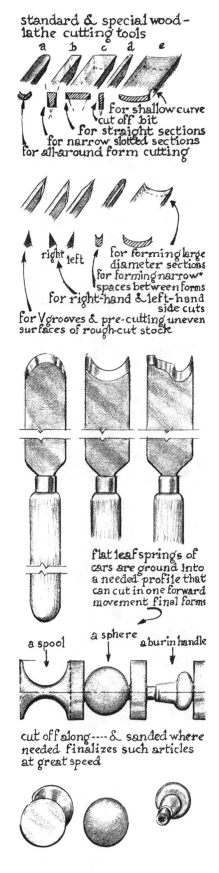

standard & special wood-lathe cutting tools

a b c d e

for shallow curve
cut off bit
for straight sections
for narrow slotted sections
for all-around form cutting

right left

for forming large
diameter sections
for forming narrow
spaces between forms
for right-hand & left-hand
side cuts
for V grooves & pre-cutting uneven
surfaces of rough-cut stock

flat leaf springs of
cars are ground into
a needed profile that
can cut in one forward
movement final forms

a spool   a sphere   a burin handle

cut off along ---- & sanded where
needed finalizes such articles
at great speed

The tailstock center can have a protruding cone at its work end or have a protruding central pin surrounded by a wide sharp crater rim. The first type of center insert holds the workpiece by the tip of the cone only. In the other, the lead pin penetrates the wood entirely until the polished crater's cutting edge becomes imbedded in the wood, thus holding the workpiece more securely than would the single point at the cone end.

The total surface of the crater rim that is imbedded into the wood should be greased somewhat to reduce friction during turning of the workpiece. The odor of scorched wood will soon tell you that the tailstock needs attention. If the wood should scorch around the crater cup, a little candle wax rubbed over the groove created by the crater edge often is sufficient to stop the scorching.

Tailstock Centers

If you have a metal-turning lathe you can make a variety of tailstock centers that have welcome features.

One type is the *open cup center*. It will hold a piece of wood that may not be entirely round. The workpiece will seek its own center and form its own seating in such a cup. Under friction an uneven wood ending would wear somewhat under compression into an even fit. After a little lubrication, cutting can start right away.

This system bypasses the need for an exact centering of the workpiece with a predrilled hole. When you use a deep cup center, the workpiece cannot be pushed out of its seating during the full run of the turning operation. The cup leaves no central hole, but some end trimming may be required afterward.

The *ball bearing center* is a very useful design you can make on the metal-turning lathe. First forge onto the stock end a wide head, which can be turned into an exact seating socket to hold a ball bearing very tightly. Next, turn a small center insert with a little protruding seating pin that fits snugly in the ball bearing hole. This system eliminates all tail center friction during wood cutting. I find myself using this type of center more than any other.

Another design has a socket with a locking screw, as illustrated. It can take a small solid cone insert or a little crater center or a small wide cup center with a leading pin. If the pin is left out entirely the wide cup can take many irregular workpieces in a self-centering action.

Other types of center arrangements are shown also; they are all variations of the same principle and serve to hold the workpiece in alignment with the headstock drive during the cutting action.

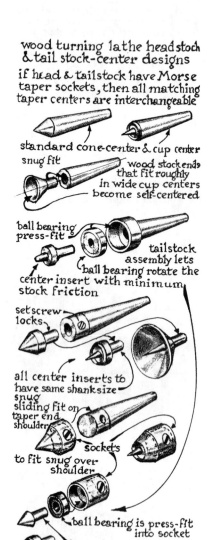

wood turning lathe head stock & tail stock-center designs

if head & tailstock have Morse taper sockets, then all matching taper centers are interchangeable

standard cone-center & cup center

snug fit

wood stock ends that fit roughly in wide cup centers become self-centered

ball bearing press-fit

tailstock assembly lets ball bearing rotate the center insert with minimum stock friction

set screw locks

all center inserts to have same shank size

snug sliding fit on taper end shoulder

sockets to fit snug over shoulder

ball bearing is press-fit into socket

interchangeable inserts

if headstock socket has Morse taper, then center inserts don't need set screw locking

The Headstock Center

This must hold the workpiece firmly anchored to the headstock spindle. If the headstock spindle has no Morse taper socket, then a cylindrical socket must seat a cylindrical-centered shank as exactly as possible and be held in position by a set screw.

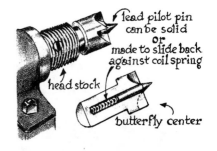

The butterfly, or four-wing, design of the center and its leading pin will leave the wood workpiece marked with a small central hole and four deep indentations, because the wing teeth must be pressed deep enough into the wood to drive the workpiece. You can improve this design with a sliding pilot pin that retracts on a coiled spring so that the pin slides back after its point has been placed on a previously marked center of the workpiece. When the workpiece is pressed onto the butterfly center, the pilot pin will be completely depressed, leaving hardly any mark in the wood.

This refinement may seem to be gilding the lily, since the little markings do no harm to a tool's functioning, and, after some hammering, the compacted wood closes up the remnants of such shallow markings. All the same, any handle, and a tool handle especially, looks better without markings of that sort.

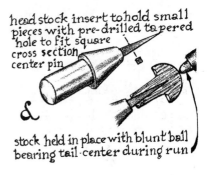

Inventiveness leads to the making of specialized tools. Not everyone cares to spend the time to make them if they are not used often. If, however, I intend to make large quantities of duplicate workpieces, I can justify designing a special tool that will improve the appearance of the pieces over and above their usefulness.

When making many small artifacts, such as small burin handles, for instance, one can design a jig as described in Chapter 16. Another special tool can be designed to hold a small handle while its marked ends are cleaned up. A long, square cross section pin holds the workpiece while a blunt ball bearing tail center keeps it from sliding off the pin during the run.

Adjustable Cutters

I have found it useful to reinforce heavy-duty tool handles with steel ferrules to prevent the wood from splitting. To get the right ferrule-seating dimension, you can make a cutter gauge. Its first cut establishes the needed dimension. Other tools can follow up the first cut to the finished ferrule seating. In this way it is not necessary to stop and start the lathe in order to measure sizes with calipers. The gauge assures the correct diameter for a press fit of ferrule on wood under great force. Since the gauge can be adjusted to whatever size you need, the saving of time and effort when many handles are to be made will justify making this tool.

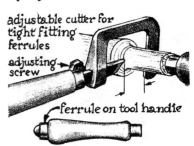

Using a Cutter Gauge

First cut each end of the workpiece stock freehand into a small cone shape. This makes it possible to slip the cutter gauge over the cone. The clearance between the cutter itself and the wood is closed in a self-adjusting action during a first cut. After this, any standard cutting tool can follow up and finish the rest.

You may be tempted to use that cutter gauge to cut the entire section you need for the ferrule seating. If so, you will have to redesign the cutter edges in order to facilitate this operation. It will be a little difficult to prevent certain uncontrolled forward movements of the tool should a false move make it "grab" the wood, thus immediately cutting the wood diameter smaller than you intended. Or, it may leave a deep local groove that reduces the hold of the ferrule seating and the handle's strength where it is most needed. After some exploration, you may decide to leave well enough alone and simply use this adjustable cutter as a dimension indicator for a first cut, to be followed up with freehand cutting.

Tailored Tool Rests

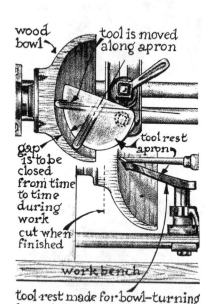

tool rest made for bowl-turning is contoured to reduce the gap between wood & apron edge

Besides straight tool rest aprons of various lengths, you may need an extra long one. For such a one it is best to make supporting columns, one at each end, fitting in corresponding tool sockets. If such an extra long tool rest only has one central post, support its ends with a wedge arrangement between their ends and the lathe bed. This will eliminate possible downward vibrations of the tool rest during the turning of the wood. Downward vibrations in a "springy" apron can wreak havoc with the workpiece since any sudden interference with a steady cut makes the tool end grab the wood, leaving a deep groove or torn wood surface.

A tailored tool rest apron is called for if you plan to turn deep, hollow parts, such as a wooden bowl. The illustrations are self-explanatory and do not need a lengthy text. What may require further description is the manner of attaching a workpiece (such as a wooden bowl) to a faceplate, which can be done without wood screws by *gluing* the piece directly to the faceplate.

The best way to do this is to first screw onto the faceplate a piece of plywood, which is then turned to the needed dimension. On the surface of this plywood, glue, with a strong adhesive, a piece of cardboard, to which in turn the workpiece is glued. The glue may be a contact cement or any other type of strong adhesive. The holding force in this arrangement is very reliable if application is done correctly. When the turning is finished, detach the piece by splitting the cardboard in two with a broad, thin chisel. The separation comes quite easily, leaving the base of the workpiece unmarked by screw seatings.

Plumbing Parts as Lathe Accessories

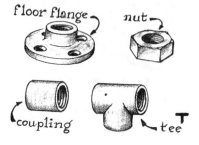

The most used is the *floor flange*. Because the threaded part has been tapered so that it will clamp itself on a plumbing pipe ending, the narrow part of the floor flange thread must be widened until it can easily be screwed on the headstock without binding. A cylindrical threading tap can be used to accomplish the widening. Or, use the standard taper tap to open the narrow end of the flange from the other side until the flange is free of binding on the headstock.

If you decide to make such a tap, use an adjustable standard pipe-threading die and cut the thread on a bar of annealed high-carbon steel. Grind the tap flutings with an abrasive cutoff wheel and finally harden the tap by quenching in oil at *cherry red* heat glow, which gives it the hardness of a cold chisel edge.

Final fitting of the floor flange on the headstock calls for refining the contact area between the flange hub and the headstock shoulder face. If these two seat exactly after tightening, the flange "facing" can be turned in position for accuracy with a hand-held *carbon-tipped* lathe tool. Hold this tool firmly on the steady rest, which has been placed with its apron edge as close as possible to, but free from the flange rotation. Holding down the tool firmly so it has the least possible chance of wandering, cut the metal in a scraping or shaving action. After all inaccurate parts have been cut away, the flange is ready to receive, with four wood screws, a plywood facing, as described before.

Any workpiece used *without* that plywood disc can be screwed directly onto the floor flange.

Abrasive Discs and Side Grinders

Making several such floor flange attachments means you can fasten to them plywood discs of various sizes to which can be attached abrasive papers for the many wood surface sanding jobs you may have to do. A disc cut from salvaged offset printing rubber blankets can be glued on first to soften the abrasive action when finally abrasive sheets are glued onto the rubber-sided discs.

The plywood disc, after it has been accurately faced, may have glued to it an abrasive cutoff disc to make a side-grinder. Such grinders, as described before, can grind accurate flat surfaces on metal workpieces.

Wood-Turning Lathe Headstocks as Grinding Arbors

Grinding wheels with large lead bores sometimes fit directly on headstocks. It so happens that a thread on a 3/4-inch plumbing pipe has an outside dimension of 1 inch, which coincides with many large grinding wheel bores.

As a rule, such large surplus or salvaged but worn wheels can be purchased at flea markets at little cost. The advantage becomes apparent if you can slip them onto the headstock ends without having to make adapters. The wheel's lead hole walls cannot harm the thread on the headstock. Therefore, if the wheel is locked up between two washers and a plumbing pipe nut, you have achieved an uncomplicated and sturdy arrangement.

In addition, the step pulley allows you to change the wheel's rpm to the one you feel is the most effective and safe. Adjust the regular tool rest so that no gap remains between it and the wheel after the wheel has been dressed to accuracy.

standard plumbing pipe fittings having same thread size of lathe head stock, can serve to fasten most variety of stock & accessories

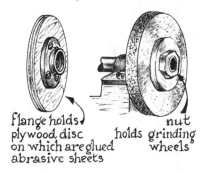

flange holds plywood disc on which are glued abrasive sheets

nut holds grinding wheels

WARNING: Any grinding with a lathe requires that the lathe be protected from *abrasive residues* that fly off the wheel during work and may creep between critical bearing surfaces or enter parts that slide together. Place paper or drop-cloth covers over endangered surfaces during grinding. After the job is done, clean carefully to remove the last trace of abrasive powders. If you have an air compressor, a forced air blast removes the abrasive dust best.

7. Tempering High-Carbon Steel

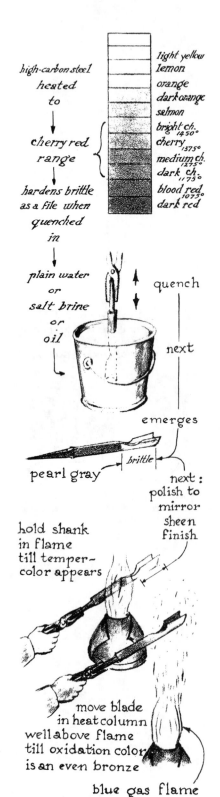

high-carbon steel
heated
to
↓
cherry red
range
↓
hardens brittle
as a file when
quenched
in
↓
plain water
or
salt brine
or
oil
↓

light yellow
lemon
orange
dark orange
salmon
bright ch.
1450°
cherry
1375°
medium ch.
1275°
dark ch.
1175°
blood red
1075°
dark red

quench ↕

next

emerges

pearl gray — brittle

next :
polish to
mirror
sheen
finish

hold shank
in flame
till temper-
color appears

move blade
in heat column
well above flame
till oxidation color
is an even bronze

blue gas flame

In shop practice, it is preferable to quench hot high-carbon steel first to *file hardness*, which is *brittle hardness*.

To make high-carbon steel brittle hard, it must have a visible heat glow of *light cherry red* to *dark yellow* at the moment of quench. After this quench it is reheated over a blue flame to subtract excess hardness until it reaches the hardness of your choice. This process is called *tempering*.

A faster method, but one with greater risk of misjudgment, involves quenching hot steel directly to its final hardness. This is called *hardening* and does not require tempering afterward.

The *quenching liquid*, as a rule, is water, but may be brine, oil, fat, or other suitable coolant. In the blacksmith shop the coolants all assume room temperature. At moment of quench, the shock impact between the heat of the hot tool and the room temperature liquid is the same regardless of the liquid used. After the moment of impact, however, the submerged steel cools at different rates in different liquids, depending on the boiling point of each; water boils at 100°C. (212°F.), oils and fats at 300°C. (572°F.).

After the quench the brittle part is made shiny by grinding and polishing. That part is next reheated over a blue flame (the kitchen stove gas flame or the flame of a blow torch will do). At first the shiny part of the tool is kept out of the flame so that the softer part of the tool is heated first, allowing the heat to travel by conduction to the shiny part. In due time *oxidation colors* will appear on the shiny steel surface, creating a color spectrum that travels, through conduction, outward to the tool's cutting edge as the heat is continually increased.

The first color to appear in this spectrum will be *faint straw*, followed by *dark straw*; *bronze*; *peacock*; *purple*; *full blue*; *light blue*, and finally *faint blue resembling blue gray*. The drawing of these colors can be arrested at any time you choose by quenching in water.

Faint straw indicates the correct hardness for wood-engraving burins, razors, scribes, and all tools that are not strained much, but are given the hardest edges that stay sharp longest.

Straw is the hardness for hatchets, axes, and comparable tools.

Dark straw is good for cold chisels, stone carving tools, punches, dies, and thick-bladed woodcutting tools.

Bronze is best for sturdy-bladed woodcutting tools that are used with hammers on medium and hard woods.

Peacock, which is a blend of bronze and purple, is best for most thin-bladed wood-carving gouges.

Purple matches the hardness for springs, spatulas, table knives.

Full blue is for gun barrels, machine parts such as bearing housings, steel brackets.

Light blue is too soft for cutting tools but may serve for handles, levers, parts that do not require springiness.

Faint blue resembling gray, the last color in the spectrum, means that the high-carbon steel has been annealed to its softest and can be turned on the lathe or drilled, milled or filed, as one can do with mild steel.

In hardening high-carbon steel files and rasps, no colors need to be drawn and brittle quenching only is required. High-carbon steel emerges from the quenching bath with a pearl gray hue. Most temperable steels found in the scrap pile can be tempered in this way.

The process of tempering high-carbon steel is in time less complicated as it becomes understood clearly. The variables involved give the smith-toolmaker a welcome chance to be inventive in choosing the combination of steps he must follow for best results.

For further information, the reader can find extended descriptions on the art of tempering in my two previous books, *The Making of Tools* and *The Modern Blacksmith.*

through experience how to foresee possible errors and avoid repeating former mistakes. (See page 230, *Notes on Drilling.*)

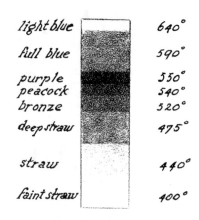

light blue	640°
full blue	590°
purple	550°
peacock	540°
bronze	520°
deep straw	475°
straw	440°
faint straw	400°

oxidation color spectrum
on all steel surfaces
when heated gradually

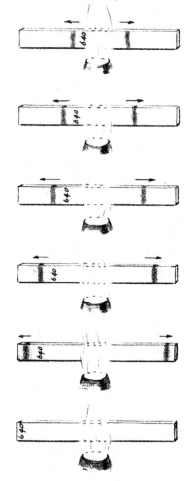

oxidation colors
during
heat-treating of
high-carbon steel

8. Making Carbon-Tipped Tools for Wood- and Metal-Turning Lathes

steel may be turned on a wood-turning lathe if tools are carbon tipped —

use tips silver soldered onto mild steel bars that are hand held

use same tips, silver soldered on high-carbon steel bar, if used in tool post of metal-turning lathe

steel turned in wood lathe

at low 800-1000 rpm, steel bar held by chuck and tail center in a wood lathe can be cut by hand-held carbon tip tools

a face plate & dog are used if no chuck is available

Carbon-tipped steel cutting tools that are sold today to take the place of the high-carbon steel ones have two advantages: (1) their hardness will not diminish when the tool heats up and (2) their hardness is so great that they can cut even hardened high-carbon steel. But the disadvantages are that any undue shock on a carbon tip will, as a rule, break it. Also, if the sharp edge gets lodged during a cutting action it is almost impossible to extract the tool without breaking its edge.

Tool bars for metal-turning should be made of high-carbon steel measuring approximately 3/8 inch by 3/4 to 1 inch. A bar is clamped into the tool post of a fairly heavy-duty metal-turning lathe and a carbon tip can then be soldered onto it. Because of the great strains that often occur during metal turning, *silver solder* is used instead of tin solder to bond the tips to the bars.

Sections of mild or high-carbon steel bars can be made into wood-turning tools with tangs at one end to take long wooden handles. Or, they can be kept short without tangs, to be clamped in tool posts for metal-turning.

Wood-turning seldom requires carbon-tipped lathe tools. Occasionally, however, you will have to trim a metal part in a wood-turning project and will need a carbon-tipped tool to do it. This happens, for example, when a finished tool handle with metal ferrules on each end needs some trimming on the lathe. Used as a wood-turning tool, the carbon tip actually scrapes or shaves the metal, since the depth of the cut is always very slight.

If you do not have a metal-turning lathe it is possible in an emergency to cut steel on a wood-turning lathe. There must, however, be a way to fasten the steel workpiece in the wood lathe headstock and reduce the rpm to about 500.

Adjust the tool post height so that the cutting edge level of the carbon tip will be at the center level of the workpiece. Keep the gap between the tool post apron and the workpiece as small as possible. In action the tool must be held down extra firmly to prevent it from being caught between the workpiece and the tool post apron. Steel bars can be reduced in this way—little by little to the wanted sizes—with hand-held carbon-tipped tools. This method was practiced extensively in the ninteenth century, when hardened high-carbon steel cutting edges were used to cut metal. A very low rpm of the lathe headstock was maintained to keep the tool edge from heating up.

SOLDERING METALS TOGETHER

Soldering metals resembles somewhat the technique of welding. In both instances it is important that the oxygen in the air not reach the melt. This can be accomplished by surrounding the melting area with a substance that keeps the oxygen out but at the same time does not interfere with the fusing of the solder on the surfaces to be bonded together.

In silver soldering a borax flux keeps the oxygen away from the metals. Zinc-saturated muriatic acid applied to the metal surfaces degreases all the pores, which will then, as a rule, accept the bonding metal at the point of its melt; the zinc element acts as a catalyst.

In silver soldering a borox flux keeps the oxygen away from the surface as well as from the melted solder. At the moment of melt the flux will be pushed aside by the solder flow. At the point of bonding it seems as though the two metal surfaces were drawn together automatically.

Failures usually can be traced to oxygen interference, which introduces oxidation scales between the surfaces. After a failure, you must scrape surfaces completely clean before attempting soldering again.

brush standard silver solder flux over clean & accurate surface of bar & heat it in forge fire cherry red as seen in semi dark room

next: silver solder wire end is held on fluxed surface and its "melt" evenly distributed over the whole surface

at the moment of coverage, place the fluxed surface of the partly heated carbon tip on the melted solder of the bar, which will "take" the carbon tip in a complete bond.

after cooling, dress the bit on stone wheels specially made to grind carbon bits.

It is often puzzling to the beginner why the best soldering job is the one for which only a minimum of solder is used. Excessive solder between two unmatched surfaces weakens the joint markedly.

Although brazing and welding pieces of steel together are not covered in this book, the same principles apply.

METAL-TURNING WITH CARBON-TIPPED TOOLS, HIGH-SPEED TOOLS, AND HIGH-CARBON TOOLS

In spite of the advantages that carbon-tipped tools and high-speed steel cutting tips have, they do not replace entirely correctly hardened high-carbon steel cutting tools in the home workshop.

If it proves too difficult to silver-solder carbon tips to bar stock, or if you do not have high-speed cutting tips, you can always fall back on the much easier way of making metal-turning cutting tools from good quality high-carbon steel. In my experience it is quite sufficient to have high-carbon steel tools without special alloys. They serve excellently if you don't speed the cutting. (Excessive speed overheats and ruins the temper of the tool.) It is also important to remember that the workpiece must be *softer* than the cutting tool.

The desirable thing is to purchase high-speed steel lathe cutting tips, if possible. These tips, though a little less hard than well-hardened high-carbon tips, or well-hardened high-carbon steel edges, have the advantage that they will not be dulled appreciably by excessive heat created during the cutting of metal at high speed.

Remember, however, that the well-tempered edge of a high-carbon steel cutting bit will, as a rule, fill most of your needs. It is encouraging to know you can do without all of the fancy modern alloy steels, so indispensible to a sophisticated modern industry, as long as you are *not in a hurry, the cutting speed of the lathe is low*, and you have *easy access to high-carbon steel.* You will find to your pleasure that in all respects the scrap pile and metal-turning in a home shop go hand in hand.

One item sometimes found in the scrap pile is like a gem. It is the *magnet.* The early automobiles had V-shaped magnets in their generators. Old Fords also used to have them. These have proven to be excellent stock from which to make tools.

Magnets will retain their magnetism at their best if the steel they are made from has the highest possible carbon content and has been made as hard as possible. Be careful in quenching a magnet, however, since there is a danger that the steel could crack at the moment it is quenched in cold or freezing liquids. Several magnets that I tried to forge into tools turned out to be worthless since they were full of cracks. But I did find other perfect ones and was able to make the very best lathe cutting tools out of them. Take the chance if this type of magnet should come your way; if they are without cracks, the time you spend making lathe cutting tools from them will not be wasted.

9. How to Drill Square Holes

Admittedly this knowledge is of limited use. There are several ways to make square holes without drilling them and they are preferred by industry. Drilling a square hole actually combines *drilling* with *milling*, during which the tool is guided by a brittle-hard, square-holed jig attached to the workpiece.

In woodworking, square holes are made with drills that rotate within a square, sharp-ended housing that acts as a chisel in the downward thrust. This instrument is a combination drill and broach. In metalwork, a broach is the instrument that shears the steel in a downward thrust. The four corners around a predrilled hole are cut this way without the use of the drill.

DESIGN PRINCIPLE OF THE CUTTER

As the illustrations show, the sides of the equilateral triangle placed in the square have the same length as the sides of the square and thus can move within the square without binding. If that triangle is modified into a configuration with curved, instead of straight, sides, it will then move within the limits of the square with three of its sides and one of its points always touching and sliding along the square's sides. The points of this configuration, however, meeting at an angle a little larger than 90 degrees, cannot fit into the square jig's 90-degree corners and must bypass them somewhat. It therefore describes a square with slightly rounded corners.

If the drill is made with sharp corners, straight sides, and a flat cutting-face, it will not act like a regular drill with the cone-shaped profile of its self-centering cutting edges. A cone-shaped cutting profile would cause the drill to seek its center of rotation automatically. With the cutting edges in a flat plane, however, the drill is allowed to wander in whatever direction the cutting forces dictate. It is here that the brittle-hard square jig, when fastened to the workpiece, is used to confine the drill's motions to a wobbling one, forcing it to slide along the square's sides as dictated by the jig, at the same time as it cuts away the steel below it.

It is best to make the drill shank as long as possible in order to cut the sides of the hole as parallel as possible to the axis of rotation. Note that the wobbling action of the milling end is made possible by the springiness of the long drill shank.

an equilateral triangle in a square as shown can rotate in it without binding

modified with curved sides it rotates in square tangentially

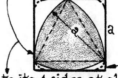

to its 4 sides at all times but bypassing its corners

a drill with cutting edges aligned in a flat plane allow the

sharp

cutters to wander freely

side edges are dulled

flat plane

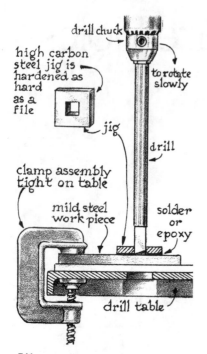

drill chuck

high carbon
steel jig is
hardened as
hard
as a
file

to rotate
slowly

jig

drill

clamp assembly
tight on table

mild steel
work piece

solder
or
epoxy

drill table

The workpiece and jig can be clamped together on the table of the drill press. An alternate arrangement, clamping workpiece and jig in the chuck of a metal-turning lathe that has a long bed, renders the cone effect of the hole negligible. The rod extension of the shank with a sleeve coupling gives us a maximum shank length. Thus the tailstock is placed at a great distance from the rotating headstock while the drill is fed by the tailstock.

1st METHOD
clamp workpiece and jig in
head stock chuck & place
drill extension in tailstock
chuck

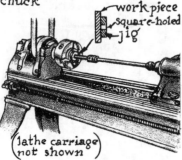

work piece
square-holed
jig

(lathe carriage
not shown)

head stock is shifted into
back gear to slowest speed

sleeve-coupling + set screws
connect drill & extension

hand feed the
tailstock

Another method uses a short stubby drill. The combined jig and workpiece are placed on a roller swivel-bearing or on a stack of well-lubricated flat, smooth discs that act as a swivel bearing.

It is amusing to demonstrate the drilling of square holes to unbelievers who are not acquainted with this procedure. I first saw it done on my last day in school in 1923 in Holland, where I was trained to be a marine engineer. The teacher in the machine shop demonstrated how to drill a square hole, thus enlivening the occasion of graduation. But on a later occasion, when I served in the Netherlands East Indies conscript army as a private in the ordnance plant, I was challenged to prove my claim that I could drill square holes. I was given the opportunity to do so, and after witnessing this feat my superiors saw fit to qualify me, then and there, as sergeant instead of private.

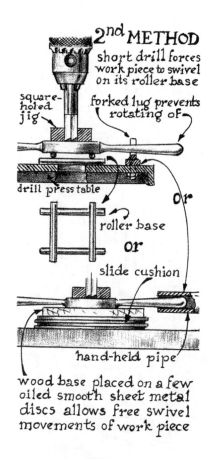

2nd METHOD
short drill forces work piece to swivel on its roller base

square-holed jig

forked lug prevents rotating of

drill press table

or

roller base

or

slide cushion

hand-held pipe

wood base placed on a few oiled smooth sheet metal discs allows free swivel movements of work piece

10. Making Hand-Held Punches

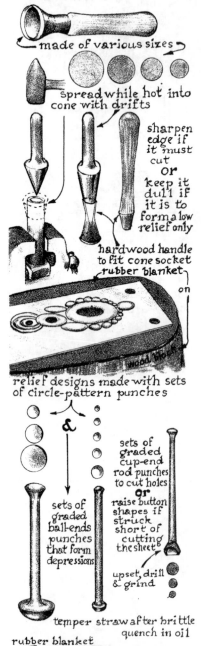

circle-pattern punches
made from high-carbon
steel rods & seamless
tubular steel

made of various sizes

spread while hot into
cone with drifts

sharpen
edge if
it must
cut
or
keep it
dull if
it is to
form a low
relief only

hardwood handle
to fit cone socket

rubber blanket
on

relief designs made with sets
of circle-pattern punches

&

sets of
graded
cup-end
rod punches
to cut holes
or
raise button
shapes if
struck
short of
cutting
the sheet

sets of
graded
ball-ends
punches
that form
depressions

upset, drill
& grind

temper straw after brittle
quench in oil

rubber blanket

hardwood block

First use ball-end punches from
below sheet before cutting a disc
free to prevent buckling of
edges

The punches are used to cut or indent thin sheets of metal or other sheet material into various shapes and designs. Simple patterns can be cut using steel tubes with sharpened ends. If the tubes are made from high-carbon steel, they will stand up better than those made from mild steel, the edges of which are too soft. Mild steel tubes have to be case-hardened at the cutting edge, which will last well when not abused.

Once in a while one finds on the scrap pile a seamless high-carbon steel tube. Seam-welded mild steel tubes, used in the plumbing trade and for electrical conduit tubes, are more readily available.

The illustrations show how a seamless tube, heated in the forge fire, can be flared out with a cutting edge at one end and a handle socket at the other. Be sure to harden all punch ends to the hardness of a cold chisel.

To use a punch designed as shown, prepare a wood stump with a smooth surface. Glue onto it a piece of rubber blanket salvaged from an offset printing shop. As a base upon which to punch out designs from sheet metal and other thin material, it resists being cut or marred. One can generally purchase for very little money, also from such a printing shop, discarded aluminum sheets.

Placed on the stump, the soft metal can be cut with one fairly hard blow of the hammer on the punch, making a clean hole through the metal. Or, a lighter blow will leave an uncut indentation with a low relief effect. The design surface can be raised still more, if desired, with a punch ending in a partial sphere.

These two types of punches are rather easy to make, and it is a good idea to make several sizes in order to extend the design possibilities. The large range of punch designs shown provides the opportunity for inventing whatever shapes you may need.

USING SPRING STEEL FOR MAKING PUNCHES

Leaf springs and coil springs from cars are ideal for making punches. Engine push rods, many linkage rods, and most car parts—because they are inherently resilient—are, as a rule, made from temperable steel. This makes it possible to harden the cutting edge of tools made from them.

Special Punches for Decorative Designs

For a 2-inch wide punch in the shape of a heart, there is an advantage in making the pattern in two halves—one left- and one right-hand punch, each the opposite of the other. Combined they make a heart profile; individually they can be used to make many other designs.

Take a 6- to 8-inch long section of a leaf spring of a car. Heat it on one end and peen it out to about 4 inches wide. Heat the other end and upset it into a head. Sharpen the wide, flat end and grind it with an outside bevel so that the head position, aligned with the center of gravity of the cutting pattern, will spread the force of the hammer-blow evenly over the full cutting edge.

Make the reverse punch with a template cut by the first punch. When this template is turned over it becomes its opposite.

Another punch design can be forged from a similar blank. Heat the blade and shape it over rod ends that act as bicks. The edge of the punch will form a three-quarter circle. In the same way, other punch endings can be shaped with curve designs hammered out on props made especially for them.

A series of punches using similar blanks make variations of three-quarter, half, and quarter moons. The punches described so far are especially suited for making interconnected designs without cutting them away from the sheet. Closed circular punches that are made to cut all the way through can be held slanted and used to cut partially through to form half circles.

Punches can be shaped to cut out squares, parallelograms, rosettes, stars—whatever figures the design calls for.

punches to cut & form designs in sheet metal, leather, fabric, paper, cardboard, plastics, etc.

use high-carbon steel as

leaf & coil springs of cars
engine valve push rods

forge punch blanks into cutting ends that follow design outlines

stock
forged

peened at yellow heat & bent to keep shank aligned with center of projected design pattern

average punch length 6"

peen out & fold hot over rod sections

forge dozens of blanks to shape various contoured cutting edges

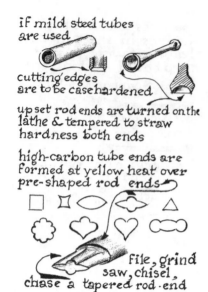

if mild steel tubes
are used

cutting edges
are to be case hardened

upset rod ends are turned on the
lathe & tempered to straw
hardness both ends

high-carbon tube ends are
formed at yellow heat over
pre-shaped rod ends

file, grind
saw, chisel,
chase a tapered rod-end

You may be tempted to use plumbing pipes to make such punches with case-hardened edges. Some of them, if not large enough, can be made larger by upsetting or spreading the end. For heavy work, however, use only high-carbon steel pipe.

WARNING: The galvanized surfaces of plumbing pipes around the working ends of tools must be ground off. Heating the punch end with the zinc coating left on will ruin the quality of the steel and make case hardening ineffective. I have successfully made cutting punches after grinding away the zinc coating, then beveling the edges to razor sharpness before case hardening them.

Remember that the cutting edges of all mild steel tools that have been case hardened should never be sharpened on *both* sides since the case-hardening compound does not penetrate very deeply. Sharpening both sides will leave a soft core edge, whereas sharpening one side only will leave the other sharp edge hard.

The working ends of all high-carbon steel tools should be annealed before they are filed into the endless variety of design contours, as the examples show.

Even if your punches seem strong and well-tempered, do not use them on a steel that may be too hard.

Using Car Axle Ends as Stock

Chapter 26 in this book explains how to free a car axle from its flange and ball bearing. You will, as a rule, have between two and three inches diameter of solid material from its end to work with.

Cut a length that can be conveniently hand held as a punch. Anneal it in the fire and turn the end into a *flat facing* on the lathe. Proceed as shown in the illustration or, instead of gluing the paper design onto the facing, blacken it with ink or a mixture of lampblack and grease. Wipe it off, leaving a little black residue.

Next, draw the design on this black facing using a pair of dividers in measured-off center-punch marks. Whichever method of design transfer you use, the correct placing of center-punch marks will enable you to place the drill depressions evenly.

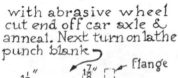

with abrasive wheel
cut end off car axle &
anneal. Next turn on lathe
punch blank

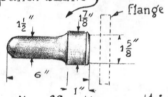

scribe off pattern on thin
paper & glue on end &

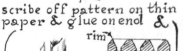

mark with
center punch
drill placement

first drill ⅛" hole
¾" deep &
follow up with
½" drill until
it reaches full
size at hole rim

file to sharpen
rims of holes

Place the shank of the punch in the vise and file away all superfluous material along the design outline until the bevel edge is *almost* as sharp as you finally want it. The very final sharpness can be reached easily with the aid of a small, rubber-backed abrasive disc rotating at high speed in a fixed grinding arbor. You can then grind off the resulting fine burr or feather edge on a tripoli-impregnated cotton buffer. This gives the punch the cutting edge it needs to easily cut out a small rosette with one hard hammerblow. Or, using a lesser blow, you can simply indent a rosette design, and then make it more pronounced by turning it over and, using a spherical tool, hammering it repoussé style to accent the raised relief.

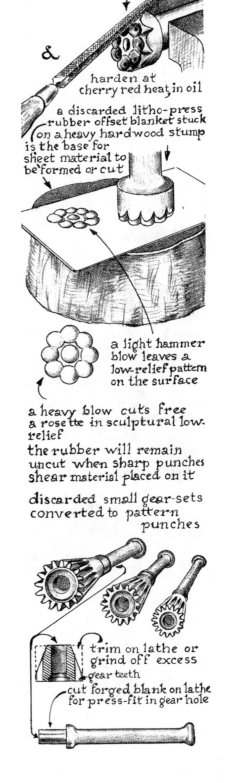

next

&

harden at cherry red heat in oil

a discarded litho-press rubber offset blanket stuck on a heavy hardwood stump is the base for sheet material to be formed or cut

a light hammer blow leaves a low-relief pattern on the surface

a heavy blow cuts free a rosette in sculptural low-relief

the rubber will remain uncut when sharp punches shear material placed on it

discarded small gear-sets converted to pattern punches

trim on lathe or grind off excess gear teeth

cut forged blank on lathe for press-fit in gear hole

Punches Made from Odds and Ends of Scrap

The little pinion gears fastened to the ends of rods are often found at scrap steel yards, and can be used to make punches. Many odd machine parts from dismantled equipment can also be converted into punches to leave indentations, as shown.

A somewhat more elaborate combination, not too difficult to make, uses a cluster of engine push rods with cup-shaped endings. Used singly, these endings can serve as individual punches, but clustered they can punch out rosette designs. In many combinations these punches can make very decorative patterns, which can be accentuated by relief techniques.

More complicated machine parts may take a little more time to fabricate, but they are well within reach of the craftsman who has come this far in his ability to make just about every tool he needs from waste material. The illustrations here are self-explanatory.

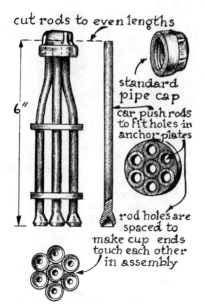

cut rods to even lengths

6"

standard
pipe cap

car push rods
to fit holes in
anchor plates

rod holes are
spaced to
make cup ends
touch each other
in assembly

after assembly heat the 7
rod ends yellow hot & force
bundled rod ends tightly in
pipe-cap

next: reheat pipe cap end
& forge it to fill all remaining
clearance between cap & rods

unsharpened cup rims, when
punched on soft sheet metal,
will leave sculptural relief
patterns, provided

the sheet rests on a rubber
blanket glued to the smooth flat
top of a heavy hardwood stump

How to Make the Star Punch

Punch the cold, hard male die into the soft *yellow-hot* seating, which yields like clay. During hammering the seating begins to cool, but at the same time it begins to heat up the cold male die. Do not continue hammering the hot seating after it has cooled off too much.

Remove the die and cool it in the quenching liquid for a few seconds. This in turn allows the seating (should there be enough reserve heat) to reheat the cooled off portion to forgeable heat. Carefully reseat the male die and deliver once more a few heavy blows before the seat again has become too cold for continued hammering.

Resist the temptation to keep on hammering; reheat the seating part whenever necessary. Cool off the male die before the next round of shaping the blank.

That same die can shape various sizes of star punches. Simply increase the depth of penetration into the female head, leaving progressively larger or smaller impressions as you wish. Having a choice of sizes of star punches makes it possible to create the design shown in the illustration. They can be used to *imprint* a low-relief decorative design on the sheet or to *cut out* a medallion-type rosette form.

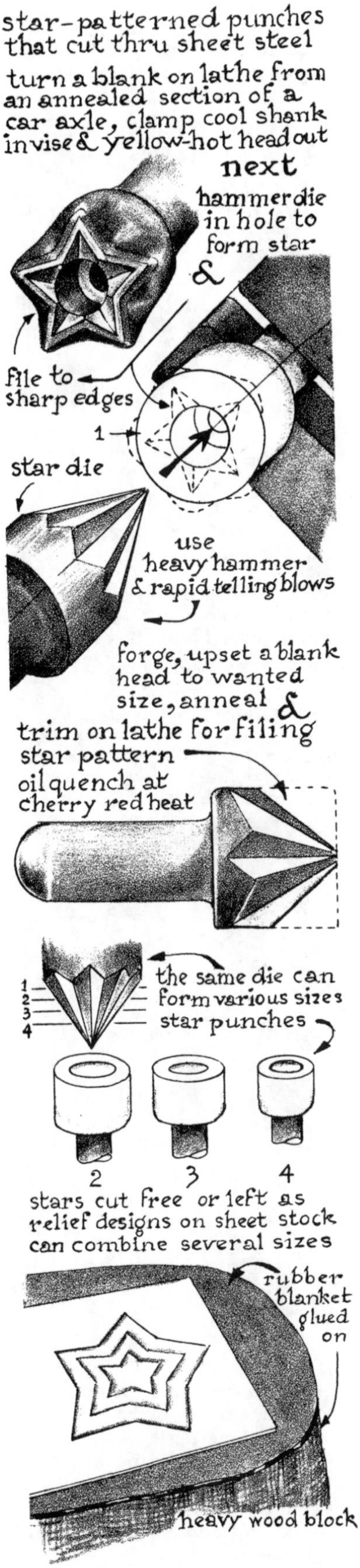

star-patterned punches that cut thru sheet steel

turn a blank on lathe from an annealed section of a car axle, clamp cool shank in vise & yellow-hot head out

next

hammer die in hole to form star &

file to sharp edges

1

star die

use heavy hammer & rapid telling blows

forge, upset a blank head to wanted size, anneal &

trim on lathe for filing star pattern

oil quench at cherry red heat

1
2
3
4

the same die can form various sizes star punches

2   3   4

stars cut free or left as relief designs on sheet stock can combine several sizes

rubber blanket glued on

heavy wood block

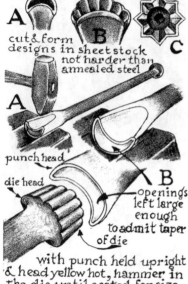

various patterned dies when
forced into hot pre-shaped
punch end openings & next
sharpened & hardened, will

cut & form
designs in sheet stock
not harder than
annealed steel

punch head

die head

openings
left large
enough
to admit taper
of die

with punch held upright
& head yellow hot, hammer in
the die until seated for size

next: reheat punch head &

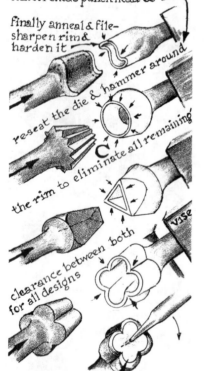

finally anneal & file-
sharpen rim &
harden it

reseat the die & hammer around
the rim to eliminate all remaining

clearance between both
for all designs

if punched-out discs get stuck
in punch heads, pry them out
with a pointed probe
       or
press punch rim in candle wax
beforehand, causing release of
waste disc automatically, if
inside cutting edge is polished
mirror smooth

How to Forge Other Decorative Punches

If you want to add still more varieties of punches, using forging techniques rather than machining them will simplify the work.

Design A, for example, can be made by upsetting a head on a rod and drilling it out to a certain depth. Then place the hollow end in a swage with the hammer peen folding a portion of it, as shown. To finish the tool, simply sharpen the edge and harden it.

In *design B*, the die head is fluted by filing. Hammer the die head into the heated end of B, which has been shaped as in A. Once the end is heated, many light hammerblows on the outside of the punch will spread the heated steel and close the remaining gaps between the inner and outer die surfaces. It may take a few heatings to accomplish this. Be careful when you insert the male die that it is correctly seated before additional hammering begins. If the seating should be incorrect, an overlapping of indentations would ruin all the work that went before.

Design C is similar to the previous star pattern but, instead of having a thick wall of material, it relies on the spreading of the thin wall. Careful hammering with a narrow peen into the groove areas of the design will close up the spaces between them.

After finishing the shaping, sharpen the forged blanks, first with smooth files, then with rotating abrasive discs. If the die has exact and smooth surfaces the blank will not require machine- or hand-filing to improve it further. Temper the punch edges to the hardness of a cold chisel.

NOTE: Sharpening razor-sharp edges on punches that are meant to cut through metal follows the same principles that apply to cutting edges of wood-carving tools. Surprisingly, the sharpness of the cutting edge is more critical if *soft* materials are to be cut. The edges can be somewhat less precise if harder material is carved. If this is confusing to you, think of trying to cut through a cork with chisel and hammer. The cork simply *bends* under the tool without being cut. Cutting material involves a shearing action, not a tearing and bending action. This explanation should be a rule of thumb regarding tool cutting edges.

Punches to Cut through Soft Material

The illustration shows a machining technique for making the punch used to cut a clover-leaf design in soft material (felt, cloth, paper, for example). The interior of the punch can be drilled out.

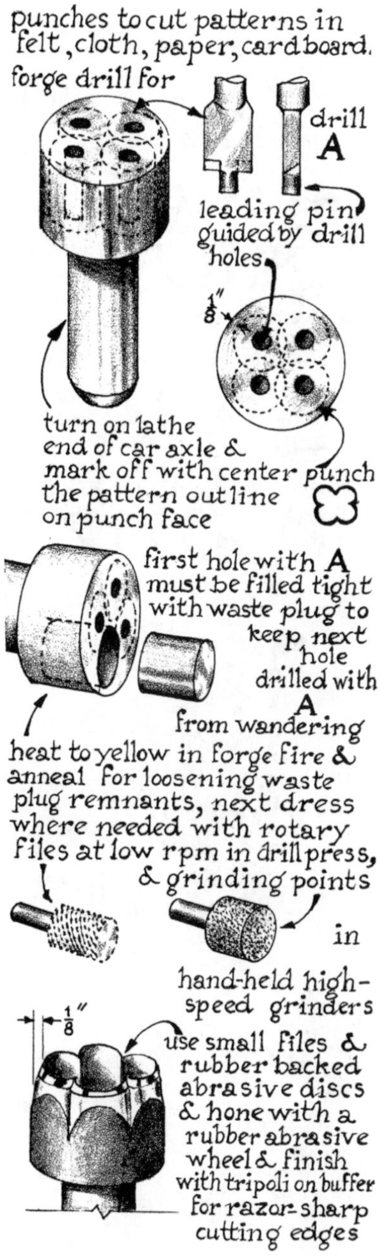

When a beginner sees an expert doing the drilling he gets the impression that he, too, can do it easily. But the expert has learned through experience how to foresee possible errors and avoid repeating former mistakes. (See page 48, *Notes on Drilling*.)

The punch in the illustration must be drilled with a *leading pin* that prevents the drill from wandering. Slow down the rpm to prevent the drill from heating up.

First anneal the high-carbon punch head to its softest. Drill each new hole after filling the previous hole with a waste plug. Ram the plug into the finished hole so that the drilling of the next one will not be affected by the previous one. If the plug is not seated very tightly it may be necessary to drive deep punch marks around the plug edge to anchor it.

If successful in this operation, proceed with refining by grinding, filing, and sharpening as has been described in the making of other punches.

punches to cut patterns in felt, cloth, paper, cardboard.

forge drill for

drill A

leading pin guided by drill holes.

$\frac{1}{8}$"

turn on lathe end of car axle & mark off with center punch the pattern outline on punch face

first hole with A must be filled tight with waste plug to keep next hole drilled with A from wandering

heat to yellow in forge fire & anneal for loosening waste plug remnants, next dress where needed with rotary files at low rpm in drill press, & grinding points

in

hand-held high-speed grinders

$\frac{1}{8}$"

use small files & rubber backed abrasive discs & hone with a rubber abrasive wheel & finish with tripoli on buffer for razor-sharp cutting edges

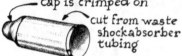

if pattern is a
circle, use high-carbon tubes
to be hardened at edge

bevel edge
is quenched
at cherry red
& tube held
horizontally
while dipping in
cut from car drive   oil
shaft

cap is crimped on

cut from waste
shockabsorber
tubing

mild steel tubes are case
hardened at pre-sharpened
cutting edges

Using Odds and Ends of Tubing for Punches

I have found that the drive shaft of a car, between the transmission and the differential, is made from a high-carbon steel tube welded at the seam. Because such drive shafts are quite large in diameter (in this case 2 5/8 inches), they are well suited for our purpose.

After cutting off an end section and leaving the stub shaft attached, all I needed to do was to refine the cutting edge on the lathe, sharpen it and temper it.

In tempering such a tool, see to it that 1/4 inch at the rim *gradually and evenly becomes cherry red*. At that point, holding the punch blank *horizontally*, quench it deeply and directly in oil, being sure to keep that horizontal position to avoid any danger of oil fume combustion. The danger is real if the piece is quenched in a vertical position, because if the air in the punch body is shut off from its surrounding the locked-up fumes of oil can become a combustible mixture with the trapped oxygen residue. The cherry red heat at the rim could possibly trigger an explosion just as a spark in a car engine explodes a combustible mixture.

Salvaged shock absorber tubing is another possible scrap source for punches. Unfortunately they are often made of mild steel, so it is necessary to case harden the cutting rims. Also a cap must fit tightly on the end of the tube to receive the hammerblows.

NOTES ON DRILLING

A drill press can feed a drill gently or forcefully; at low speed or high speed; using a small drill, a medium one, or a large one. These variables must be brought together in the correct combination if one is to succeed in a drilling job. In addition, it is important to know how *hard* the material to be drilled is.

Modern industry demands higher and higher speeds to cut most material, including steel. This naturally calls for cutting tools that can withstand breakage, overheating, wrong positioning of the tool, and a host of other factors that threaten to cause accidents or unsatisfactory results.

In my own shop I have reduced setbacks by *slowing down* all machinery in order to work with less danger, less heating-up of steel, and greater flexibility in tool alignment during these operations. In short, I propose that in learning to make do with what we have, we rely more on a clear understanding of what is involved during drilling, lathe turning, blacksmithing, and other shop activities and less on data geared to mass-production and precision fabrication.

Important Things to Remember about Drills

The standard twist drill has beveled cutting edges with both sides ground the same length and under the same angle so that the cutting edges automatically meet and seek the center of drill rotation.

Hand sharpening twist drills without the aid of a guiding instrument may leave cutting edges unsymmetrical. This will pull the drill outward, making it wander instead of seeking its own center of rotation. Once it tends to wander, eccentric forces are created, which often strain the drill to its breaking point.

From time to time during the sharpening, hold the drill before you in silhouette against a light background. Sight it to be sure that the sides are as near equal length as your skill can tell you.

11. Christmas Tree Candle Holders and Decorations

Punches can be used to make Christmas decorations from thin sheet metal. Tin cans that have a gold or silver sheen make good ornaments and, when kept indoors, will remain shiny year after year. Mine have lasted over thirty years and are as bright today as when I made them.

CHRISTMAS TREE CANDLE HOLDERS

These keep an upright position when hooked over a tree branch because they are weighted down by a little steel plug at the bottom. The farther the decorative element extends *below* the candle the more stable the upright position will be. Of course, the weighted part must stand free from surrounding branches.

The simplest way to fasten the weight plug into its seating hole is to strike the inserted plug with a heavy hammer on an anvil. The diameter then expands into the hole, locking the plug in place. This is a fastening method that comes in handy on many other occasions because it does away with precision work to keep plugs from falling out if the hole has only a moderate fit to start with.

After punching out the various elements for the candle holder, assemble them in a tight fit, soldering all the critical parts with regular tin solder. Since this is not much additional work, I would recommend that you do so for permanency.

christmas tree ornaments & candle holders made from flattened food cans & waste industrial sheet steel strips

matched holes

bend

bend to fit candle

to fit holes

crease & bend drip cup

baling wire

lace parts together & cinch tightly &

bend to hang on tree branch

lace wire thru tin & cinch

punch hole thru tin to match size rod

& insert,

place on anvil to

weight keeps candle upright

anvil

expand with hammer tight in hole

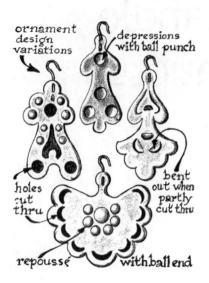

ornament
design
variations

depressions
with ball punch

holes
cut
thru

bent
out when
partly
cut thru

repoussé

with ball end

OTHER DECORATIONS

These can be made with the same tin can material or any other sheet metal you may find appealing. Keep in mind that punches for cutting through steel do not need to be absolutely razor sharp. If they are slightly dull, it introduces an element of surprise when the metal is punched on the rubber blanket glued on a stump. If the punch is slightly dull it merely *indents* the metal, leaving an interesting *low-relief* design. To make of this a *high-relief* pattern, turn the decoration over and use the repoussé technique of hammering a rounded punch into the depressions. These push the relief further outward.

Punches can also be used to cut only *partway* through the sheet. The cut section, bent outward, will catch and reflect light, adding sparkle to the decorations.

If you have enjoyed making the great variety of punches described in Chapter 10, you may want to gather friends together for a Christmas decoration "punching spree," as I have often done. The participants, sitting in front of the wooden stumps on the floor, can create beautiful designs as they share the variety of punches. The skill needed to use those tools can be acquired in minutes. And the decorations that result are well worth keeping.

12. Making Design Layouts for Punches

sample design cut & raised in low relief from discarded litho-offset plates

TRANSFERRING DESIGNS ONTO METAL

1) The design patterns for decorative articles can be drawn on paper and traced onto the metal with carbon paper.

2) Or, the paper on which the design is drawn can be rubbed with beeswax on the reverse side, then rubbed down onto the metal with a smooth, shallowly curved object (such as a tablespoon). If the metal is thin and soft, be careful not to bear down too hard. You may use a fingernail if the metal is fairly hard or heavy gauge. Following the outlines, use the various punches to cut through the paper and metal to free the piece. Finally, pull off the paper remnants.

3) Or, the pattern can be punched out of a transparent material, such as a sheet of acetate, using all the variety of punches you have. This sheet is then used as a template with which to draw the design onto the metal. In this way, a complete design can be laid out first and then cut with the same punches.

These methods make it unnecessary to scratch in the design with a sharp-pointed steel scribe, which can damage the metal if you miss the mark.

I offer here as examples several designs, but they are not necessarily to be copied. You will no doubt enjoy following your own imagination, no matter how elaborate the design may be. The wider your variety of punches, the greater will be your design possibilities.

Whether your design is elaborate or simple, you may want to combine the silhouettes with relief work made with the repoussé technique. Turning the sheet over, push the metal out, or make depressions on the original side, using the rounded tool ends made for that purpose. These hollow designs are very effective, looking like so many shining buttons. The combined raised and depressed patterns will give your design a three-dimensional character.

It is the leather worker, as a rule, who uses such repoussé and chasing punches on his material. For metalwork, the rubber blanket on the stump (as described in Chapter 10) acts as a good base when treating the metal in the same way.

template for design outline is made with the same pattern punches, in transparent acetate sheet

13. How to Make Miniature Chisels and Punches

heavy-gauge industrial high-speed steel hack saw blades used as stock for making small chisels

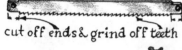

cut off ends & grind off teeth

with abrasive cutoff wheel cut 3 strips ⅜" wide & cut 1 strip into six even lengths

clamp the 6 pieces in vise grip & grind one end slightly tapered

round off taper-end edges but leave sides untouched

side   ⅜"   edge

⅜"

1"   1"

Miniature chisels and punches are needed when making a matrix and many small workpieces, for example, an escutcheon.

Used hacksaw blades can be very useful items that you may find discarded by industrial plants. They are generally made of high-speed tool steel. They are, as a rule, thick enough to be used as stock for small cutting tools such as miniature chisels and punches.

MAKING THE CHISEL BLADES

To cut up the blade, clamp a wood fence on the table of the abrasive cutoff wheel. Space it to cut six strips 3/8 inch wide from the blade. Hold the stack of six small pieces together with the visegrip plier.

Grind, as accurately as you can, a tapering end on this stack so that each is exactly the same as the others. From these make miniature chisels with different cutting edges. Fit the tapered end of each into a chisel-holder socket; it should be easy to insert or knock out the chisels so you can replace one with another when you want a different cutting edge.

MAKING THE CUTTING BLADE SOCKET

The holder is an annealed high-carbon steel bar or rod as shown. Drill a small hole in the end of the holder. Heat the holder end to a light *cherry red* and flatten it somewhat to match the width and thickness of the chisel's taper ending.

Heat the hollow flattened rod ending *yellow hot* and hammer in the taper end of the chisel, causing the hot steel to yield and seat the chisel taper. Then hammer the hot metal around the hard cold taper of the chisel until all surfaces contact one another exactly. The socket will now fit on such a taper as a lathe tailstock fits a Morse taper center.

SHAPING THE CUTTING ENDS

It is up to you to decide what shapes you wish to give the chisel cutting ends. The shaping is done in a simple forging action.

Note that high-speed tool steel, when heated *yellow hot*, can be bent into curves of your choice. But, unlike hot-forged high-carbon steel or mild steel, it will resist peening or upsetting. High-speed tool steel has been designed to maintain considerable hardness even when heated; it is only at *light yellow* heat that it will yield to bending. And once it is cooled without quenching or tempering it will return to its original hardness.

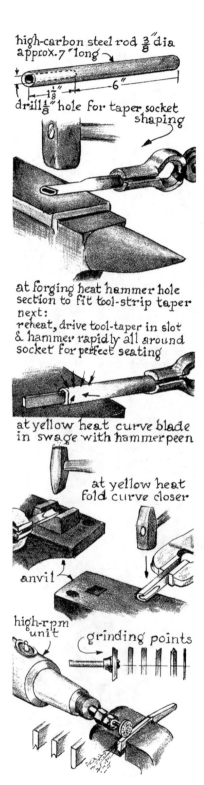

high-carbon steel rod 3/8" dia approx. 7" long

drill 1/8" hole for taper socket shaping

at forging heat hammer hole section to fit tool-strip taper
next:
reheat, drive tool-taper in slot & hammer rapidly all around socket for perfect seating

at yellow heat curve blade in swage with hammer peen

at yellow heat fold curve closer

anvil

high-rpm unit   grinding points

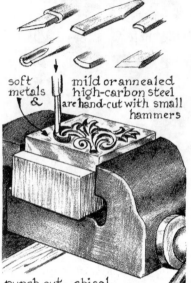

useful cutting ends shaped
with hand-held high-rpm.unit

soft
metals
&   mild or annealed
high-carbon steel
are hand-cut with small
hammers

I have always gone by this rule of thumb: After heating and then cooling, high-speed tool steel will cut any partially annealed regular high-carbon steel satisfactorily and any well annealed high-carbon steel and mild steel easily.

punch out, chisel,
& cut key slot decoration
on a block   of endgrain
pear wood   held in vise

sheet metal

MAKING AN ESCUTCHEON

Transfer the design onto a piece of sheet metal of your choice. It should be annealed steel or annealed silver, brass, aluminum, or any other metal of comparable softness.

The self-explanatory illustrations show step by step drilling, shearing, punching, filing, grinding, buffing.

when using steel ½" thick
or over, cut pattern with
abrasive cut-off wheel

Cutting with chisels and punches is best done on a base of endgrain fruitwood clamped in the vise, as shown. The sharper the tool edges, the fewer difficulties you will meet.

When keyholes require exact patterned contours, smooth files, especially in small sizes such as locksmiths use, will serve best.

It is in these small projects that you can apply your acquired skill in making your own files, especially if you must reach small spaces for the cross section design you need.

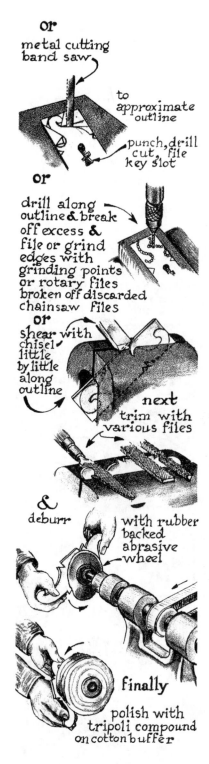

or

metal cutting band saw,

to approximate outline

punch, drill cut, file key slot

or

drill along outline & break off excess & file or grind edges with grinding points or rotary files broken off discarded chainsaw files

or

shear with chisel little by little along outline

next trim with various files

& deburr

with rubber backed abrasive wheel

finally

polish with tripoli compound on cotton buffer

14. A Punch to Cut Small Washers from a Metal Strip

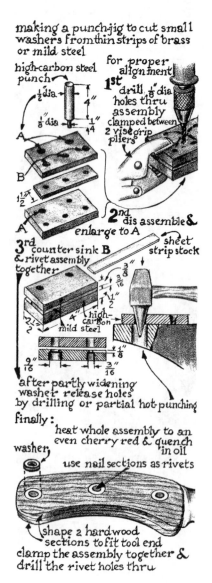

making a punch-jig to cut small washers from thin strips of brass or mild steel

high-carbon steel punch

for proper alignment

1st drill ⅛" dia holes thru assembly clamped between 2 vise-grip pliers

½" dia
4"
⅛" dia
¼"

A

B

1½

A

2nd dis assemble & enlarge to A

3rd countersink B & rivet assembly together

sheet strip stock

⅜"
³⁄₁₆"

high-carbon mild steel

2¼
9⁄₁₆"
½"
⅛"
³⁄₁₆"

after partly widening washer release holes by drilling or partial hot punching

finally:
heat whole assembly to an even cherry red & quench in oil

washer

use nail sections as rivets

shape 2 hardwood sections to fit tool end clamp the assembly together & drill the rivet holes thru

Brass washers can be used to rivet wood onto the steel handles of knives, cleavers, wood-carving tools, fireplace tools, and others. The color of the wood, the color of the washer, and the color of the rivet will all be different, enhancing the appearance of the handles.

Turn the punch on a metal-turning lathe and temper it to the hardness of a cold chisel. Follow the sequence in the illustrations, being careful to maintain correct alignment of punch and guide hole in the riveted assembly. If all has been carried out successfully, the brass stock fed into the jig and then punch-cut will easily produce small, clean-cut brass washers.

ASSEMBLING A HANDLE WITH THE WASHERS

If you do not want the washers and the rivet heads to protrude from the wood surface, you must first countersink the washers themselves before imbedding them into countersunk depressions in the wood. The washer countersinking is done with a center punch. This may be a novelty to some craftsmen but is much simpler than doing it with a countersink drill.

Just place the washer on an endgrain wood stump, put the center-punch in the washer hole, and hammer it down. This leaves a slightly bent cup shape in the washer. When all parts are assembled for riveting, the rivet head will fill the washer cup depression completely.

The washers are seated with a small countersink cutter that has a pilot pin to keep it from wandering.

Use files, abrasive discs, and polishers to smooth the surface of the wooden handle. This will bring out the contrasting colors of wood, washer, and rivet head.

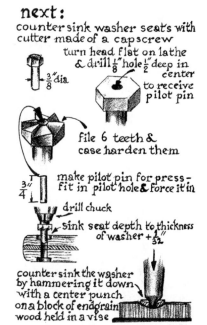

next:
countersink washer seats with cutter made of a capscrew

turn head flat on lathe & drill ⅛" hole ½" deep in center to receive pilot pin

←⅜" dia

file 6 teeth & case harden them

¾" make pilot pin for press-fit in pilot hole & force it in

drill chuck

sink seat depth to thickness of washer +$\frac{1}{32}$"

countersink the washer by hammering it down with a center punch on a block of endgrain wood held in a vise

15. Makeshift Bearings

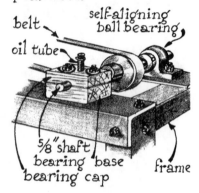

emergency bearings made of hard, close-grained pearwood

belt

oil tube

self-aligning ball bearing

5/8" shaft
bearing base
bearing cap

frame

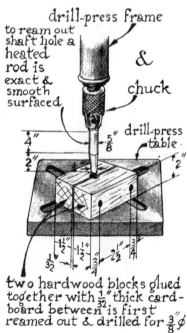

to ream out shaft hole a heated rod is exact & smooth surfaced

drill-press frame

&

chuck

5/8" drill-press table

4"

5/8"

2"

2"

1 1/2" 1 1/2" 3/4" 2 1/2" 3/4"

1/32" 1/2" 3/4"

two hardwood blocks glued together with 1/32" thick cardboard between is first reamed out & drilled for 3/8 ⌀

MAKING A BEARING OUT OF WOOD

If a bearing breaks down in the shop and if you are located very far from stores where they are sold, you can make one out of wood.

First Step

Choose a dry branch of fruitwood from orchard prunings or some comparable fine-grained wood that has the closest grain possible. As a rule, pearwood is best. Cut from it two blocks and glue them together with thin cardboard in between. Bolt this block onto the location of the bearing that is to be replaced. Mark off the exact center of the hole to be drilled through the wood to fit the shaft.

Drill the hole *slightly undersized* at first but as well aligned with the axis of rotation of the shaft as possible.

Second Step

Use a steel rod the size of the shaft and taper it gently so that the small size of the taper fits in the predrilled, undersized hole. Place the block on the table of the drill press and put the tapered rod in the drill chuck to test the setup. When you are satisfied that all is ready, remove the tapered rod and heat it in the forge fire to the point that it will scorch wood.

Quickly transfer the rod to the drill chuck and burn it through the wood, making sure that the heat is not so great that the smoke combusts into flame. The heating does two things: momentarily it *softens* the wood; and, while it scorches, makes the final surface of the hole exact, smooth, and hard.

If all has gone well and the block barely fits onto the shaft, grease the hole and let the shaft run into it with its tight fit. The grease, upon heating, will penetrate the wood.

Third Step

After the bolt holes have been drilled, split the block in two along the cardboard. The top part of the block is the *bearing cap*.

Drill the bolt holes with some clearance, so that you can make a slight adjustment in the placement of the bearing if necessary. If, in assembling, you find that the shaft hole is still too tight, placing thin shims between the bearing cap and bearing pillow will eliminate the friction. Provide the bearing with a lubrication hole and groove.

When installing the bearing, spin the shaft freehand and you will notice immediately whether or not the bearing is well aligned and if adjustments are called for.

Such a bearing will stand up under regular use for a long time. I have used one in my own shop for years.

Adjustments

If the bearing is too low, you may have to place below it strips of thin metal in the pattern of its footing, with the holes corresponding to the bolts. Or if, on the other hand, it is too high, plane or sand off enough wood to lower the bearing to the correct height.

The thickness of the cardboard must be made up for with two shims of wood or metal of that same thickness. If that is a little too thick, try progressively thinner shims until the bolted-down cap does not bind onto the rotating shaft. The shim's edge should just touch the rotating shaft.

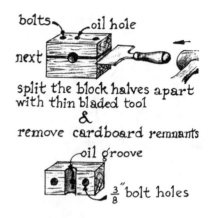

next

split the block halves apart
with thin bladed tool
&
remove cardboard remnants

cut shims of appropriate
thickness for exact fit
of shaft in bearing

USING DISCARDED BALL BEARINGS

Electrical motor and automotive repair shops quite often discard noisy ball bearings; some are from burned-out motors and are hardly worn at all. If proprietors of such shops are sympathetic with your efforts, they will let you rummage through their scrap barrels. You will be surprised how often you will find unexpected treasures in addition to the things you are looking for.

Bearing shaft holes that are slightly too large

Often you can locate a thin-walled tube that fits over the shaft but may be a little too thick to fit within the bearing hole. Slip the tube section onto a tight-fitting wooden rod and clamp it in the lathe chuck or the drill press chuck. Turn or grind down a small section of that outer tube surface, testing the bearing over it for a fit. If that section is left to taper outward somewhat it will act as a wedge between the bearing opening and the shaft diameter. In assembly, you can tap it into the space it is to fill between bearing and shaft.

After assembly, clamp a little collar onto the shaft to hold the protruding section of the tube shim in position.

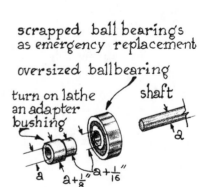

scrapped ball bearings
as emergency replacement

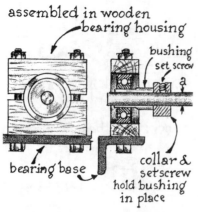

assembled in wooden bearing housing

bushing set screw

a

bearing base

collar & set screw hold bushing in place

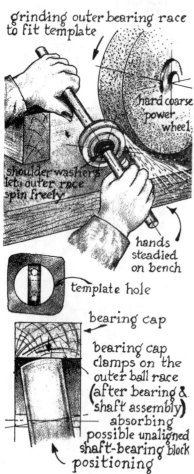

grinding outer bearing race to fit template

hard coarse power wheel

shoulder washers let outer race spin freely

hands steadied on bench

template hole

bearing cap

bearing cap clamps on the outer ball race (after bearing & shaft assembly) absorbing possible unaligned shaft-bearing block positioning

How to install the ball bearing

The illustration shows the wooden pillow block that holds down the ball bearing. Since the wood in this instance is not used as a bearing itself but simply as a seat for the ball bearing, the wood can be hard-grained; it need not be close-grained. Oak, maple, or ash will do.

After the height of the bearing has been determined and the whole assembly has been positioned, do not cinch the hold-down bolts until you are certain that the ball bearing itself is perfectly aligned with the shaft. This will protect the ball races from wearing at their sides.

NOTE: The outer ball races in self-aligning ball bearings have spherical outer surfaces representing the central section of a sphere measuring the width of such an outer ball race. Examine a self-aligning ball bearing in a ball bearing supply store, if the three-dimensional aspect of the spherical outer surface of the bearing is too difficult to visualize from drawn illustration.

In the case of these you do not need to make the surface of the bearing seat spherical. The illustration shows that a two-part *cylindrical seating* can be clamped onto the approximate spherical outer race in a test run. This allows the spherical surface of the outer race to seek its own aligned position just before the bearing seating is clamped down onto the ball bearing.

Grinding the Spherical Surface of the Outer Ball Race

Cut a template with a circular opening the size of the outer diameter of the outer ball race. Fit a wooden rod tightly into the ball bearing. Cut two cardboard discs with central holes to fit snugly on the wooden rod. These discs are to keep abrasive dust from entering the vital parts of the ball bearing; they are placed on each side of the ball bearing with thick grease in between. The discs do not interfere with the spinning of the *outer* race.

Holding the assembly in both hands, move the outer race into contact with a rotating hard and coarse grinding wheel. The ball bearing race, spun by the wheel, is at the same time being ground down. Aim to grind the steel while it rotates at maximum effectiveness by slanting it, as shown.

The grinding job can be called finished when the outer bearing race can slip through that hole, evenly touching the portion of the circle in the template.

NOTE: The foregoing methods of improvising makeshift bearings to replace broken ones are only two of the many possibilities one can inventively resort to in an emergency.

16. Making Accessory Tools for the Wood-Turning Lathe

If your shop has a wood-turning lathe, and also an assortment of machine tools, you can easily use this lathe to drive implements designed for other uses. Grinding wheels or sanding discs can be mounted on the lathe headstock and driven by the lathe. This is an ideal adaptation because the step pulley on the lathe headstock allows for various speeds; you should use the fastest speed for small wheels, the slowest speed for as big a wheel as the lathe will take.

A long grinding arbor makes room for such hand-held instruments as a reverse lathe unit, which facilitates the regrinding of engine valves. But it also provides much needed room when awkward-shaped articles have to be ground. Extra sturdy, precisely fitting arbors must be securely clamped in the headstock to prevent the vibration of high rpm small grinding wheels.

Very small, worn-down abrasive cutoff wheels work ideally on the wood-turning lathe when you need to do an engine valve regrinding job. Use the method shown in the illustration.

Hand hold a small reverse lathe unit, keeping both hands steadied against the lathe bed. Align the bevel of the valve with the side surface of the cutoff wheel remnant; the grinding job is finished in a fraction of a minute. In this way twelve valves of a six-cylinder car engine can be reground to perfection in no time.

Note that the illustrations show an arrangement made in my own shop to suit my particular needs. Inventiveness and resourcefulness play a major role in using recycled material from an often maligned junk pile. This points up once again how much sense it makes to build up a stock of scrap steel and salvaged objects: you will become better and better equipped as time goes on. Thus "making tools to make tools" should become more and more the norm of daily activity.

If you live in or near rural areas you probably have access to the fruit tree prunings stacked around an orchard. These fruitwood branches make excellent stock from which to make small tool handles. The farmer in my area is generally sympathetic with my needs when I explain to him that I am a practicing artist-craftsman. When I offer him some picture postcards of my work, he usually permits me in exchange to help myself to his prunings.

I have designed a jig to round off the end of a wooden handle and at the same time to leave a seating hole in the center to receive the tool's tang.

wood lathe accessories & their use

shatterproof thin cut-off disc
lathe headstock
mandrel holds grinding disc on shoulder

abrasive cut-off wheel worn to smallest diameter is then used at 3000 to 5000 rpm on wood lathe head stock for special grinding jobs

special application

engine valve grinding

hand-held reverse lathe is positioned at angle of valve bevel in alignment with wheel's side, which, upon contact, spin-grinds precisely a new & smooth final bevel surface texture

a handle cutter-head fastened in wood lathe head-stock is to spin at approx. 1750 rpm to cut hand-held branch

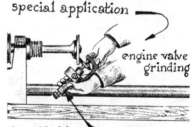

cutter is made on metal-turning lathe from a car axle section

bevel

drill hole for leading pin for a snug fit

bevel

grind one side of pin flat to cut the wood while pin's round side keeps wood from wandering

bevel, pin & filed-in teeth are to be tempered dark bronze

pin can be knocked out when needed

pin makes the preliminary seat for the tool tang, which is to be burned in

turning wood handles for small tools in a cutter designed for limited quantity production using a simple homeshop lathe

handle for burins

lathe head

replaceable cutter bit **D**

holes to release wood shavings

A E B D

slot fits cutter bit D

pipe ring guides & cuts wood stock to size

case harden teeth

C

a-a   a→| |←b   b-b

D

sharp   a→| |←b   sharp

locking ring B press fit on A

Without a guiding pin, there is a danger that the stock ending may wander at the point of engagement with the cutter's teeth. This will happen if you have extracted the guide pin because you do not want a tang hole in a wood ending. In this case, bevel the wood edge so that it will slip inside the jig about a quarter of an inch. The stock can then be safely fed freehand with the tailscrew.

AN ADJUSTABLE CUTTER TO MAKE SMALL WOODEN HANDLES

I contoured cutter element D in such a fashion that I could slip it into a slot while the ring-guide B and cutter C would hold it snugly in the jig assembly.

Examine the illustration carefully. It clearly shows the locked-up position of D during the cutting action. Element C, though fitting snugly on A, does not fit so tightly that it could not be knocked off should you wish to make D interchangeable with a cutter element of a different profile.

The side holes in both jig body A and cutter ring C simply serve as escape gates for the wood pulp that is thrown out centrifugally during the high-speed cutting action.

Element E is simply a cutting pin, which can be inserted tightly in the main body A so that a tang hole can be cut into the small handle. At the same time, that pin will act as an alignment guide to prevent any wandering of the workpiece during the cutting action.

The illustrations show the jig and wood stock held and fed by the tailstock center and kept from turning by the clamped-on visegrip pliers.

The craftsman who has become increasingly skilled in making things to fill his needs is offered in the foregoing an approach to designing and making a new tool. In my own experience, I have found that dexterous people seem to be endowed also with a good measure of inventiveness, which enables them to design and construct whatever accessories may be useful to them. In this way they gain ever greater independence and freedom to earn their livelihood in a unique and individual manner.

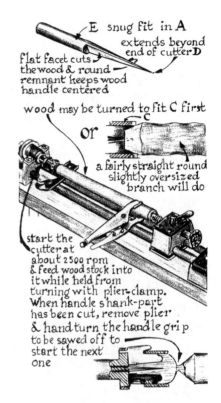

E snug fit in A
extends beyond end of cutter D
flat facet cuts the wood & round remnant keeps wood handle centered

wood may be turned to fit C first
or
a fairly straight round slightly oversized branch will do

start the cutter at about 2500 rpm & feed wood stock into it while held from turning with plier-clamp. When handle shank-part has been cut, remove plier & hand turn the handle grip to be sawed off to start the next one

17. Wire-Straightening Tools

when straight annealed wire is bent only a little it will spring back straight

}A

when bent too much some of it stays bent

B

when kinked wire is pulled through a tube with three staggered A bends, it is straightened accurately

hardwood jig to bend tube

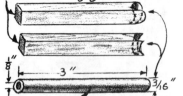

⅛"   3"   ³⁄₁₆"

flexible copper tube ⅛ inch inside diameter

hammer down

end of wire is strung thru before tube is bent between jigs

vise jaws

view

each bend is to stagger 60 degrees between bends in head on view

clamp wire end in vise

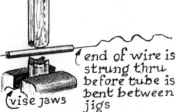

&

pull bent tube along kinks in wire

a less accurate way is to pull kinked wire

back & forth over a

vise

1-inch smooth round rod

Coils of old kinky wire can often be quickly and easily straightened with a jig.

THE FIRST JIG

Slip the end of the twisted wire through a small copper tube bent as shown in the illustrations. Clamp a short protruding end of the wire between the vise jaws, or loop it around a fixed anchor point, and pull the bent tube along the full length of the wire. This will straighten it completely. The diagrams explain what happens.

It is awkward to reuse the tube because it is difficult to manipulate another wire ending through it *after* it has been bent. For this reason, a second type of jig is more versatile.

THE SECOND JIG

Cut three sections from a hexagonal bar and drill a small hole through each one. It is important to make the hole just barely large enough to let the wire slide through.

Round off the holes at both ends to allow the bent wire to slip by without being cut on its way through. Ample lubrication during the pulling will also smooth the process of unbending the wire.

The jig housing is made as illustrated to hold the three hexagonal sections.

After the wire ending has been laced through each section separately, insert them in the housing in a staggered position. This easy assembly eliminates the problem of threading the wire through as must be done in the first method. The grip latch on the housing makes it convenient for the instrument to be pulled with great force, should that be necessary.

Keep in mind that straightening wire can only be done in this way if the wire is not too thick or if it is not made of spring steel.

It should go without saying that a tool such as this one is useful only if the wire to be recycled is in good enough condition to be salvaged at all.

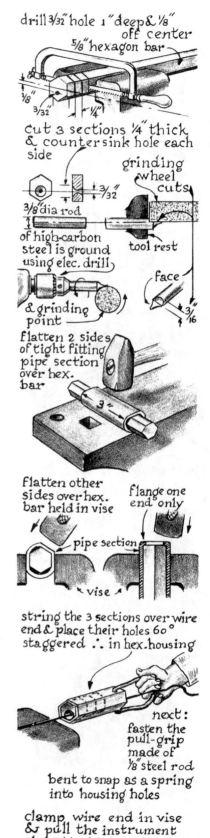

drill 3/32" hole 1" deep & 1/8" off center
5/8" hexagon bar
1/8"
3/32"
1/4"

cut 3 sections 1/4" thick & countersink hole each side

grinding wheel cuts

3/8" dia rod
of high-carbon steel is ground using elec. drill
tool rest

& grinding point

face
3/16"

flatten 2 sides of tight fitting pipe section over hex. bar

flatten other sides over hex. bar held in vise

flange one end only

pipe section

vise

string the 3 sections over wire end & place their holes 60° staggered ∴ in hex. housing

next: fasten the pull-grip made of 1/8" steel rod
bent to snap as a spring into housing holes

clamp wire end in vise & pull the instrument along the kinky wire. It straightens wire accurately

247

18. Flat Filing and Drilling

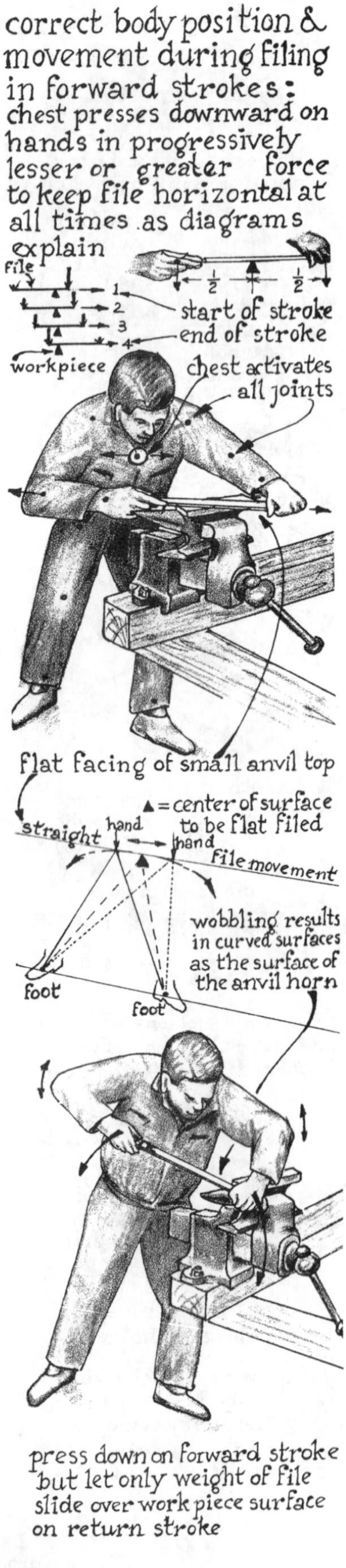

correct body position &
movement during filing
in forward strokes:
chest presses downward on
hands in progressively
lesser or greater force
to keep file horizontal at
all times as diagrams
explain

file

1
2
3
4

start of stroke
end of stroke

workpiece

chest activates
all joints

flat facing of small anvil top

▲ = center of surface
to be flat filed

straight hand

hand

File movement

wobbling results
in curved surfaces
as the surface of
the anvil horn

foot

foot

press down on forward stroke
but let only weight of file
slide over work piece surface
on return stroke

Leonardo da Vinci designed a machine to cut file serrations automatically. This made it possible for others to improve filing techniques in the centuries that followed; during the age of steam engines, craftsmen learned to file with a precision that matched the machine-tool accuracy of later years. The stress was on learning to file flat surfaces to an absolute perfection. One could say that the craft of *flat filing* is the hallmark of the completion of the machinist's training.

THE PROPER USE OF THE FILE

The illustration shows the proper stance for flat filing. In this case, a small, well-annealed anvil face is to be filed perfectly flat.

The worker's leading foot is placed in the direction of the file stroke and the other foot is placed far enough away to steady his position. This makes it possible for him to move his body backward and forward through the hip, knee and ankle joints just enough to keep pace with the backward and forward movements of the hands holding the file.

The beginner should aim to hold the file in a position perfectly parallel with the horizontal surface to be flat filed. He must learn to bear down on the forward stroke of the file and release all downward pressure on the return stroke.

Soon he will become aware that it is the movements of his *whole body*, not just of his arm and hand, that must combine for accuracy and effectiveness. From then on repeated practice is required to eliminate the initial wobbling motion that every beginner experiences. Examine the diagrams carefully to visualize what actually takes place.

EQUIPMENT

The items listed are the tools I was provided with during my first year at the college of Marine Engineering in Groningen, Holland, in 1916. They are still the basic ones used today. Half the course time was devoted to shop training. The first six months were spent learning the art of flat filing. Next came blacksmithing. The remaining time was spent on applying what had been learned to make actual steam engine parts. Hand skills were combined with learning how to use the basic machine tools: the drill press, metal-turning lathe, and shaper.

In this school, all equipment was old and fairly dilapidated. That required still more skill on the part of the student if he was to carry out his assignments and meet the final test requirements for accuracy. In those years it was absolutely necessary in the event of a mid-ocean breakdown that the marine engineer on board ship be able to repair, in the ship's machine shop, the engine with sufficient skill and know-how to allow the ship at least to limp into port under its own steam for more permanent repairs. The skills I acquired in those early years have been a boon to me all my life. Many times have hand skills at their best saved the situation when only a minimum of equipment was available.

needed equipment to carry out assignments in learning flat filing

a 2-lb ball peen hammer

a 35-to 50-lb machinist vise

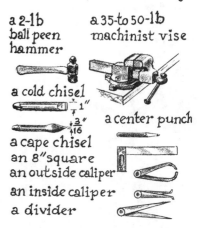

a cold chisel $\frac{1}{4}\times 1''$

a cape chisel $\frac{3}{16}''$

a center punch

an 8" square
an outside caliper
an inside caliper
a divider

types of files
one of each

12"x 1¼" flat   bastard
12"x 1¼"   "   smooth
12"x 1¼" half round   b.
12"x1¼"   "   s.
10"x $\frac{7}{16}$" round   b.
10"x $\frac{7}{16}$"   "   s.
10"x $\frac{3}{8}$" square   b.
10"x $\frac{3}{8}$"   "   s.
10"x $\frac{1}{2}$" triangular   b.
10"x $\frac{1}{2}$"   "   s.
8"x $\frac{3}{8}$"   "   s.
8"x $\frac{1}{4}$" square   s.
8"x $\frac{1}{4}$" round   s.
8"x $\frac{1}{2}$" half round   s.
8"x$\frac{1}{2}$"x$\frac{1}{8}$ thin flat   s.

a surface gauge or a part that has been machined flat as a drill press table or a circular saw,- planer table

some soot & shellac to blacken surfaces for scribe marking and some salvaged litho-offset ink for testing the accuracy of flat filed surfaces on surface gauges

Flat filing exercises
1st assignment
File all sides of a bar 1/16" less

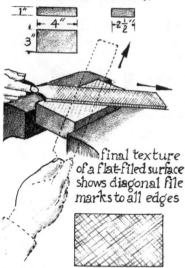

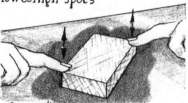

final texture of a flat-filed surface shows diagonal file marks to all edges

test for accuracy: jiggle bar on machined table gauge that has been blackened a little & next move bar over table to reveal low & high spots

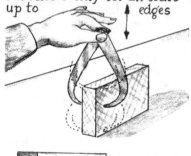

after 1st accurate flat surface file its opposite flat & parallel. calipers must lightly "feel" total surface evenly for all sides up to edges

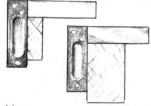

all sides must be filed 90° to each other, checked on surface gauge & with calipers for parallelity

HOW TO LEARN FLAT FILING

From the easiest assignment to the most difficult one, the student will gradually develop increasing skill in flat filing. Cumulatively this enables him to compete with the machine in accuracy. Perseverance and self-discipline will carry him through. It did so for the thousands before us who had to make do with a minimum of machine help.

First Assignment

Cut from a hot-rolled mild steel bar a piece 3 inches wide, 1 inch thick, and 4 inches long. Clamp it in the vise in a horizontal position and proceed as shown. Sometimes a new piece of hot-rolled steel has a hard scale that could possibly damage the sharp file teeth. If you find that this is the case, it would be wise to hack off the scale with a sharp chisel. Cut off only a thin sliver, being careful that at no time does the chisel cut deeper than necessary. Once the surface has been scaled, the file can be used without the danger of becoming dulled.

As soon as the cold chisel marks have been filed away, put the surface to the first test for flatness on a surface gauge. A flat machined part in the shop will serve well for a gauge.

Rub into its surface a little lampblack and grease mixture, wiping off all excess. Sliding the filed surface over it will show, by the black spots left on it, the degree of accuracy that has been reached and will indicate where corrections must be made.

After total surface accuracy has been achieved and demonstrated with the surface gauge, replace the *bastard file* with the smooth file. The final smooth surface must pass the test for absolute accuracy.

It is at this point that, with gained skill, you can file the opposite side, which then must be tested with the caliper instead of the surface gauge. The caliper, finely adjusted to barely touch the steel, will make it possible for you to *feel* where corrective filing is needed.

After tests show accuracy, the remaining sides of the block can be filed in the same way. One additional test must then be made. The square will show whether all sides meet at exactly 90 degrees.

Place the square over the edges of the block and hold it up against the light. Any inaccuracy will show up as a light leak, indicating where corrective filing must be done.

When the workpiece has been completed, there is no doubt that you will have quite a few blisters on your hands to show for your efforts, but you can now face more complicated assignments with greater confidence.

The steps described in the first assignment are all applicable in what is to follow. Each successive project is designed to exercise and utilize your acquired abilities.

Second Assignment

Once the 1 by 1 by 4-inch bar has been filed, mark off with a scribe a slot location and accent it with small center-punch marks about 1/4 inch apart placed on the scribed lines. This ensures that the slot location will remain visible even when extended work is done on that surface.

Use a cape chisel to precut the slot within the marked lines to a depth just under the given measurement. A follow-up filing into the slot can be done accurately without danger of overshooting our aim. It is necessary, nonetheless, to file cautiously and accurately.

250

Making a key to fit the slot. Forge a key blank (an outmoded design) somewhat larger than would fit the slot. File it down to correct size. Probing will indicate where corrective filing must be done. During corrective filing one must learn how to hold or slant the file in such a way that local areas only can be filed without touching surrounding areas in order to correct inaccuracies. Often using the tip of the file only will reduce local spots to needed size.

Third Assignment

This assignment repeats the second one with the added complication that each of the four slots measured with the caliper must show that its opposite slot bottom runs parallel and that all are equidistant. All four slots must furthermore fit the same key evenly when the key is tapped snugly through the slots with the back of a hammer stem. The key should still hold the slot in the last 1/8 inch of its travel. When the piece is held up against the light for that last 1/8 inch fit, a slight glimmer, if any, should show evenly between contacting sides.

Fourth Assignment: Drilling

The drill press is used in this project. First check the proper sized drill for accurate alignment. Clamp it in the drill chuck and rotate it momentarily under power to test it for perfect central alignment. Since most modern drill presses have three-jawed chucks, see to it that the drill shank has been inserted as far as it will go in order to intercept any possible wobble.

Old-fashioned drills have Morse taper endings that fit Morse taper drill seatings. That system eliminated all possible wobble of the drill. Modern three-jawed drill chucks, however, have been machine perfected to such an extent that cylindrical drill shanks can be accurately aligned and held fast effectively.

Since it is cheaper to manufacture drills with cylindrical shanks, and the three-jawed chuck can take all sizes of drills, the modern system has replaced almost entirely the old-fashioned taper shank drill.

Adjust the drill press table to its desired height. Place the workpiece on it and lower the drill so that its point fits exactly in the previously made centerpunch mark on the workpiece. Hold it in that position momentarily and, with whatever means is available, clamp the workpiece to the drill press table securely.

Position the drill up and down to make sure that the correct centering and aligning has not gone awry during this "make ready." (The term *make ready* is used in all trades that require a dry run before actual cutting, punching, printing begins.) Even then the first real cut with the drill must be closely watched so that a corrective step can be applied immediately.

Sometimes the corrective step involves giving the drill press table a gentle sideways tap with the back of a hammer stem, or the workpiece on the drill press table might require a gentle tap in whatever direction to ensure that all is centrally secured. Such adjustments will make the drill cut correctly before it goes much deeper.

In this progressive fashion, it is possible to drill holes precisely in marked-off locations. A test with the caliper afterward will check on the accuracy of the end result.

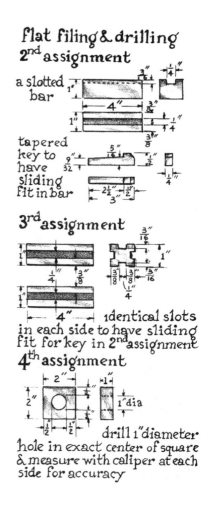

flat filing & drilling
2nd assignment

a slotted bar

tapered key to have sliding fit in bar

3rd assignment

4" identical slots in each side to have sliding fit for key in 2nd assignment

4th assignment

drill 1" diameter hole in exact center of square & measure with caliper at each side for accuracy

WARNING: If a workpiece has been clamped on the drill press table so that the drill will be aligned with the hole in the table, the final cut will make the thin steel remaining at the bottom bend outward the moment the drill begins to emerge below. The thin steel then bends away and, instead of being cut by the drill, the edges begin to bind under progressively increasing strain. It is therefore necessary that you develop the ability to "feel" and "hear" telltale signs of binding. Once you notice them, immediately reduce the downward pressure of the drill to relieve the strain on it should it be slightly dull. Otherwise its outside cutting edge will begin to *rub* instead of *cut*, generating heat and endangering the temper hardness of a high-carbon steel drill.

All this can be prevented if you clamp the workpiece on a flat piece of waste steel. This allows the drill to cut cleanly all the way through solid steel.

Fifth Assignment

Scribe off a hexagon around the hole and tangent to it. Two of its sides should run parallel to the sides of the workpiece. Make precise center-punch marks on each corner of the scribed hexagon. Duplicate this procedure on the other side of the workpiece.

Next, file each side of the hexagon with a bastard file, keeping the hexagon slightly undersized. Constant checking of each side with calipers will reveal well in advance what corrective filing will be required. Follow up with a smooth file to *refine* the surface textures, stopping short of an absolute final dimension.

When you reach this point, stop and switch to making a hexagonal bar, as illustrated in the sixth assignment. This bar must be made to fit in the hexagonal hole. Once made, return to where you left off in the fifth assignment and test the fit between the hexagonal bar and the hole. This will indicate whether the 120-degree angles at which the sides must meet are exact. (A template can be cut with a single exact 120-degree angle. It will serve during the making and testing of the two workpieces.)

As the final test check that the bar has an even sliding fit from the moment it is inserted into the hole until the last 1/8 inch of travel. At that point only a small glimmer of light should be visible at the sides of the hexagon.

Remove the bar and tumble it 60 degrees. Insert it in the hole again and test it as before; repeat the 60-degree turn until the original position has been reestablished.

If you have successfully passed this test, the seventh and eighth assignments are within reach of your ability. The illustrations of these assignments should now be self-explanatory.

No additional assignments will be described although my past training carried me through many more. This gave me the ability to do three-dimensional projects with perfect sliding fits.

Trained in this way, any craftsman could fabricate from the ground up most parts of a steam engine; a complete crankshaft bearing head; piston rod bearing ends; sliding steam valves; speed regulator hinges; and similar projects.

I hope that not only eager students, but also dedicated teachers will have the desire to revive old-fashioned methods in teaching handcrafts, including flat filing at its best.

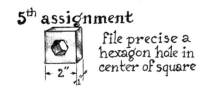

5th assignment
File precise a hexagon hole in center of square
2"

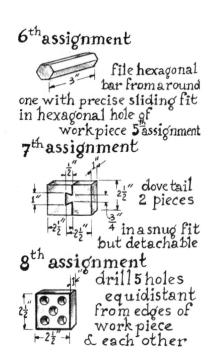

6th assignment
file hexagonal bar from a round one with precise sliding fit in hexagonal hole of workpiece 5th assignment
3"

7th assignment
dovetail 2 pieces
1" 1/2"
1"
2 1/2"
2 1/2" 2 1/2" 3/4"
in a snug fit but detachable

8th assignment
drill 5 holes equidistant from edges of work piece & each other
1"
2 1/2"
2 1/2"

19. Files, Rasps, and Grindstones

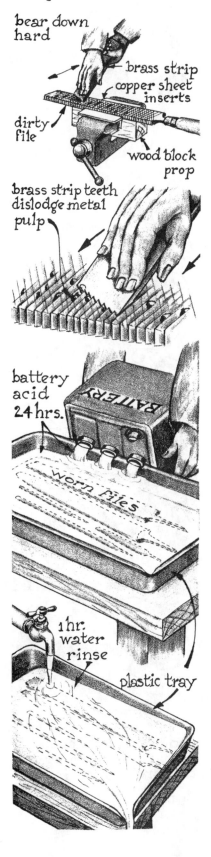

cleaning & reconditioning dirty & worn files

bear down hard

brass strip

copper sheet inserts

dirty file

wood block prop

brass strip teeth dislodge metal pulp

battery acid 24 hrs.

worn files

1 hr. water rinse

plastic tray

Old files, thrown out because they are too worn, frequently come our way. Their excellent high-carbon tool steel can be used as stock for making high-quality gouges and other cutting tools.

Before forging files into tool blanks, grind off the teeth. This will eliminate the danger of the steel cracking where serrations remain.

HOW TO CLEAN DIRTY BUT SHARP FILES

If metal pulp clogs the file's serrations, it can be dislodged with steel wire brushes made for that purpose. A disadvantage in using a brush is that its hard steel tines compete with the file's own hardness. This tends to reduce the sharpness of the file, as well as its lifetime. If you do use a brush, it should be one with *brass* tines, which will not wear down the steel teeth of the file appreciably.

My preference is to use a strip of brass or even mild steel measuring 1 inch by 8 inches by 1/16 inch, instead of a brush. Press it on the file surface, moving it parallel to the serrations. The file will automatically cut teeth in the brass strip, and these in turn can reach the bottom of the file grooves to dislodge the metal pulp.

ACID SHARPENING

If the file teeth are not too worn, they can be sharpened somewhat by immersing the files in a bath of acid. The acid seems to bite into the walls of the tooth serration, rather than into its edge, and thus sharpens that edge to some degree.

I have successfully used the acid from old car batteries for this purpose. The proprietor of the nearest scrap yard handles turned-in batteries, which he sells for the worth of the lead cells in them and not for the acid. I bring along an acid-proof container and he is glad to let me have all the acid I may need to recondition my files.

To hold the files, use an acid-proof tray such as those used in photographic darkrooms. Carefully pour the battery acid into it. Immerse the files, leaving them in the bath for twelve to twenty-four hours. At once you will see little gas bubbles rising from the files; this indicates that the etching of the steel is taking place. When you think the steel has been etched back enough, fish out one of the files with a pair of pliers, taking care that no acid drippings fall on skin or clothes. Rinse it in running water for a few minutes and dry it. Put a handle on it and try it out on a piece of steel.

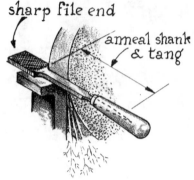

reshaping dull files by grinding on large coarse & hard motor-driven wheels

sharp file end

anneal shank & tang

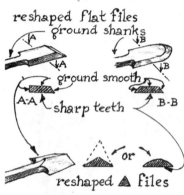

reshaped flat files

ground shanks

ground smooth

sharp teeth

or

reshaped ▲ files

knock off sharp file end

&

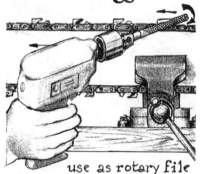

use as rotary file in chuck of electric drill to sharpen chain saws that are clamped in a vise

If you are satisfied with the sharpness, rinse all the files for long enough to leach out the last trace of acid (an hour or more). The principle is the same as the rinsing of photoprints to leach out the last trace of hypo so that no aftereffect of the chemical will spoil the prints in the future. Any rust discoloration that shows on the files later will not affect their use.

Over the years I have restored dull files in this way at no cost except the time spent.

NOTE: Etching metal with acid need not be dangerous if safety precautions are taken. The main thing to keep in mind is never to handle the liquid with bare skin or breathe the fumes carelessly.

Keep your eyes well protected against splattering from files as they are slid in the acid bath. Also, be aware that the gas emitted by the bubbles is more or less explosive if lighted by flame.

If you have no photographic tray, plastic basins used for household purposes are, as a rule, acid-proof. You can also make a tray from plywood. Coat the wood with beeswax or candlewax, heated over a flame, to make it leak- and acid-proof.

HOW TO KEEP FILES UNCLOGGED

Strange to say, the file's worst enemy, oil or grease, will also keep it unclogged if a minimum coating is used on the serration walls only, not on the sharp edges. A lightly oiled file may not bite into the steel for the first few strokes, but if you bear down hard, and the steel filed upon is soft mild steel, the file *edges* will soon be scraped free of the lubricant and will bite into the steel better than before, at the same time rejecting all filing pulp.

Another approach is to rub into the file from time to time a *chalk powder*, which prevents the metal pulp from getting lodged in the serration grooves.

A file that accidentally gets oily in the shop will not be ruined, as uninformed people claim. Remove the excess oil with a solvent and bear down hard on the file while cutting on mild steel. A dozen strokes or so will restore it if the teeth were sharp to begin with. More than once I have received as a gift such a supposedly "ruined" file. I would have felt almost guilty about accepting it were it not for the fact that the giver insisted positively that it was no longer usable.

RESHAPING OLD WORN FILES

This can be done on a power grinder, but it carries with it the danger of overheating, which will quickly ruin the temper of the steel. Grinding, therefore, must be done very carefully, with frequent cooling of the steel. The illustrations show how to do it.

ROUND FILES

The round files used to resharpen chain saws are discarded when their middle sections have become too dull. Their end sections, which are not used, remain sharp. These can be used as rotary files when clamped in the chuck of a small hand-held electric drill. In this way, chain saws can be sharpened in a fraction of the time that hand-filing would take. A brand-new chain saw file could conceivably be broken up into small rotary file sections to get still more use out of one new file.

When. many years ago, I showed my method to a friend who made his living at cutting wood, his gratitude was unbounded: doing it my way eliminated a major part of his chores.

GRINDING WHEELS

The surfaces of grinding wheels cut in the same way as files do. When using soft, fine-grit wheels that go very slowly, you must oil-saturate them to prevent the metal pulp from clogging the stone's pores. Do it the same way as you would when sharpening tools by hand. Saturate with thin oil and wipe off the metal pulp from time to time before adding more oil.

If a wheel rotates fast, the centrifugal force will throw off excess oil. After that use a toothed steel dressing wheel to sharpen the cutting surface somewhat. You will find that an oil-treated wheel grinds the softer metals with much less tendency to clogging than an unoiled one does.

Should mild or high-carbon steel be ground on very coarse, hard, dry wheels running at 1700 rpm or more, no lubrication against clogging is needed since the metal pulp will be thrown off by centrifugal force.

MAKING FILES AND RASPS

It is not difficult to make a fairly crude file, but handmade files cannot compete with machine-made industrial ones. In any critical filing job, machine-made files should be used. If, however, you are in emergency need of a particular file and the nearest supply house is too far away, you can make one without too much difficulty.

Forge the file blank from high-carbon steel stock. Anneal it slowly to its softest. Fasten the blank on an end-grain piece of wood clamped in the vise.

Any sharp-edged cold chisel, held in a fairly upright position and struck with a heavy hammer, can raise a sharp cutting edge like those of a file. The illustrations give an idea of what happens when the angles at which the bevels of cold chisels are ground cut the steel in various ways. They can form fine or blunt teeth depending on the slant of the chisel to the surface of the file blank.

It is true that any sharp, raised part of a piece of steel will, when hardened correctly, cut steel like a file. But centuries of experience in file-making have taught manufacturers which angle and which combination of spacing and cross-cutting of the teeth will give the best results. These findings have been applied in designing file-making machines that cut file teeth at their most effective and uniform. Readers will find that hand-cut file teeth can produce a usable file even though it may be less refined than machine-cut files.

use cold chisel edge to cut steel surfaces or raise file & rasp teeth

steel surface is cut in a hacking action

serrating action raises cutting edges in files & rasps

metal is raised by wedge action

bastard file teeth are cut deep & spaced farther apart smooth files are cut shallow & more finely serrated

bastard files are more raised spaced &

smooth files are less spaced & raised

chisel direction raises teeth in a wedge action 15°

chisel facet keeps groove bottom open

wedge forces
for best results:
space between teeth equal
depth of teeth equal
each hammer blow to 1 tooth
all blows of the same force
groove bottoms somewhat rounded

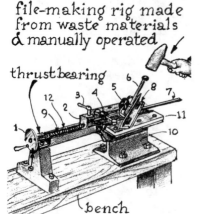

file-making rig made from waste materials & manually operated

thrust bearing

bench

1 disc settings measure spacing between file teeth
2-3 clamps file stock 7 which is moved by threaded rod 9
4-5 rubber rollers press file stock & chisel 6 in their tracks
8 frame holds 6-5 & is bolted onto 10
10 is I beam section onto which are riveted bridge-plates 11

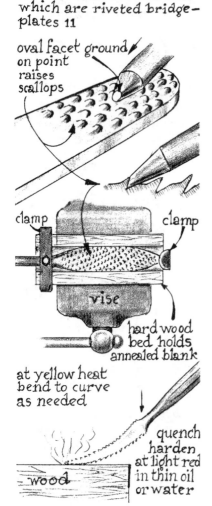

oval facet ground on point raises scallops

clamp   clamp

vise

hard wood bed holds annealed blank

at yellow heat bend to curve as needed

quench harden at light red in thin oil or water

wood

A File-Cutting Jig

You can put together a file-cutting jig in your home shop if you wish to spend the time.

Examine the illustrated detail of an enlarged file tooth. Analyze what takes place. A cold chisel can be used to raise a sharp edge on the file blank a few hundred times. If the teeth are as evenly spaced as possible and uniformly struck by hand, a moderately useful file will result. Only correct hardening is needed to finish the job.

A File Made from Mild Steel

File teeth can be cut in a short time without danger of dulling the cold chisel. Once all teeth are cut, heat the whole piece evenly to *yellow heat* and sprinkle case-hardening compound over the toothed surface of the file.

When the workpiece is replaced in the forge fire, the hardening compound will bubble and melt until all of it has been consumed. At that moment, brittle quench the file in brine or water. Since the teeth edges are very thin, the hardening effect of the powder easily penetrates all the way.

It is remarkable how well such a tool will stand up in an emergency, and the same can be said for a case-hardened mild steel rasp.

Making Rasps

Instead of a cold chisel, a pointed tool is used to form the teeth. The cutting point is created by grinding a small slanted oval facet on the end of a one-point stone-carving tool. Hold the tool at a slant to the prepared surface of the rasp blank and strike it with a one-pound hammer. A single tooth is thus bent outward from the surface.

The next tooth can be placed as you choose. The illustration shows a pattern with staggered teeth. With practice, you will be able to cut teeth as regular as those seen in skillfully handmade Italian rasps. It is an inspiration for a naturally dexterous person to learn how to make good rasps by hand.

NOTE: Files and rasps made of high-carbon tool steel should be heated a *light cherry red* and quenched in thin oil or brine. This gives the teeth a hardness to cut mild steel or annealed high-carbon steel. At the same time the softer core of the file makes it resistant to breaking under strain.

If a piece from the scrap pile proves to be barely temperable, you may have to heat it a little *lighter* than light cherry red in order to obtain the hardness needed. Quenching in water will then be most effective. Be careful, however, that the visible heat glow stops where the tool's tang begins so that the tang will not break off at the handle after quenching.

Rotary Files and Rasps

Blanks made for rotary files and rasps can first be turned on a lathe. The profile can be cylindrical or any shape you like. The shanks are finished to fit the chuck of a small hand-held electric drill or a table-mounted arbor.

Place its working part in a bed carved into the end-grain surface of a block of wood. (Pear, maple, or a comparable fine-grained wood is preferable.) Cut teeth serrations into its upper exposed surface, as in a file, or individual scalloped teeth, as in a rasp.

After all the teeth have been cut in the exposed surface, lift the blank out of its bed and turn it over, seating it with the textured surface on the wood below. In this way the raised teeth will not be damaged during the cutting of the rest of the teeth.

After the rotary files and rasps have been hardened, it is wise to use a propane torch to locally anneal their tang stems with a pinpoint flame. A protective shield made of asbestos or metal can be slipped over the teeth to keep the flame away from the working end.

Making a Cylindrical Rotary Rasp

Although high-carbon steel is preferred for rotary rasps, they can also be made from a section of plumbing pipe or fitting. When using plumbing material, grind off the outside galvanization or cut it off on the lathe. This eliminates any interference with later case-hardening treatment.

Cut a 2-inch length from a standard 1-inch galvanized plumbing pipe. Next turn from a well-seasoned piece of hardwood a core that fits tightly within the pipe and clamp the assembly in the lathe chuck.

With a 3/8-inch diameter drill in the tailstock chuck, drill a hole through the core to receive an arbor. When the two have been assembled, clamp the arbor in the lathe chuck. The outside of the galvanized tube can now be turned to clean off every trace of the galvanized outer skin while aligning the piece as well. The unit is now ready to be held in a hardwood bed where the teeth can be cut and, finally, case hardened.

There is no limit to the number and variety of files and rasps that can be made once the craftsman has seen how quick and simple it is to make them in his own shop, from scrap material. The skill is useful when odd shaped tools that cannot be bought in stores are required to fit specific needs.

257

20. The Reverse Lathe

stationary lathe drives the workpiece

headstock center drives

hand-held tool moves to cut

However

when process is reversed a movable lathe is hand held & guided to effect cutting on contact with rotary cutter

circular saw acts as a rotary cutter

& workpiece is allowed to spin freely between two ball-bearing centers

profiled rails guide pilot pins during movement of lathe.

saw guard   saw teeth

stop

adjustable center spacings

ferrules to fit tightly

countersink to seat centers tightly

press-fit self-sealing ball bearing & center press-fit in it

lathe frame   handle to lock

workpiece   split

saw table

added support plate

split sleeve is press-fit into lathe-center   frame

can accomodate various length workpieces

The reverse lathe is the opposite of the conventional lathe, which has a headstock that drives the workpiece while the hand-held cutting tool, placed on a steady rest, cuts the driven workpiece. In a reverse lathe, a small movable lathe is hand held and guides the workpiece against a driven cutter. On contact, the workpiece is spun and cut at the same time.

The illustrations make clear the advantages of the setup. Instead of using a circular saw, any kind of stationary rotary cutter would be suitable.

Such a reverse lathe must have two adjustable tail centers to allow a workpiece to spin freely when it is brought into contact with the driven cutter element. The workpiece will start to spin instantly, but its inertia will cause the speed to be slightly less than the speed of the moving cutter teeth or blades; that is why a cutting action takes place. By moving the workpiece back and forth in front of the cutter, you will be able to shear off the wood that is in the way of the cutter.

The illustrated example shows stationary profile rails along which pilot pins slide to direct the movement of the reverse lathe assembly. After repeated back-and-forth movements have cut all material in the cutter's path, these pilot pins contact the profile rail at every point, showing that the job has been completed.

The whole operation, from start to finish, can be done in a few seconds for such simple projects as wood handles for carving tools, files, and other hand-held tools.

To refine the wood surface, transfer the reverse lathe and its workpiece to a stationary rotating sanding disc with a rubber backing. The spinning workpiece, in contact with the rotating sanding disc, can be smoothed in a few seconds.

Any further refining of wood surfaces can then be done first with shellac fillers and then with waxed cotton buffers. Being able to hand-manipulate the reverse lathe assembly against any sander or buffer so easily makes the setup very practical.

APPLICATIONS FOR THE REVERSE LATHE

The illustrations show an application of the reverse lathe principle: how to grind engine valves in a fraction of the time that it would take on more complex grinding machinery.

With a machinist-blacksmith shop, it is fairly simple to make both instruments. I have done this with results that could not be improved upon. It must be remembered, however, that, as with all material presented in this book, each craftsman who aims to make things from scrap material will come up with an instrument that looks different, but which works in the same way as the ones shown here.

EXTENDED USE OF THE REVERSE LATHE

Once you have tried out the reverse lathe, you will be able to visualize many other applications that use a variety of cutting elements. When the forms of the workpieces have been established roughly, the surfaces can be refined with stationary anchored rotary files, rasps, grindstones, or rubber-backed flexible sanding discs. Sand-blasting tips can also act as combination drivers and cutters.

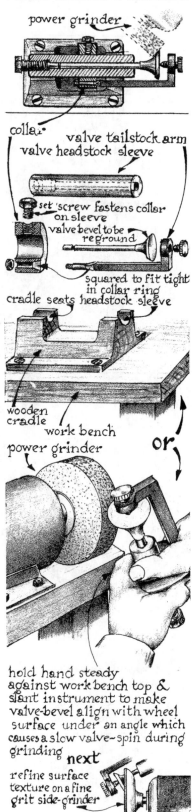

engine valve grinding-jig is hand held in wooden cradle which is fastened to workbench

power grinder

collar
valve tailstock arm
valve headstock sleeve
set screw fastens collar on sleeve
valve bevel to be reground
squared to fit tight in collar ring

cradle seats headstock sleeve

wooden cradle
work bench
power grinder

or

hold hand steady against work bench top & slant instrument to make valve-bevel align with wheel surface under an angle which causes a slow valve-spin during grinding **next**

refine surface texture on a fine grit side-grinder

the Metal Turning Lathe

1 headstock step pulley
2 drive belt
3 main ball bearing
4 backgear guard
5 top part of unit 9
6 back gear
7 main drive pinion gear
8 intermediate gear
9 bracket holding 8
10 intermediate gear
11 bracket socket

12 interchange gear
13 interchange gear
14 adjustable frame for 12 & 13
15 headstock chuck
16 tailstock locking handle
17 tool post locking screw
18 tailstock
19 tailstock feeding wheel
20 tool bit holder
21 tool post bed & tool feed handle
22 lathe bed

23 lead screw bearing
24 leadscrew spindle
25 carriage cross feed handle
26 handle for longitudinal feed
27 carriage unit
28 angle positioning lock screw
29 headstock assembly
30 carriage locking screw

not shown: automatic cross-
feed & longitudinal feed units

21. How to Recycle and Operate a Metal-Turning Lathe

In the year 1880 in Holland, my father was taught how to operate a machinist's lathe as part of his training to become a marine engineer in the East Indies navy.

The difference between his training and mine, in Holland in 1915 to 1923, made me realize that the lathe had come a long way since the hand-held cutting tools used for turning metal in his time. By my time, they had been completely eliminated.

The lathe that I used in technical school in 1919, though old and worn, seemed a fine piece of machinery. Overhead transmission shafts connected all shop machines with idling and live pulley units from which a belt reached a machine tool below it. The machine tools were driven by a one-cylinder, 25 hp, large bore horizontal steam engine with a huge flywheel.

Standing at the lathe one could easily reach overhead to a wooden handle ending in a fork with which to move the belt from the idler to the live pulley, thus engaging the lathe. Once connected, it took a split second before the back-geared lathe headstock would move. This was because wear and tear had created so much play in the machine parts that in back gearing it took a moment for the slack to be taken up. But once the headstock was moving in the well-oiled and accurate sleeve bearings, remarkably precise work could be done on the old lathe.

Visualize, then, a lathe of that vintage to be found in many home shops today. They are still being sold by families whose fathers and grandfathers ran independent shops or used such a lathe to pursue their hobbies in retirement years.

With the nationwide revival of handcrafts today, many modern craftsmen enhance their shops with these old machine tools. Once dusted off, cleaned, oiled, and renovated, the old, sturdy metal-turning lathe will serve its present owner extremely well as a recycled machine tool.

I myself came by such a lathe, and a few weeks of repair work on it enabled me to do all the lathe work needed in my shop, where I combine blacksmithing and machinist-type activities. My experiences may help you learn how to recycle and use old, worn machine tools so that you can get by with what you have.

My lathe is a 1919 South Bend all-purpose model. The bed measures 6 feet in length and I can cut a 15-inch circle. Missing were all the gears for thread-cutting as well as the one gear that linked the headstock pinion gear with the gear that drives the tool carriage longitudinal lead screw to make automatic feed possible. It had a three-jaw self-centering chuck, but was missing a four-jaw chuck, a faceplate, and a lathe dog. I easily forged the lathe dog and turned an adaptable faceplate to be clamped in the three-jaw chuck.

As luck would have it, a friend had given me, twenty years before, a box of small gears that had belonged to a smaller lathe of unknown

recycling an old metal-turning lathe

forged bracket **A** takes stub-shaft **B** to anchor the transmission gear **b**

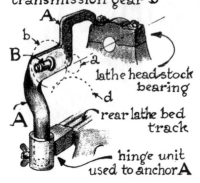

lathe headstock bearing

rear lathe bed track

hinge unit used to anchor **A**

A takes stub shaft **B** to place gear **b** to transmit drive from **a** to **d** to the tool carriage screw spindle

Or

screw spindle can be driven by belt transmission if gears are missing

with motor & step pulleys

activator cam starts or stops tool carriage automatic drive

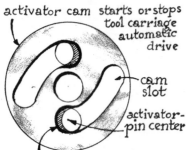

cam slot

activator-pin center

slot extension filed out to bring pins closer together for better grip of split nut on screw spindle

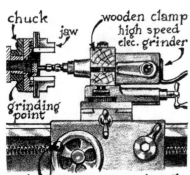

chuck

jaw

wooden clamp

high speed elec. grinder

grinding point

tool carriage activator handle

vintage. Lo and behold, it turned out that the pitch of the teeth on these narrower gears was the same as on the wider ones of my old lathe! All I had to do was to turn adapter sleeves to make up for the difference in shaft diameters. Once the keys, dowels, and idling sleeves were installed, almost any lathe work could be handled very well, with one exception: the cutting of all the different threads. I had to juggle my narrower little gears around to do this, which would take quite some time.

But in practice I found myself more often cutting thread by placing a tap in the headstock and, with the lathe in back gearing, pushing the workpiece into alignment against it with the tailstock center feeding-screw.

In a similar way, one can easily cut a thread on bars clamped tightly in the headstock using one of the many threading dies one seems to accumulate over the years. Keep the die from moving by clamping a visegrip plier on the housing handle and push the die forward with a suitable adapter placed in the tailstock. Let the visegrip or die-housing handle follow by sliding over the tool post bed.

The major gain these narrow compromise gears gave me was to permit the tool carriage to move automatically again. For this I had to forge a complicated bracket in order to bolt to it a steel stub shaft. This shaft holds a transmission gear that connects the headstock pinion gear with a large idling gear, which in turn moves the lead screw spindle in the carriage. Thus I can cut long pieces of work automatically and evenly.

The need for an automatic carriage drive must have been felt by the previous owner of this machine tool. Not having the gears, he had installed on a bracket bolted to the lathe leg a quarter-horse electric motor. With a step pulley on the motor and one on the lathe lead screw, he could vary the moving speed of the tool carriage at will, thus bypassing a gear drive he did not have.

This points up the fact that, if worse comes to worse, belt-driven transmission arrangements will do well in a home shop. Do not hesitate to be inventive in these matters. There is a choice of the widest variety among the many kinds of transmission systems in salvaged machinery, including automobiles. For instance, chain transmissions of timing gears in cars would be ideal, since the parts can be exposed and center-to-center distances can be made adjustable with an idler roller.

Another worn element in this lathe that had to be reactivated was the split lead screw housing, which engages or disengages the tool carriage's longitudinal feeding movement. This screw housing had worn to the point that it skipped a beat. Taking apart its eccentric coupling, I simply extended its cam slot with a rotary file. That allowed the housing activators to move a bit further so that the split lead screw housing could clamp onto the screw a little more tightly, engaging it effectively. It was the brass screw housing, not the lead screw spindle, that had worn.

Next came the inaccurate three-jaw chuck, which had to be made accurate again. If an accurate round workpiece, free from a tailstock center, is clamped into a three-jaw chuck, its end generally shows a little wobble, especially when the workpiece is a fairly long rod. Often this is a sign that the chuck jaws have worn at one end more than the other. This telltale sign means that regrinding the jaws is called for if, after several tries, it continues to show an end wobble.

GRINDING CHUCK JAWS FOR ACCURACY

As the illustration shows, I made a wooden mount for my electric grinder so that it could be clamped onto the tool post bed. By aligning the grinder unit with the chuck center, I ensured that its grinding point could enter the opening between the chuck jaws and could be made to move freely back and forth between them over their full length.

Rotating the headstock at slow rpm in direct drive, I cross-fed the grinding point gently outward toward me until the first spark showed contact between the point and the steel of the chuck jaw. Next, moving the lathe carriage back and forth evenly by hand, I brought the grinding point into contact with the full length of each of the jaws. From time to time I would increase the bite of the grinding point by 1/1000 of an inch. I could tell upon examination if all three jaws had been resurfaced by the grinding points, thus correcting them. Any further grinding was unnecessary; it would only reduce the lifetime of the jaws.

Wear and tear in the chuck body itself must also be met. It would show up after forceful tightening of the chuck jaws onto an accurately round workpiece. If the jaws grab the workpiece at the jaw base ends and not at the jaw tip ends, it will produce some wobble at the free end of the workpiece. To compensate for such chuck wear, the grinding must be graduated from jaw tip to jaw base in order to space the jaw base ends 1/1000 or 2/1000 of an inch farther outward than the tips. This makes it possible for all jaws to be clamped on the workpiece over their full length with even bearing instead of on one end only.

Another improvement for an old, worn, three-jaw chuck that can no longer clamp very small diameter work is as follows:

Remove the jaws, but first make certain that each jaw is marked with a stamped-in number or code corresponding to the proper location on the chuck body. This assures correct replacement, and most chucks do have such matching marks stamped in.

Regrinding each chuck jaw as shown in the illustration reduces the width of the contact area with the workpiece so that all three jaws can come centrally closer together. Since each jaw has been hardened to hold fast to mild steel workpieces, it is the workpiece that will wear down under jaw slippage. But if the piece should be *harder* than the jaws, such slippage will wear down the jaws.

I suggest that the contact groove width between jaws and workpiece not be smaller than 3/16 inch for a lathe of a 12-inch workpiece diameter or over.

CLAMPING OUT-OF-ROUND OR ECCENTRIC WORKPIECES IN THREE-JAW CHUCKS

It is necessary to position such a piece by using a filler shim or strip between one of the jaws and the workpiece. The filler strip should be the length of the jaw itself so that at no point can an uneven clamping action take place. A simple probe between tool post and rotating workpiece will easily show if the piece has been centered correctly, where such centering is required. (In industry a finely sensitive dial indicator probe is used to measure off-center movements.)

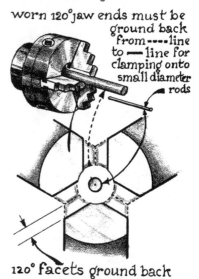

repairing parts of a worn metal-turning lathe

worn 120° jaw ends must be ground back from ---- line to — line for clamping onto small diameter rods

120° facets ground back

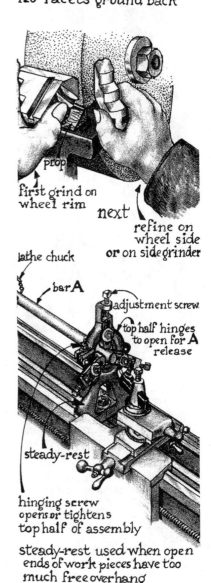

first grind on wheel rim

next

refine on wheel side or on side grinder

lathe chuck

bar A

adjustment screw

top half hinges to open for A release

steady-rest

hinging screw opens or tightens top half of assembly

steady-rest used when open ends of work pieces have too much free overhang

position of cutter held in tool post of lathe to cut most metals with minimum friction

basic position cutting point at center — for all lathe turning jobs

swivel base slants tool up or down to adjust tip elevation

schematic plan for cutter movements & positionings

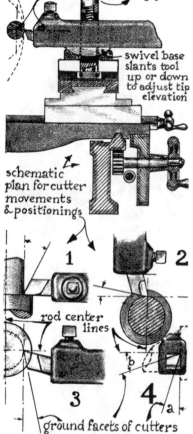

rod center lines

ground facets of cutters

1 top view shows clearance from longitudinal straight rod if cutter tip at height of rod center

2 & 3 clearance from vertical tangent plane to rod circle

4 angle a' clears cutting tip from binding against side faces, & angle b helps the cut metal to ease off the tip unbroken

TURNING OPEN-ENDED WORKPIECES

This means that no tailstock is used when turning a long piece that is clamped in the three-jaw chuck. First place the tool bit elevation position at dead center so that no workpiece center remnant will force the tool bit below it during rotation. Otherwise great strains might pull the workpiece out of center in the chuck or break the tool bit at its point. There is a real danger when using carbon-tipped tools that such brittle bits will then break.

Another make-ready method is to center-punch the bar ending and position it onto the tailstock center before tightening its other end in the headstock chuck.

The Lathe Steady-rest

If the free end of a long bar must be faced or bored or shaped, a steady-rest can be placed at its end.

Clamp the bar in the chuck and test run it without the steady-rest. If the end of the workpiece shows no significant wobble, position the steady-rest assembly, as shown, for a cutting action to come.

In this aligned position, with the workpiece at rest, hand screw the loosened steadying sliding sleeves, one after the other, onto the bar end until each gently touches its surface. Next oil the contact area between sleeve ends and workpiece and run the headstock in direct drive. Now move each sleeve inward somewhat with the wrench and temporarily secure each one at the moment that a little dragging of the workpiece surface between the sleeves takes place. At this point, minutely retract the sleeves just enough to stop the dragging, then tighten them in their slots to anchor them securely. In this way all elements seek their own best positions without binding after being locked in place.

You may now begin cutting without worrying about the workpiece moving out of center, its outward face freely exposed to the cutting tip.

TOOL BIT SHARPENING

The beginner, taking his first steps in working at the machinist's lathe, must soon learn the design principles of shaping and sharpening tool bits.

Tool bit profiles must slant *away* from the workpiece to make them bite into the steel effectively. The illustrations show what is involved. Professional lathe operators offer much advice on how to angle the tool bits to make them cut at their best, and there is no doubt that experimentation and scientific analysis have established the exact profile angle for lathe cutting bits for every imaginable set of circumstances. They must be tailored to cut workpieces of different diameters, hardness, types of metal, etc. For all-around purposes, however, and especially in the shop geared to a one-man operation, one soon learns to "feel" how a cutting bit should be shaped and sharpened once one understands what takes place when metal is cut. I have found that such an empirical approach gives me excellent results in most situations.

It is very important to keep in mind that the grinding angles should not weaken the bit unnecessarily. They should *just clear* the workpiece below the actual cutting edge to avoid binding or dragging. A cutting bit should be positioned in such a way that the steel will not only be cut easily, but that the track it leaves on the steel surface becomes almost polished at the same time. It is good practice, if the cutting bit seems to cut properly, to lubricate it as well to insure smoothness of operation and to keep workpieces and cutters cool. There are many of us who do not care to purchase special lubrication liquids for lathe operation. Used motor oil will do; a small stiff brush in a can of oil kept within arm's reach serves very well.

DIFFERENCES BETWEEN CUTTING TOOLS FOR WOOD-TURNING AND METAL-TURNING

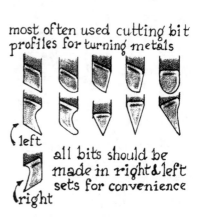

most often used cutting bit profiles for turning metals

(left

(right

all bits should be made in right & left sets for convenience

Because wood is softer than steel, wood-turning tools can have thin, sharp-angled edges. But tools for cutting steel must have edges that are thicker and blunt-angled. The thicker edges for steel cutting, therefore, do not slope away from the workpiece as much as do the edges of wood-cutting tools. (See Chapter 6, page 209.)

I suggest that, when tailoring lathe tool cutting edges, the beginner not hesitate to experiment, even at the expense of a few misjudgments. He can gain from mistakes and thereby learn to do the right thing at the right time.

THE HEADSTOCK FACEPLATE AND THE LATHE DOG

If no faceplate is available you can duplicate its action by holding the workpiece between two lathe center points. First turn a center with a cylindrical shank and clamp it in the three-jaw chuck; clamp a lathe dog onto the workpiece. Such a dog can be forged so that its arm will reach between the chuck jaw's walls, thus driving the workpiece with it. This duplicates the faceplate-lathe dog setup.

Advantages of Using the Lathe Dog Method

The greatest value in this method is that the tailstock body can be adjusted horizontally under right angles with the lathe bed, creating an offset between centers of headstock and tailstock. Such an offset makes it possible to cut a cylindrical workpiece into a tapered one with the cutter travelling parallel to the lathe bed. Another advantage is that inaccuracies of the chuck can be completely ignored as long as the center inset clamped in the chuck is made to move in true center.

This center, as well as the one in the tailstock, acts in reality like a universal joint to the workpiece center seatings, driving it in perfectly centralized rotation free from any possible chuck-induced tensions. See to it that the head- and tailstock centers are kept well lubricated. An additional advantage of this method is that the workpiece can easily be removed from the lathe whenever needed and replaced between centers again without risking any variation in the positioning of the workpiece.

HOW TO CUT A TAPER

If we wish to make a duplicate tailstock center shank with a Morse taper we can do so without any measuring instruments at all, using sighting methods only.

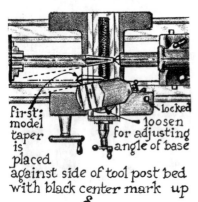

cutting tapers on a lathe
1ˢᵗ method

first: model taper is placed against side of tool post bed with black center mark up & turn tool post base until black center line on model is sighted parallel to lathe bed lock it in this position & move carriage to reach workpiece for cutting action & lock it too. feed cutter with tool bed screw only

locked
loosen for adjusting angle of base

First Method

Keep the tailstock center aligned to the headstock center. The tool post body, being adjustable, can be rotated horizontally to any angle you wish and locked at that new angle. Feeding the cutter with the tool post body only will then cut the workpiece at that new angle while the lathe carriage is securely locked to the lathe bed. The Morse taper, however, happens to be measured, not in degrees, but by offset and distance—in this case 5/8 inch to the foot. The measure is an approximation because Mr. Morse, the inventor of this tool's design, found that the required taper angles had to differ a bit when these center inserts were used to hold fast to large or small diameter workpieces. The design of the center's taper shank is intended to give it a holding capacity, as a clutch does, without freezing into its seat and making it difficult to extract.

Using your own center inset as a model, place it on a blackened, flat, machined surface so that a black line rubs off on the inset. This mark will enable you to hold the inset tangentially against the side of the adjustable tool post assembly with the horizontal black line facing vertically upward. Rotate the loosened tool post assembly while sighting as accurately as possible until the black line runs parallel to the lathe bed below. At that angle, lock the tool post bed position.

You can now begin cutting at the start of the cutter's travel span and the beginning of the tail-center end of the workpiece. After several cuts have brought the workpiece to a size that can be tested for a fit of its taper angle in the tailstock center sleeve, stop the lathe, slide back the tailstock assembly, and remove the center from the tailstock.

Next, remove the center's sleeve and slip it over the as yet unfinished lathe-turned end that remains in the chuck. If you find that only the smallest or largest diameter of the partially turned taper shank contacts the sleeve, it means that the angle adjustment of the tool post assembly must be corrected. This time, you could hold the model tangentially against the workpiece in a reverse directed taper position. Its center black marking line should run parallel to the lathe bed. If it doesn't, the tool post bed position should be corrected accordingly.

Still another way is to estimate the amount of play by rocking the sleeve at deepest insert of the unfinished taper shank and to adjust the angle setting by feel.

After adjustment, make a new cut, which, as a rule, will prove to be so nearly accurate that only the tiniest adjustment, if any, is needed. Tapping sideways with a small hammer or mallet on the end of the slightly loosened tool post body will reposition the tool post bed almost imperceptibly. If the test (sleeve over taper) shows the angle to be correct, proceed to duplicate the model for size and length, after which the sleeve should show a perfect seating when placed over the new taper.

NOTE: In most lathes of this nature the limited tool travel requires more than one full run. To reach the final taper shank length the lathe carriage must be loosened, moved up a little, locked on the lathe bed once more, and the cutting continued where it was left off.

After the final test, refine the taper shank's texture further if needed with a fine-grit, rubber-backed abrasive disc held in a small electric drill rotating against the surface of the taper as it rotates in the headstock. When the surface texture shows even and smooth, cut off the needed part of the workpiece. This is the blank out of which the head of the tailstock center piece can be turned to meet the design planned for it.

Second Method

This method uses the faceplate and dog. A suitable size rod (not longer than necessary) is first provided with deep drilled center seatings at both ends so that it can be held snugly between headstock and tailstock centers. Again, the surface of the model center inset is marked with a black line. With that horizontal line facing upward, hold the model horizontally and tangentially against the rod (workpiece stock).

Next, turn the adjusting screws on the tailstock base to an offset position until you sight the black line on the model as parallel to the lathe bed below it. In this way, you can determine the tailstock center position. Secure it in that position.

Begin the cutting with a shallow cut. Feed it with the lathe carriage back and forth along the rotating dog-driven rod. As soon as enough of the length of the taper has been turned for an angle test, remove the workpiece and slip it into the emptied tailstock sleeve. If necessary, adjust the position of the tailstock setting with the two offset screws to meet whatever inaccuracy you have observed, and restore the setup for the next cut. After that cut has been made, test it again and continue this procedure until a perfect fit has been established.

When the taper is finished, return the tailstock center to lathe center with the tailstock offset adjusting screws.

Whatever type head you wish to turn on a new blank for a tailstock center can now be done in the normal way.

HOW TO MARK THE TOOL POST BODY ON THE TOOL POST BED FOR A MORSE TAPER ANGLE

On the end of a flat-ended punch, grind a right angle cutting edge. Place it at a chosen spot and match the right angles between tool post body and tool post bed. Strike a light but telling blow on the punch. This leaves the two meeting lines indented in the upright and horizontal metal parts, indicating the correct position of the tool post body for cutting a Morse taper. It allows instant and accurate positioning of the tool post assembly for future occasions. No such mark can be made on the tailstock body since the needed offset differs with the length of the workpiece held between lathe centers.

NOTE: Trial and error methods in lathe-turning versus sophisticated industry methods have been adequate when making tools out of worn and discarded machines and parts. Instead of using refined measuring instruments, I would need occasionally a regular inside and outside caliper, a depth gauge, and such ordinary tools as a square, a straight edge, a pair of dividers.

HARDENING TAIL-CENTER TIPS

Since the central seatings of workpieces turn around tail-center tips in the same way that shafts and bearings rotate around one another, the center tip surface must be hard, polished, and lubricated in order to stand up under the great strains and friction created by heavy cuts made with the tool bit.

Harden the tips by drawing a *straw color* after brittle quenching. Be cautioned that the tip must be heated *very slowly* in the forge until that tip begins to glow locally to a *light cherry red*. At that time the slow heating has spread the heat gradually, through the steel's con-

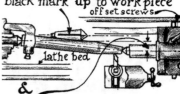

2ⁿᵈ method

place model taper with center black mark up to work piece
offset screws
lathe bed

&

offset tailstock center until black line is sighted parallel to lathe bed
next: position carriage & tool post assembly for cutting action & feeding tool parallel to lathe bed with either one

marking punch

90°

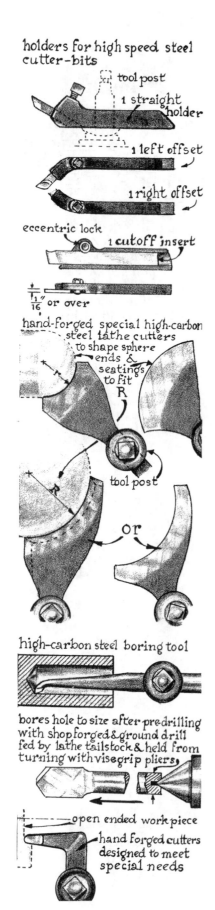

holders for high speed steel cutter-bits

tool post

1 straight holder

1 left offset

1 right offset

eccentric lock

1 cutoff insert

$\frac{1}{16}$" or over

hand-forged special high-carbon steel lathe cutters to shape sphere ends & seatings to fit R

tool post

R

or

high-carbon steel boring tool

bores hole to size after pre-drilling with shop forged & ground drill fed by lathe tailstock & held from turning with visegrip pliers,

open ended workpiece

hand forged cutters designed to meet special needs

ductivity, while not heating up the shank part of the taper to a critical degree. A sudden quench will, therefore, keep that shank fairly soft, which will give it a better holding ability than a hardened, highly polished shank would. After all, taper and sleeve act more or less as a clutch does to keep its opposite member from slipping while letting it go easily when necessary.

STANDARD AND SPECIAL METAL-TURNING LATHE CUTTING TOOLS

Almost all metal-turning lathes are provided with an adjustable tool post. The slot in the tool post can hold the standard lathe cutting-tip holders. It is advisable to have in one's collection a straight holder, and a lefthanded and righthanded one. A holder that can contain a cut-off blade insert is recommended. The same tool post can hold handmade cutters forged from high-carbon tool steel and ground to suit. In this instance the correct dimensions of a sphere, or a seating in which such a sphere might fit, can be used to make cup shapes as shown in Chapter 26. The preliminary shaping of such spheres and sphere seatings with standard lathe cutting tools can be followed up with the special tools for a last refined cut.

These special tools are very convenient and, as a rule, any machinist who is a blacksmith as well can easily and quickly make all lathe cutters from scrap steel. Boring tools for small hole diameters are especially easy and quick to forge. That goes for the forging of any size drill as well. If a workpiece can be clamped in the headstock chuck and such a drill fed with the tailstock center at slow rpm, drilling large holes becomes a simple matter.

Another tool, the right-angle cutter, designed to cut open-faced projects on the lathe, is readily forged. It can reach close quarters in complicated setups, simplifying make-readies.

USING THE LATHE IN PREFERENCE TO THE DRILL PRESS

It is difficult to drill holes larger than 1/2 inch in diameter on a standard modern drill press because its slowest speed is often too fast. If you have no means to reduce the speed, the metal-turning lathe is the solution. Everyone who owns a metal-turning lathe should acquire a drill chuck insert with a Morse taper shank that fits the lathe tailstock, or should adapt one by having it welded onto a Morse taper shank.

The drill, placed in the chuck, can thus be fed by the tailstock screw if a workpiece lends itself to being clamped in the headstock chuck. The illustration shows the setup in which the drill is placed just before touching the workpiece. It is cradled in a V-slot ground into the end of a standard lathe cutter-holder. In this position, the slot may be moved sideways without pushing the drill forward. Thus the drill can slide snugly along the slot walls, which keep the drill from wandering once it engages the workpiece. You will quickly learn how to meet the slightest deviation in position of all elements involved in this simple method of securing exact drill alignment with the workpiece rotation.

Sometimes the drill tip must be steadied further in the slot by pushing the drill tip a little more sideways. The cutting edge of the drill tip then acts as a lathe cutter-bit placed in that position until enough of a bite has been made in the workpiece to create a central seating for the drill tip. Gradually backing off the tool-holder guide while feeding the drill with the tailstock will soon establish correct alignment so that the tool-holder guide can be moved out of the way.

This system is especially convenient should you wish to drill very small holes at high speed without having to make a centerpunch mark in the workpiece beforehand. Such small holes are needed before using a very large drill with a leading pin. The leading pin fits into that small hole and is fed into the workpiece at the slowest speed of the lathe in back-gearing. A common difficulty is that drilling large holes will make the drill shank slip in the chuck, which often cannot be tightened sufficiently to prevent it from doing so. In that case, a visegrip plier clamped firmly onto a small, flat facet ground on the drill shank will keep the shank from slipping under such strains.

Instead of a drill chuck, the tailstock center inset can hold the drill. In this setup, the drill shank should have a deep center seating to hold the drill onto the tip of the tailstock center inset. A small, flat facet should be ground onto the shank so that a wrench or visegrip plier can be clamped over it to keep the drill from turning.

CAUTION: There is danger that the drill shank will slip off the tailstock center tip if the feeding of the drill is slackened during drilling. This could make the drill "grab" the workpiece sideways, thereby breaking it.

It is remarkable how effectively such important drilling jobs can be carried out, especially if one has to bore large holes in workpieces that can be clamped into the headstock chuck. Predrilling of the large hole will simplify a much larger diameter boring job to be done later.

Using the Headstock Chuck to Hold the Drill

This must be done if the workpiece cannot be clamped into the headstock chuck. The illustration shows a setup in which part of the workpiece rests on the top of the tool post assembly to slide along it during drilling. In this instance, the tailstock center is shown removed and the sleeve pushes against a wooden block, which allows the hole to be drilled all the way through instead of partway.

It is easy to see that endless combinations of drilling problems can be solved using a lathe with different setups. The main thing is to stay within the range of possibilities that the tools offer and not to overreach ourselves to the point that we break tools needlessly. Many tools are broken and ruined by overloading or speeding.

USING THE LATHE AS A MILLING MACHINE

It should be recognized at the outset that a milling action can be successful only when all machine parts involved are very sturdy and steady. It must be possible for the actual cutting to be done without the slightest vibration caused by overloading or speeding the machine or allowing excessive tolerances between moving machine parts.

The illustration shows only one method to give an idea of what can be done successfully should the need for such an operation become important in the shop.

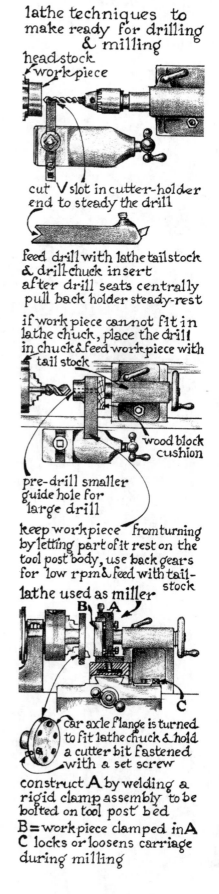

lathe techniques to make ready for drilling & milling

headstock
workpiece

cut V slot in cutter-holder end to steady the drill

feed drill with lathe tailstock & drill-chuck insert after drill seats centrally pull back holder steady-rest

if workpiece cannot fit in lathe chuck, place the drill in chuck & feed workpiece with tail stock

wood block cushion

pre-drill smaller guide hole for large drill

keep workpiece from turning by letting part of it rest on the tool post body, use back gears for low rpm & feed with tail-stock

lathe used as miller
B A

C

car axle flange is turned to fit lathe chuck & hold a cutter bit fastened with a set screw

construct A by welding a rigid clamp assembly to be bolted on tool post bed
B = workpiece clamped in A
C locks or loosens carriage during milling

269

22. The Trip-Hammer and Its Use

Readers who have not seen a trip-hammer in action should examine carefully the full-page illustration of the one shown here.

Observe:

1) How the motions are activated

2) How the rise and fall of the hammer element comes about

3) How to raise or lower the hammer position when the hammer is at rest

4) How the hammer weight is balanced by the flywheel-crank combination

5) The slack or tight belt adjustment, which makes the clutch engage or stop the hammer action to speed it up or slow it down

6) The location of the brake mechanism that releases or stops the lower pulley on the hammer drive shaft

The text and accompanying illustrations are intended to further clarify the workings of the trip-hammer.

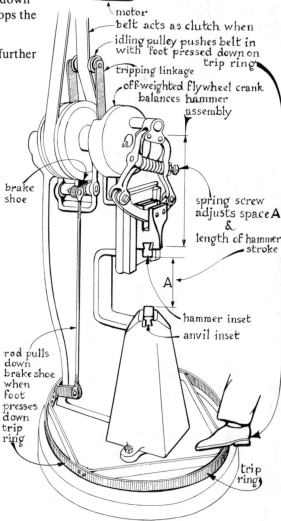

motor
belt acts as clutch when
idling pulley pushes belt in
with foot pressed down on
trip ring
tripping linkage
off-weighted flywheel crank
balances hammer
assembly

brake
shoe

spring screw
adjusts space A
&
length of hammer
stroke

A

hammer inset
anvil inset

rod pulls
down
brake shoe
when
foot
presses
down
trip
ring

trip
ring

looser spring lowers hammer
& tighter spring raises
hammer
when spring is looser, hammer
speed must be slowed down.
when spring is tightened, the
hammer can run faster at a
maximum hammer lift

top
inset

highest
lift when
spring is
tightest

bottom
inset

wedge size

when top wedge is placed front
& bottom wedge is back,
top inset is aligned
with bottom inset
wedges
insets   (1)

when top wedge is placed front
& bottom wedge is front
then top inset=$\frac{1}{4}$"in
$\frac{1}{4}$" front of bottom inset
(2)

when top wedge is placed back
& bottom wedge front
then top inset=$\frac{1}{2}$"in
$\frac{1}{2}$" front of bottom inset
(3)

when top wedge is placed back
& bottom wedge is back
then top inset=$\frac{1}{4}$"in
$\frac{1}{4}$" front of bottom inset
(4)

if an inset is built excentrically
its apron can jut out
front or back
(5)

WHY THE INSET SLOTS ARE OFFSET

The slots are receptacles for the hammer and anvil insets. The offset gives the smith a choice of various inset placements.

The *first diagram* shows a center-to-center alignment of insets. It calls for placing the locking wedges in opposite positions. If the insets are symmetrical it makes the hammer action on the hot workpiece central and symmetrical.

In the *second diagram* the insets are set *unaligned*, with both wedges placed nearest to the smith. In this placement, the workpiece will be indented at the bottom but not at the top when its end is kept flush with the edge of the hammer inset farthest from the smith.

In the *third diagram* the same takes place as in the second, but *shortens* the indentation on the workpiece a little. This wedge positioning gives the smith a chance to forge a local lip at one side of the bar end.

In the *fourth diagram* both wedges are placed on the *far* side of the smith. The insets' positions, in relation to one another, is the same as in the second diagram.

The *fifth diagram* shows that if the smith wants to offset one side of the workpiece longitudinally, he will need an inset with an *extended apron* so that the hammer will strike that portion of the workpiece that rests on the spot directly below the hammer path. The apron then serves as a convenient steady-rest while keeping the bottom of the workpiece unaffected by the hammerblows.

The five diagrams for inset positioning should make it clear that the smith can choose among many inset designs and positionings to suit his plans.

THE FORGING

The illustrations show the positioning of the workpiece, first in the top view of the anvil, and then in the side view between the two insets. Notice that when the two insets are centrally aligned, the edges nearest to the smith are curved somewhat.

Note also that in a long workpiece the portion that protrudes beyond the hammer will bypass the trip-hammer frame because the inset slots have been placed at an angle to the main frame.

If the workpiece is a bar, its heated portion can be steadily pushed forward during the uninterrupted hammering to create a flattened-out portion without leaving sharp indentations in the steel. If that section is moved backward and forward during the hammering, the flattened portions become thinner, longer, and wider.

MOST FREQUENTLY USED INSETS

These are the ones shown in diagrams 1–4. If they are positioned center-to-center and have somewhat rounded edges along the flat facings, they will leave the workpieces with flattened surfaces. Hammering then replaces sledging by a blacksmith's helper.

It is logical that when human helpers became less and less available to the smith he would reach for a trip-hammer to take the helper's place. Without a trip-hammer, sledging frequently *is* needed if the hardest blow the smith can deliver (with the heaviest hammer) is not enough to accomplish what he aims to do. Not only would the smith overtax his strength and endurance, but it would take too much time to finish a given job. When it became possible to produce trip-hammers at a reasonable cost, many a smith looked upon the acquisition of one as a new lease on life. He no longer was dependent on a helper to do sledging. Besides relieving him of the brutal part of the work, he could also make more intricate forgings by designing special insets to meet special shapes. Such insets are referred to as *dies* or *matrixes*. It is that particular aspect of trip-hammer use that proves most useful to the modern smith.

If he has the skill (or can acquire it) to make insets himself, they will help him to make multiples of a single type of workpiece and still stay short of mass-production. Making multiples often holds out the promise that the smith can improve his earnings by doing everything single-handedly.

USING FLAT-FACED INSETS WITH ROUNDED EDGES

The most frequent use of the trip-hammer is for workpieces that need to be *flattened* or *drawn out* in a stretching action. If the facings are perfectly parallel to one another, accurately parallel flat sides can be forged on a bar from thick to thin.

For example, a delicate spatula can be made in seconds and in *one heating* and without using the cross-peen. Resist the temptation, however, to keep on hammering when the visible heat glow has disappeared. The steel is then too cold to change its form further without the danger of cracking.

Remember that the insets are quite cold compared to the hot steel and that the thinner the steel becomes, the greater its surface and the more rapidly it cools despite the enormous amount of energy the heavy hammer pours into it. If the steel becomes too cold for hammering, reheat it before continuing.

One unexpected advantage of the rapid and forceful blows of a trip-hammer is that they increase the quality of the steel; this is especially true in the making of cutting tools. The improvement is caused by the *compacting* of the molecular structure, which the trip-hammer's heavier blows do better than a lighter hammer blow would.

During the final steps in the forging of a carving tool blank, the blade can be compacted deliberately by letting the last blows, before the blade gets too cool, be the heaviest and fastest. This is referred to as *packing* the steel.

anvil inset slot is angled with column top to allow long workpieces to clear the frame upright

anvil column

workpiece

frame upright

inset alignment dowel is anchored to column

aligned flat-faced insets are used

for most work-pieces that must be flattened while lengthened & widened

MAKING WEDGES

To make a heavy wedge, the stock of a salvaged car axle will do very well.

First cut off the hub and flange part of the axle. The remaining length, about 24 inches, can then conveniently be hand held during the forging.

The first skill to acquire in using the trip-hammer is holding the workpiece in exact alignment with the anvil inset facings. As a rule, this is at right angles with the path of the hammer. Accurate holding of the workpiece prevents its bending out of line during forging. It also prevents the often painful jerk in the hands that hold the end of the workpiece. This happens at its worst when heavy-gauge stock is held inaccurately and when it becomes a little too cold to continue forging. (If the part to be forged is at *yellow heat*, inaccurate holding bends it out of line without much strain on the hands.)

Begin forging the wedge with *medium blows*, increasing them in force while gradually drawing the workpiece toward you. When you have reached the very tip, commence all over again where that section is thickest, hammering with medium blows and increasing them gradually to *forceful* ones, again drawing the workpiece toward you to its very tip. In this way you forge the graduated thickness required for a wedge.

By working steadily and quickly with forceful blows, you will be able to keep the workpiece at forging heat. Much of the heat loss is compensated for by heavy hammerblows converted into heat, thus slowing the rate at which the workpiece cools off. Keep in mind with any forging you do that forceful hammerblows *add heat* to the workpiece.

When making wedges, the thinner part of the wedge becomes broader in time, making it necessary to forge its sides together intermittently if a parallel wedge is wanted. If the slope of the insets is not too steep, you can forge the sides of the wedge approximately without having to change the insets. Or you can parallel the sides from time to time by hand-forging on the anvil.

HAMMER TEXTURE ON THE WORKPIECE

If the workpiece must be smooth and free from indentations left by the somewhat rounded edges of the insets, the inset design will have to be changed somewhat. This is done by *sloping* upward the hammer inset face nearest the smith and rounding it slightly. If the anvil inset remains horizontal, no local indentations will result.

The workpiece is held the same way as before, at right angles with the hammer path. If only the anvil inset face slopes downward toward the smith, he should lower his hands somewhat, holding the workpiece to even out the effect of each inset facing on it during hammering. If *both* inset faces slope outward toward the smith, he can again position the workpiece at right angles with the hammer path as was done when both insets had facings parallel to each other.

In whatever combination the insets are arranged, the workpiece should be pushed back and forth gradually during the hammering while the tripping-foot control directs harder or lighter blows, slower or faster, as the smith judges is required.

CORRECTIVE HAMMERING

Watching closely how the *yellow hot* steel shapes up during the hammering tells you immediately where and how you must place the heated bar to get what you are after, or to correct what you did wrong. In spite of the accuracy of machine action, some differences do show up if the workpiece is held a little off dead center of the hammer path. If the center of the workpiece is not in the exact center of the hammer inset it seems as though the entire machine structure *gives* somewhat, one way or the other.

If the trip-hammer is an old, worn one, these differences will show up more, but, strange to say, good use can be made of such flexibility if you wish to use only the right or the left part of the hammer during corrective hammering. Having worked with worn trip-hammers a great deal, I find that their loose-jointed action becomes almost an advantage, giving me an additional choice of workpiece placement.

MAKING INSETS TO FORM TAPER ENDINGS ON RODS

Clamp two inset blanks together and drill a hole at the center where the two faces meet. Next grind each inset separately, as shown, to form an approximate cone shape. It must end up well-rounded at the end edges. Both insets are centrally aligned to receive the round stock at the larger part of the cone that faces the smith.

During the hammering, feed the *yellow hot* stock forward gradually between the insets, rotating the rod continuously. This squeezes the steel so that it develops a circular, rather than an oval, profile as each hammerblow counteracts somewhat the oval profile created by the previous blow. Toward the end the force of the blows should be tapered off while the piece is worked backward and forward, and rotated all the while. The result will be remarkably accurate as your manipulation becomes more and more skillful.

VARIATIONS

The illustration shows how to position a round rod, heated locally about one or two inches from the end. Continuously rotate the rod during the hammering; when it reaches the desired size and dimension, gradually push the workpiece forward to position 2, hammering all the while; end up finally with position 3, rotating the rod around its own axis all the while. A taper with a *round shoulder section* results. This shape is the first step in making a wood gouge blank.

The untouched end can be widened for the gouge blade, first with the peen of the hammer, and then flattened between the flat-faced insets of the trip-hammer. It is up to you when peening must be done; you may choose instead to spend the time changing insets for this purpose and afterwards replace the flat-faced insets and smooth the widened blade section.

Making a wide blade from a larger diameter stock with a trip-hammer eliminates the need to upset lighter stock for the blade, thus saving time and effort. It is practical to start with heavier stock, since the drawing-out and peening of steel with the trip-hammer takes only a little time and the least physical effort.

Two- and Three-step Insets

The illustrations show insets that have a choice between two seating profiles instead of one. This allows the smith to make long, slender tapers more easily. The first profile has the larger diameter seating, which permits forging of a shorter, stubbier taper. After reheating, the taper can be made more slender in the smaller profile next to it.

Special Insets

If specific blanks are needed for *quantity forgings*, insets must be made that can shape various cross sections of workpieces more easily than if they were forged by hand. It is here that the time spent on making special insets becomes justified.

The most frequent cross section of salvaged stock is *square* or *round*. With the insets shown, a round is first hammered out into a triangle bed with cross sections that are fairly deep. The resulting

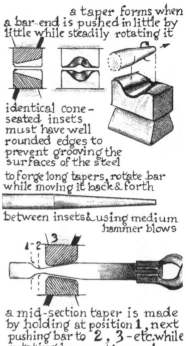

a taper forms when a bar-end is pushed in little by little while steadily rotating it

identical cone-seated insets must have well rounded edges to prevent grooving the surfaces of the steel

to forge long tapers, rotate bar while moving it back & forth between insets & using medium hammer blows

a mid-section taper is made by holding at position 1, next pushing bar to 2, 3 - etc. while rotating bar continuously

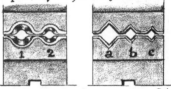

triphammer insets for forging taper, square, triangle sections

using graduated seating profiles in successive steps

at yellow heat & using rapid medium blows, form taper in seat 1 & in same heat continue in 2 if long slender taper is needed

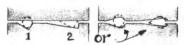

forging round rod in triangular seat curves the formed section & must be straightened before seating 2 is used

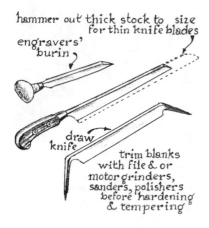

hammer out thick stock to size for thin knife blades

engravers' burin

draw knife

trim blanks with file & or motor grinders, sanders, polishers before hardening & tempering

forging, when reheated, is placed in the *next shallower* profile. It will then fill it completely, producing a triangular stock for making small knives or engraver's burins. With a little practice, you will be able to forge such stock without curving it, (curving being a natural tendency when a bar is stretched on one side more than the other).

CUTOFF INSETS

These can do what cutoff *hardies* do on the anvil. The power of the trip-hammer makes it possible to cut heavy-gauge stock without great effort. The insets must do the cutting without ruining the sharpened knife edge. To do this, each knife section ending must be flat-faced and at the same level as the knife edge. The final blow that cuts the stock in two will fall at the point where the flat-faced endings meet and the knife edges barely touch one another.

Another method is to leave the bottom inset annealed for cutting purposes. In this instance, a *hardened hammer inset* edge cuts the stock the moment it starts cutting into the annealed surface of the anvil inset.

WARNING: There is a real danger that the severed part of the stock will, at that moment, fly off at high speed. It could hit a bystander or start a fire should it land in combustible material.

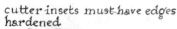

cutter insets must have edges hardened

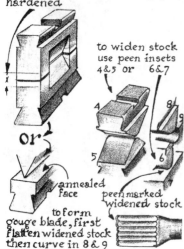

to widen stock use peen insets 4&5 or 6&7

or

annealed face

peen marked widened stock

to form gouge blade, first flatten widened stock then curve in 8 & 9

A Third Way of Cutting Stock

In this instance the cutting action can also be combined with the *shaping* of the stock. If you wish to make a cross peen hammer, for example, you can forge the sloping sides with hammer insets made with that same slope. Make these insets with a curved edge where the sloping planes meet so that they can be used for peening purposes as well.

If sections of car axles are cut with this pair of insets, keep the *yellow hot* part firmly in position while the hardest and fastest hammerblows bring the curved endings of the insets together, leaving a very thin connecting link that holds the stock together. At that point, stop the action and knock off the end piece over the anvil or pry it off with tongs. In this way, the hot piece will not fly off uncontrolled.

I practice this method for making hammer heads with good results on my own trip-hammer. It has saved much effort and time compared with hand-hammering a cross peen separately on the anvil.

SWAGES AND FULLERS

Almost all accessories for hand-hammering on an anvil can be duplicated as trip-hammer insets, functioning as top and bottom swages and top and bottom fullers.

23. Making a Pair of Insets to Forge a Gouge Blade

Most of the difficulties encountered in hand-forging a wood-carving gouge can be overcome with a trip-hammer. In the design of the gouge shown here, the shank and the bottom of the blade form a straight line. In hand-forging it is difficult to force the thicker part of the blade down to the point at which it meets the shank without distorting that part. There is the danger that the thinner areas of the steel will give way under hammerblows much sooner than the stiffer, thicker parts.

In order to curve the blade at the point where it is narrowest, yet thickest, a heavy blow must be struck on that spot with a matching curved hammer forcing it down onto the flat facing of the anvil. Failing to do so correctly causes the bottom profile of the tool blank to hump up at that thickest point. To meet this difficulty when hand-forging, place the heated blank with the bottom profile on the flat anvil face. Hold the ball of an appropriate size ball peen hammer on the hump and strike its face with a hammer to force the hump down flush with the anvil face. As a rule, one or two such treatments will correct the discrepancy. Only after the blank is correctly aligned can the shape of the blade ending be refined.

"Humping up" can be avoided from the start with specially designed trip-hammer insets that shape a gouge blank with the proper alignment. The *bottom inset* can be made by joining two pieces of steel as shown for insets A and B. Should you possess a piece of steel large enough to make the unit out of one piece, by all means do so. But since the inset apron falls outside the hammer's path, very little force is transplanted by the workpiece onto the apron. In fact, the apron serves more or less as a steady-rest, easing the smith's manipulation during forging and allowing him to keep the workpiece in perfect alignment, holding it firmly seated in that bottom rest during hammering.

Working the blank backward and forward a bit will permit you to feel when the most forward position has been reached, the point at which the thickest part of the blade meets the shank. A few firm hammerblows on that spot will seat it flush with the shank bottom. You can now pull the blank *toward* you again until the top inset reaches the very edge of the gouge blade ending. If the blade is still hot enough, you can end up using the most forceful hammerblows to *pack* the steel.

It may appear at first glance at the illustrations that making such an inset unit is too difficult. But if you can visualize the step-by-step procedure, you will see that making each part is actually rather simple. If you feel, therefore, that the gain outweighs any difficulty you might encounter in making the unit, I strongly recommend that you do so.

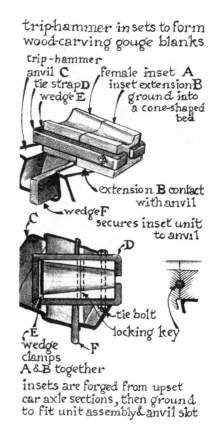

triphammer insets to form wood-carving gouge blanks

trip-hammer
anvil C
tie strap D
wedge E

female inset A
inset extension B
ground into a cone-shaped bed

extension B contact with anvil

wedge F secures inset unit to anvil

tie bolt
locking key

wedge clamps A & B together

insets are forged from upset car axle sections, then ground to fit unit assembly & anvil slot

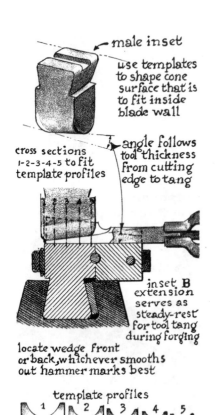

male inset

use templates to shape cone surface that is to fit inside blade wall

cross sections 1-2-3-4-5 to fit template profiles

angle follows tool thickness from cutting edge to tang

inset **B** extension serves as steady-rest for tool tang during forging

locate wedge front or back, whichever smooths out hammer marks best

template profiles

1 2 3 4 5

cut templates from tin cans & refine curves to match the previously hand-forged tool blank

Making the *top inset* is an entirely different matter. I suggest that you avoid making hammer insets with horizontal extensions. The extension can cause enormous strains on the machinery as a whole, particularly if the extreme outward part of it hits the bottom inset unintentionally with full force. The effect would be to wrench the moving parts of the machinery to the breaking point. Do not underestimate the enormous forces that come into play when a rapidly moving hammer mass releases its stored energy upon impact with the anvil inset. The hammer inset must, therefore, be allowed to distribute its energy within the range for which it is designed. It must not fall beyond the projected area of the hammer body itself.

The hammer inset should be designed to fill the inner space of the finished blade of the gouge so that, in the end, a full contact is established between the workpiece and both inset walls. The illustrations show how to accomplish this. Examine the cross sections of the hammer insets; they match the curvature of several templates you can make.

Use a hand-forged finished tool as a *model* to make the insets. First, fit the *bottom inset* depressions exactly along the curves of the templates. (As shown in the illustration as profiles 1–5.) Next, use the templates to fit the *hammer inset* curves.

If made well, the two insets will allow sufficient steel between them to shape the blank correctly at the moment the surfaces of both insets contact it. Make certain that the edges at each end of the hammer inset are rounded off so that at no time can local depressions mar the blade blank beyond repair.

After both insets have been installed, position the hammer at its lowest so that the small space between them allows you to check the center-to-center alignment. If the insets prove to be unaligned after they have been wedged in, when hot steel is hammered between them the offset hammerblow would immediately result in sideways forging. The blade would then become much thinner on one side than on the other and would also curve out of line. The location of the error itself will indicate in which direction inset positions must be corrected to establish a perfect center-to-center alignment.

Only when you have mastered correct positioning of these two insets will using them become a joy. They will cut time in half or better when you make such carving tool blanks on the trip-hammer.

It is at this point that you have a choice: Shall you leave well enough alone, having reaped the benefit of a quickly and beautifully forged blade section of a gouge, and go on to hand-forge the rest (shank, shoulder and tang)? Or should you attempt to make additional insets for the trip-hammer with which to shape the remaining parts of the tool? It is a critical decision to make. The temptation to mechanize further is great. We all fall prey to it at one time or another, overreaching ourselves without being aware that we are doing so.

Making insets for complicated workpieces is bound to take time. Sometimes we make mistakes and must do the work all over again. In general, my advice is that, with limited equipment, it is preferable to *combine* the trip-hammer with some hand-forging. The remaining sections are really not too difficult to forge by hand. You can accomplish what you set out to do: to make multiples, short of mass production, within time limits that are acceptable to the independent, self-employed smith.

24. Making Trip-Hammer Insets From Trolley Rail

Once you have learned to operate the trip-hammer using simple insets, the next step is to make additional ones from easily available varieties of scrap steel. Small-gauge trolley rail is one suitable item that can often be found. Its T-shaped cross section will need to be built up from below so that the finished inset units can be anchored in the hammer and anvil slots.

It would be overreaching to try to fill the open spaces with steel sections, especially since hardwood sections will do as well. No excessive heat ever reaches the wood, so there is no danger of it burning. Using hardwood props simplifies making trip-hammer insets from odd-profiled bulky scrap steel, especially if that material is a high-carbon steel that can be tempered to the exact hardness we want.

The step-by-step procedure illustrated shows a simple way of installing these insets. Treat them as if they were made from one solid piece of steel. The only difficulty is making the matrix parts of the insets by forging methods. It is possible that you will find in your own shop a simpler way of doing what has been suggested here. The perfect results that I have obtained do, however, justify the somewhat laborious task of making the male and female parts of these insets.

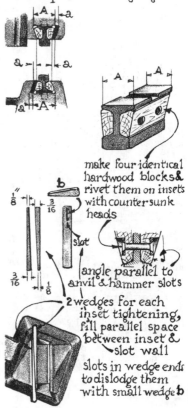

using sections of scrap rail in the making of inset units for triphammer projects
Example: a wood gouge

make four identical hardwood blocks & rivet them on insets with countersunk heads

angle parallel to anvil & hammer slots

2 wedges for each inset tightening, fill parallel space between inset & slot wall

slots in wedge ends to dislodge them with small wedge **b**

279

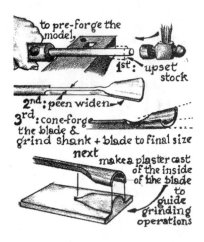

to pre-forge the model.

1st: upset stock

2nd: peen widen

3rd: cone-forge the blade & grind shank + blade to final size

next make a plaster cast of the inside of the blade to guide grinding operations

THE INSET DIES

The plan for forging a carving gouge blank is that only the tool *blade* and *shank* are to be shaped on the trip-hammer.

First, hand-forge a *model tool* to be copied in making the inset dies for the gouge blade and shank. Finish all parts of the tool until you are satisfied with the shape and design.

Making the Top Inset Die

Using the model tool, cast an impression of the *hollow part* of the blade and shank in plaster of Paris. This form can then be duplicated in steel to make the *top inset*.

NOTE: When two opposite parts, as in male and female dies, must be made, we think of one as being *positive* and the other *negative*; or one being solid and the other hollow. The opposites fit together like a pudding and the pudding mold. If we look at the parts separately, but must *visualize* the two fitting together, a confusing "mirror" effect often results in our making the forms the *opposite* of what they should be. It is the mirror that makes of our *real* right side a left side when we look into it. Be on guard against this mirror switch. Making a plaster cast is a good way to avoid the difficulty. If you duplicate the cast of the *tool* precisely, the result will be the correct *inset* shape.

To make that inset, use as tools anything in the shop that is suitable. Filing, hacking, grinding, and hammering are all permissible as long as you can reach the narrow recesses without breaking fragile tools.

The Correct Use of Grinding Points

The grinding points should be *exactly centered* and *exactly round* before a grinding operation begins. If they are eccentric, the procedure to restore accuracy is simple: Hold the rotating grinding point driven by the high-speed grinding unit (gently but steadily) against the rotating surface of the large, coarse, hard grinding wheel so that the rotation of both tends to grind away every trace of eccentricity of the small point.

While you are at it, such small points may be further shaped by the big wheel into whatever profile you may want. Small wheels and grinding points are not expensive and all of them can be reshaped for special jobs.

I find that grinding points at high rpm in hand-held, motor-driven grinders are most useful *after* the bulk of the material has been removed with larger tools.

WARNING: High-speed grinders can be dangerous instruments if overloaded because of misjudgment or lack of skill in controlling their manipulation. If the grinding point grabs the steel, it may fly apart. The danger is that sections can injure the operator or bystanders if no protective shields are used. Wear safety goggles at all times when doing this type of work.

Testing the Final Surface of the Inset for Accuracy

Coat the surface of the model with a little blacking. Place it in or over the inset matrix surface and wiggle it in that position. Lift it off and examine the surface of matrix for telltale black spots left by the

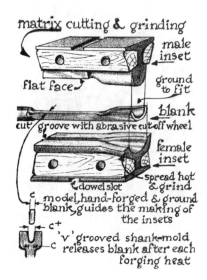

matrix cutting & grinding

male inset

flat face

ground to fit

blank

cut groove with abrasive cut-off wheel

female inset

spread hot & grind

dowel slot

model, hand-forged & ground blank, guides the making of the insets

c+

'v' grooved shank-mold releases blank after each forging heat model. These indicate where more grinding must be done. Continue to grind and test until an overall *even* marking of black contact areas is achieved. This shows that the form surface fits exactly its opposite.

Shaping the Bottom Inset

Heat the *anvil* part of the inset *yellow hot*. Clamp it into the post blacksmith vise and hammer that part of the yellow hot groove wall *outward* with a blunt-ended bar in the approximate shape of the gouge blade contour. Anneal the inset so that the steel will be at its softest.

Next, do the machining by every means available—filing, hacking, hand-grinding. Finally, test with the gouge model to see whether the job has been done accurately.

Tempering

Harden the trip-hammer insets in oil. The resulting soft core is desirable because it will cushion the hardened surfaces against breaking. And if hardened surfaces should crack, the outer part will not shatter because the softer center holds the broken parts together.

As a rule, such cracks are shallow and remain insignificant during further use of the tool. Nonetheless, the cracks may show that you have hardened the piece more than necessary or that the inset was overloaded by accident when a powerful hammerblow struck the hard anvil inset face without any soft-heated steel between the two. This also makes a good case for reheating steel that has cooled off too much, since heavy hammering of cold, hard steel has a similar effect on the insets to striking the empty anvil face itself.

TROUBLESHOOTING

There are various types and makes of trip-hammers, and their owners have to develop a skill to operate them within the limits of their designs and capability. Using trip-hammers and making insets for them sometimes introduces unforeseen difficulties. These show up in faulty end results or accidents. Following is a discussion of what are, in my experience, the most frequent mishaps and what steps can be taken to avoid trouble.

The Insets Become Loosened

During forceful hammering the insets may loosen, but this frequently is not noticed in time because the noise of the operation drowns out the telltale sounds of a loose wedge. If that wedge should drop out of the hammer slot before you can stop the hammer, it can fall free onto the lower inset. The next hammerblow by the *empty hammer body* would then hit whatever part lies below it. There is a great danger that something will be irreparably damaged by such impact.

tools used to forge & finish shapes & surfaces of matrixes

heat rail end yellow hot & clamp in heavy post vise

use heavy hammer

blunt punch end is to lessen marking surfaces

punch open the precut slot both sides

use post vise
or
heavy machinist vise

after annealing, use rotary files, drum sanders, grinding points to refine pre-forged & cut shapes

flexible grinding disc in electric drill chuck

rotary file

high speed hand-held electric grinder

on a wood lathe turn from close grain hardwood, grinding points of suitable shapes & held in drill chuck at slow rpm, refine with lapping compound all final surfaces.
next
surfaces can be smoothed further with tripoli on cotton buffers

harden at bronze temper color

The Wedges Become Loosened

Be sure to provide the seatings with a correct slant to match the wedge. If insets do not have the right wedge seatings, there is a great chance that the wedges will loosen during hammering. Do not consider that you have finished making an inset until you have ensured that the wedges anchor it securely.

Tips for Aligning Insets

Insets *must* be correctly aligned before good results can be expected. If your sighting ability is inadequate and you require measuring instruments, use calipers or any other measuring device to make certain that both insets are perfectly aligned when they have to be, and accurately offset when the design calls for that.

If the position of the insets should be misaligned sideways due to excessive clearance between the centering dowel pin and the inset's own seating slot, a little trick can often correct it. Install the male and female insets but leave the wedges slightly *loose* during the first two or three hammerblows. Hold a strip of thick, stiff cardboard, instead of hot steel, between the insets. The sideways forces during the blows will tend to automatically move the two insets into alignment. If indentations left on the cardboard by these hammerblows show an *even bearing* right and left, it means that the trick succeeded, and the wedges can then be tightened.

If the indentations show only on *one side*, the inset position must be examined and relocated somewhat. Do this by making more room to one side of the seating, using a rotary file as a milling cutter; make up for the excess clearance on the other side with a *filler strip*. If you have the use of an acetylene torch, a few blobs of melted metal will do instead of a filler strip.

Pocking of Inset Facings

Inset facings that have not been hardened correctly will in due time show pockmarked surfaces, which in turn leave uneven surfaces on the forged workpiece. Regrinding and smoothing all contact areas and then correctly hardening them should solve the problem.

25. Trip-Hammer Upsetting

Upsetting steel with the trip-hammer offers an improvement over upsetting by hand only if the machine can eliminate the human errors in hand-hammering and hand-holding.

Since hand-holding must be done in *either* case, trip-hammer manipulation can meet the problem only halfway. Cup-shaped anvil and hammer insets can be made and used to center the workpiece automatically. The illustrations show how it is accomplished.

In practice, however, even the slightest unaligned hand-holding requires that the smith continually move the workpiece around its own vertical axis in order to compensate for previous inaccurate positioning. The illustrations show how to prevent bending the workpiece during upsetting.

To prevent further sideways bending, the workpiece should be cooled by a quick immersion of 1/4 inch of each end in water for a second or more before placing the piece in vertical alignment between trip-hammer insets. Rapid hammering, combined with circular holding movements, will then upset the middle part and not the ends.

Once that midsection has become thick enough to resist outward bending, replace both cup insets with flat-faced ones. Now the piece can easily be upset into a short, thick cylinder to make, for example, a hammer blank, as shown.

Making a Hammer Head

First drill a 1/4-inch diameter hole at the center of the blank, which you made by upsetting the axle section as shown before. This acts as a guide for the cone-head fullers, which will be hammered into it to enlarge it to fit the size hammer stem you want.

CAUTION: The guide hole must run through the dead center of the blank, not to either side. Otherwise the cone fuller will spread the metal on the thinner side more than on the other, which would weaken that side as well as placing the hammer stem hole still further out of center.

If you intend to locate the center by eye, it is best to clamp the workpiece in the drill vise and, bringing your line of sight level with it, visualize the extension of the drill as it bisects the circle. After the drill has entered the steel 1/8 inch, resight the position and make whatever correction may be needed. In this way, you can establish acceptable accuracy without measuring instruments.

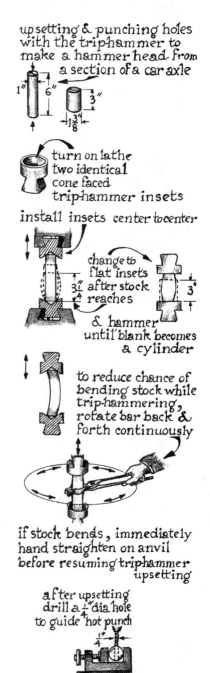

upsetting & punching holes with the triphammer to make a hammer head from a section of a car axle

turn on lathe two identical cone faced triphammer insets

install insets center to center

change to flat insets after stock reaches 3¾ & hammer until blank becomes a cylinder

to reduce chance of bending stock while trip-hammering, rotate bar back & forth continuously

if stock bends, immediately hand straighten on anvil before resuming triphammer upsetting

after upsetting drill a ¼" dia hole to guide hot punch

Correcting an Off-center Guide Hole

In the event that the guide hole is drilled off center, the recommended method of correction is as follows:

Keep the thinner side as cool as possible short of brittle hardness so that when the cone fuller is introduced the steel on the thin side will not stretch as much as the thicker, hot side will.

This is a difficult procedure and there is a danger of ruining the workpiece. Care should, therefore, be taken to drill accurately to begin with.

Finishing the Blank for Final Shaping

After the cone-head fullers have widened the hole sufficiently at top and bottom, a narrower channel is left between the two. This is desirable since it will hold the hammer stem securely at the point it most needs to be. A steel wedge driven into the top end of the hammer stem will spread the wood, locking the stem into place.

At *yellow heat* forge the blank between flat-faced hammer insets, as shown. This automatically reshapes the hole into an oval.

It is now up to the smith to decide the type of hammer head he wishes to make. He can end up with a cross peen, a ball peen, or other shapes useful to him.

The surfaces can then be refined by grinding, filing, turning on a lathe. Finally, such a hammer head should be tempered as explained in Chapter 7.

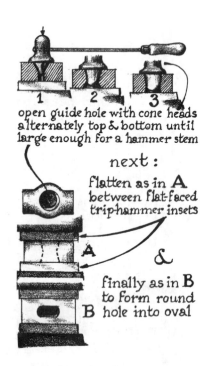

open guide hole with cone heads alternately top & bottom until large enough for a hammer stem

next :

flatten as in **A** between flat-faced triphammer insets

A

&

finally as in **B** to form round hole into oval

B

26. Insets Made from Car Axle Flange Endings

Sometimes certain types of scrap steel parts come our way in such quantity that it is a challenge to the imagination to create something from such a windfall. That was the case when I fell heir to several hundred pounds of fine quality *steel discs*, 3/8 inch thick by 2, 3, and 4 inches in diameter.

The first idea that came to my mind was to forge deep cup-shapes which, in turn, could be forged into rosettes. Large, heavy-gauge rosettes can be used as giant washers for bolt heads on the long tie rods used to keep building walls securely spaced. The same type of rosette, being heavy, can also serve as a base for a lamp or candlestick (see Chapter 3) or a column that holds an arm with a magnifying glass. In short, the discs have become valuable stock in my shop and have been put to many other uses also.

MAKING INSETS TO FORGE CUP SHAPES

Car axle ends have enough bulk to make a ball inset. First, cut off the hub section from the car axle. Next, using the lathe, cut off the flange at the hub end, since the ball bearing cannot be removed with the flange in the way.

Hold the remnant between the jaws of the heavy post vise so that the edges of the press-fit ball bearing rest on them. With a few, very heavy hammerblows on the stub shaft, drive the ball bearing off its seating. You may need to use a sledge hammer, but in that case the vise must be of a very heavy caliber, 100 pounds at least.

The end of the hub body frequently has a deep depression, which can be filled with a tight-fitting lug in order to make the best use of the greatest volume of hub steel. To make certain that the plug will not drop out during use as a hammer inset, taper the plug slightly so that the larger part of it fits the bottom of the depression in the hub end.

With the cold plug held in readiness, heat the hub body collar around the depression to a *yellow heat*, quickly insert the cold, tapered plug, and hammer the hub until all clearances around the plug are closed. Cool all slowly in an annealing action. This assembly will behave as if it had a solid head.

Next, clamp the blank into the lathe's three-jaw chuck, as shown, and turn the end into a *ball shape* with the regular cutter bits. Finally, refine it with a special cutter made for this project, as illustrated on page 268.

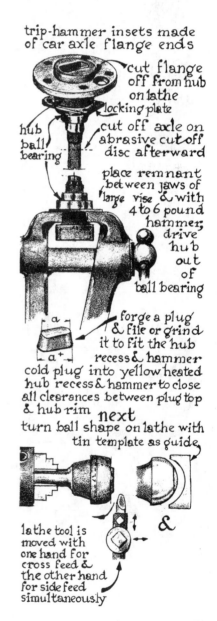

trip-hammer insets made of car axle flange ends

cut flange off from hub on lathe
locking plate

hub ball bearing

cut off axle on abrasive cutoff disc afterward

place remnant between jaws of large vise & with 4 to 6 pound hammer; drive hub out of ball bearing

forge a plug & file or grind it to fit the hub recess & hammer cold plug into yellow heated hub recess & hammer to close all clearances between plug top & hub rim

next turn ball shape on lathe with tin template as guide

lathe tool is moved with one hand for cross feed & the other hand for side feed simultaneously

&

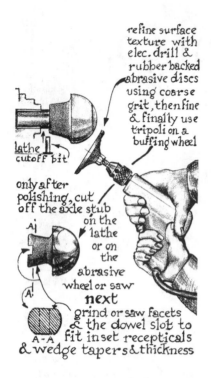

refine surface texture with elec. drill & rubber-backed abrasive discs using coarse grit, then fine & finally use tripoli on a buffing wheel

lathe cutoff bit

only after polishing, cut off the axle stub on the lathe or on the abrasive wheel or saw

next grind or saw facets & the dowel slot to fit inset recepticals & wedge tapers & thickness

A-A

making cup shapes from flat scrap steel discs, using trip-hammer with ball & bowl insets

$\frac{3}{8}$"t $9\frac{3}{4}$"

forge a hold-down tool from a $\frac{1}{4}$"x3"x10" bar   crimp ring

to be hand held in guiding hot cup centrally in insets

hot   cold

yellow hot disc kept aligned centrally   bear down

Making the Bowl-shaped Anvil Inset

Trial and error is often the quickest method when using items available from the scrap pile. If scrap articles are too light to withstand the great strains insets must take under forging, they must be reinforced to prevent distortion.

Heat a second car axle hub, from which the bowl-shaped lower inset is to be made, to *yellow heat*. Clamp it in the vise and, with the heavy ball peen hammer, enlarge the depression until it becomes round and bowl-shaped. Its small diameter below the bowl needs to rest on a 1/2-inch-thick washer to bridge the slot opening in the anvil and to cushion the inset base against distortion during hammering. The washer may have to be turned on the lathe.

Cut off the excess axle section and cut the inset base into a profile to fit the standard wedging arrangement. The illustrations show the two assemblies ready to be installed for making cup shapes out of the flat round discs.

Examine the illustration in which an inset rim is reinforced by a crimped-on rim around it.

FORGING A CUP SHAPE OUT OF A FLAT DISC

Instead of trying to hold the heated disc centered over the inset bowl with tong-type tools, you may find the system I have devised a better solution.

The Hold-down Tool

Use a bar 1/4 inch by 3 inches by 10 inches. Heat one end *yellow hot* and place it over the center of the bowl-shaped anvil inset. A few blows with the ball inset hammer will shape the end of the flat bar into a cup.

This cold hold-down tool is intended to be held on a centrally positioned, flat, hot disc over the bowl anvil inset and to be hammered into the hot disc. It forms the disc into a cup shape. Examine the illustrations carefully, noting the sequence for making these cup shapes.

The hold-down tool should be firmly pressed down by hand onto the hot disc, guiding it this way and that, slanting and correcting its position during the hammering. You will quickly develop a skill in manipulating the tool until it seems to center itself as the cup begins to fill the depression of the bowl inset.

Save the final blows for a second heating, before which you should clean the bottom of the inset bowl of accumulated oxidation scales and coal dust. The final hammering on the reheated shaped cup in the cleaned bowl inset will yield an accurate and smooth finish if the ball inset also is smooth and accurate.

To machine the outside of the cup, spread the chuck jaws in the cup that is placed over the jaw tips. Hold it in that position with the tail-center point pressed onto the cup's center. After the cup rim is machined, it can be clamped between the chuck jaws and finished further with the large, circular cutter shown in the illustration on page 268.

that is placed over the jaw tips. Hold it in that position with the tail-center point pressed onto the cup's center. After the cup rim is machined, it can be clamped between the chuck jaws and finished further with the large, circular cutter shown in the illustration on page 86.

286

MAKING LARGE ROSETTES WITH THE TRIP-HAMMER

Locate a heavy, solid piece of scrap from which to make another inset. Turn it on the lathe into a receptacle for one of the already forged blanks. Its diameter should be *half* that of the cup.

On the lathe turn an *outside ring* that can be slipped over the circular inset, as shown, to act as a guide to keep the cup blank from wandering sideways.

The next item needed is a *flat set-hammer*, made by riveting a head onto two 3/8-inch-thick scrap discs to give it the needed stiffness. This is intended to slip easily inside the guide ring and rest on top of the heated, open-faced cup blank.

With the heated cup blank in position below it, the set-hammer is then hammered down with the trip-hammer. This flattens the free portion of the cup blank onto the flat, marginal, circular face of the anvil inset.

The workpiece is now a circular base with a marginal flange and a high, curved hump in the middle. It is around this hump that deep scallops are forged to make a decorative rosette.

The scallops can be made with hand-held ball heads that are hammered down with the trip-hammer while the heated marginal flange is held with tongs between the flat-faced top and bottom inset.

The illustration also shows a tool to create toe-holds for the ball head on the flange surface. These preliminary indentations are made with the same hammerblows that make the marginal flange. You next use ball heads to create an end result that looks entirely handmade. This is because that last hand-directed placement of various sizes of ball heads removes every trace of a mass-produced machine-made article.

Surface Treatments for the Finished Rosette

The appearance of such a rosette can be enhanced through different surface finishes. It can be steel-brushed, then heated up a little before applying warm linseed oil, which will fill the tiniest pores in the surface. This produces the black finish typical of forged steel.

Another surface treatment is to move the rosette surface against a rotating rubber-backed, fine grit, abrasive disc, allowing this flexible disc to rub over the protrusions of the rosette surface but not to reach its valleys. The result is that the ridges show as silver in contrast to the black, bringing out the texture sharply.

Follow up this treatment by holding the rosette surface against a tripoli-impregnated buffer. The silver ridges become highly polished, but the black depressions are left unaffected.

It is now important to clean the wax residue from the surface with turpentine or comparable solvent. Next heat the whole over the blue flame of a gas burner; the kitchen stove burner will do. Wait for oxidation colors to appear, stopping the process by quenching the whole in water when the color is to your liking.

The rosette can be sprayed with an acrylic fixitive to extend the lifetime of the patinated surface and to protect it against erosion. Once you have become proficient at such surface-treating and coloring, your end results will have a jewel-like appearance.

The completed rosette is ideally suited for a variety of uses. For instance, it can be used as a heavy base for the column of a candlestick a lamp, or as any object to be held in a stable position. Rosettes can also, of course, be used as decorative ornaments.

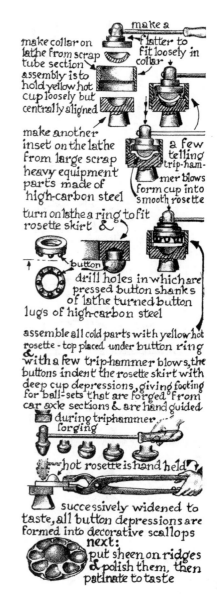

make collar on lathe from scrap tube section — make a flatter to fit loosely in collar

assembly is to hold yellow hot cup loosely but centrally aligned

make another inset on the lathe from large scrap heavy equipment parts made of high-carbon steel

a few telling trip-hammer blows form cup into smooth rosette

turn on lathe a ring to fit rosette skirt & button

drill holes in which are pressed button shanks of lathe turned button lugs of high-carbon steel

assemble all cold parts with yellow hot rosette - top placed under button ring & with a few trip-hammer blows, the buttons indent the rosette skirt with deep cup depressions, giving footing for ball-sets that are forged from car axle sections & are hand guided during trip-hammer forging

hot rosette is hand held

successively widened to taste, all button depressions are formed into decorative scallops

next: put sheen on ridges & polish them, then patinate to taste

27. Sharpening Tool Edges

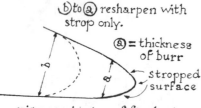

ⓑto ⓐ resharpen with strop only.

ⓐ = thickness of burr

stropped surface

microscopic view of final edge anything thinner than ⓐ will bend or break, no matter how perfect the temper

testing for sharpness
right way to detect burr

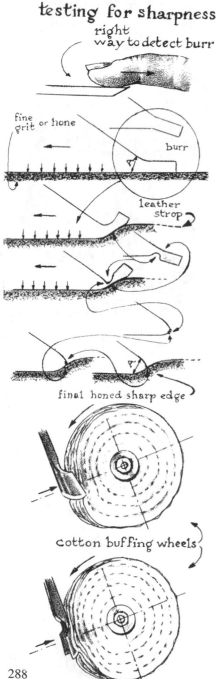

fine grit or hone

burr

leather strop

final honed sharp edge

cotton buffing wheels

It is necessary that the student first visualize what takes place at a tool's edge when it is ground for sharpness.

The angle at that edge where the two sides meet is a line of intersection of two planes. Mathematically, the dimension at that point is zero. It means that its cutting edge would have in actuality no dimension; hence, strength is absent. This visualization is at first difficult to comprehend. But in geometry, when two lines are shown intersecting *on paper*, it is easily understood.

It is the microscopic size of the sharp edge, then, that we must deal with—a size that approaches zero. At the same time it must remain strong enough to withstand the strain during the cutting action. This is our basic concern in sharpening all cutting tool edges.

Therefore, if that microscopic edge is made slightly *rounded* it will mean that its final size is *more* than absolute zero, giving it enough strength to withstand bending while cutting smoothly.

If you find it difficult to visualize this minuteness, consider the notion of engraving the Lord's Prayer on the head of a pin. Minuteness is somewhat easier to comprehend thanks to electron microscope photography. It can show us the gigantic-looking forms of hair on the surface of the almost invisible eyes of insects; it can show us body cells, bacteria, viruses, atoms.

The illustration shows the enlarged portion of a tool edge. When, during grinding, zero dimension is approached, the thin steel will bend away under the pressure of the grinding stone. Then the grit no longer cuts, but *slides* off the steel, forming a *feather edge*.

That feather or *burr* must be removed. If it were not, it would fold over during cutting, and would buckle, break or tear off, leaving a row of microscopic, jagged teeth. In use, the tool would drag and tear, instead of *shearing* the material, leaving a ruptured texture instead of a smooth one. The solution is to *strop* the edge on a leather strop like the ones barbers used to sharpen their razors.

A MODERN METHOD OF "STROPPING" TOOL EDGES

In the shop, a cotton buffing wheel replaces the strop. The rim of the buffer is rubbed with a tripoli abrasive compound and the final burr of the tool edge is cleanly removed by it. It does what the barber's strop did, but better and a thousand times faster.

Wearing down the "feather" of a freshly ground tool edge with such a buffer creates a microscopically rounded and smooth edge that will turn out to be the sharpest, smoothest, and strongest edge possible. Any future (microscopic) damage to that edge caused by prolonged use needs only this tripoli buffing to restore it to perfection.

Such occasional rebuffing may be the only resharpening required for a long time. When, in time, a dragging is felt during cutting it means that it is time to regrind the edge on *stone*. A feathered edge will, of course, result; and that will again call for the rapid removal of the burr with the tripoli-impregnated buffer as described.

No modern shop should be without a simple motor-driven buffing wheel if knifelike edges are to be sharpened at their best.

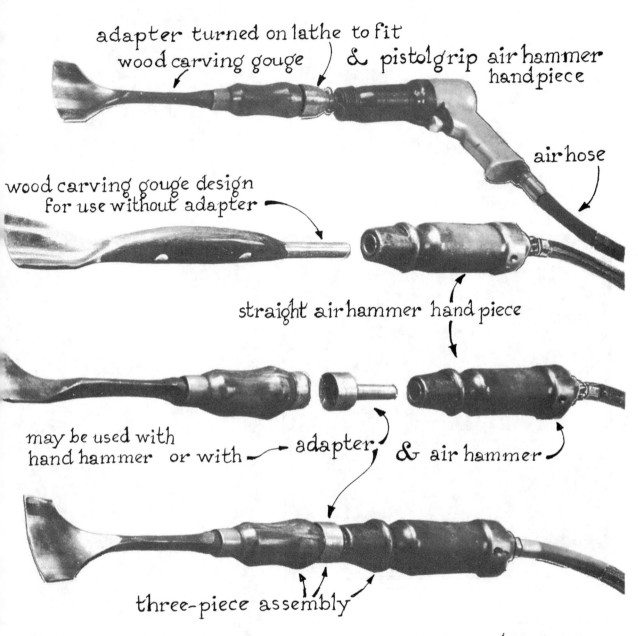

adapter turned on lathe to fit
wood carving gouge & pistol grip air hammer hand piece

air hose

wood carving gouge design
for use without adapter

straight air hammer hand piece

may be used with
hand hammer or with — adapter & air hammer

three-piece assembly

woodcarving gouges forged from salvaged car leaf or coil springs.
adapters are turned on lathe from sections of scrapped car axles.
airhammer uses from 90 to 120 psi stored in a 100 gal. tank & a
1 to 2 hp motor driving one or two compressors

above hammer handpieces used without adapters for standard
stone carving tools, steelcutting tools

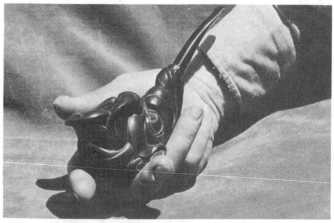

Engraver style wood-carving gouge was used to carve this pipe in manzanital burl.

Engraving burins shown on page 276 cut designs on blocks of end-grain pearwood.

The Wave, by Alexander G. Weygers, engraved with burin shown on page 276 on end-grain pearwood.

Lifesize Indiana limestone carving of garden statue by Alexander G. Weygers made with tools forged from scrap steel.

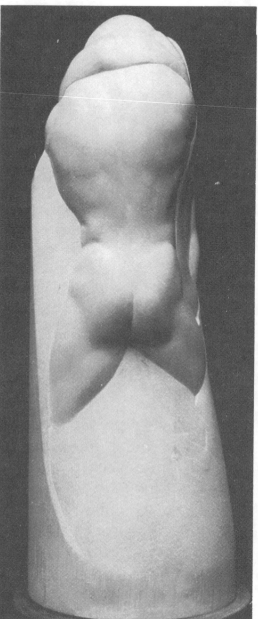

The Embrace, by Alexander G. Weygers, carved from salvaged lemonwood when orchards were destroyed to make way for housing developments. Tools used made from scrap steel of coil springs.

Hints on Using Power Tools and Other Admonitions

Use safety goggles at all times, especially since you will be using grinding wheels without wheelguards, in order to have full access to them. If you use prescription glasses, wear goggles over them to avoid their being pitted or broken. For long, sustained grinding, also use a nose filter.

Test the grinding wheel for flaws. When balanced in your hand, the stone should "ring" when tapped with a light hammer. If, instead, the sound is "dull," the stone may be cracked and should be *discarded.* Any wheel being used for the first time should be run for a full minute (while you stand aside) before you accept it as sound.

If grinding wheels are used without wheelguards and mounted directly on a ⅓ to ½ hp 1750 rpm motor, use wheels not more than 8 inches in diameter and ¾ of an inch thick, for safety; on a ⅓ to ½ hp 3600 rpm motor, use for safety reasons wheels 6 inches in diameter or less and preferably not less than ¾ of an inch thick.

Keep the gap between stone and tool rest closed. When the tool rest is bolted to the table beneath the stone, as shown in this book, adjust it by tapping it toward the stone until contact produces a few sparks.

The wheel should rotate so that the upper surface revolves toward the operator. Thus, the steel being ground will be held pressed on the tool rest.

If you use the *sides of the wheel,* do it sparingly, with very light pressure, to prevent grooving. If the side of the wheel has to be re-dressed too frequently, it becomes thinner and may fly apart.

If the side of a wheel must be used extensively, use the specially designed side grinder described in *The Making of Tools,* Chapter 6.

Always move steel back and forth against the stone, during grinding, to avoid cutting grooves into the stone.

Use hard *coarse-grain stones* (1 inch thick or more) for removing large quantities of steel. Use *fine-grain stones* for delicate, more precise steel removal — and for refining texture preparatory to honing and polishing.

Remember that *motor power* should be utilized to its fullest if you plan to remove quantities of steel effectively. Therefore, bear down hard on the stone while moving steel back and forth (short of slowing down the motor), if steel has not yet been tempered or if it is mild steel.

When grinding tempered tools, always take care that the steel does not lose its hardness through *overheating* during grinding. Cool the steel frequently by quenching. (Mild-steel and unfinished tools, of course, may be overheated since hardening and tempering are still to follow.)

Keep a one-gallon can of water next to your grinder at all times, for cooling workpieces.

Remember that should *oxidation colors* accidentally appear on the cutting part of a tool, the color spectrum will indicate the degree of temper loss: *Yellow* (when cooled instantaneously) means the tool still remains sufficiently hard; *bronze* is a little less hard, but still safe; *dark purple* is almost too soft for cutting steel; *light blue* means you have lost your tool hardness at that spot.

If hardness remains unaffected behind the softened cutting edge, grind the tool back to the point of remaining hardness. This means losing some of the lifetime of the tool, but no retempering is needed.

Be careful, this time, to cool the tool frequently in order not to overheat again. The new cutting edge should be ground entirely free of the soft portion.

Should accidental overheating anneal too much of the blade, it is best to retemper the tool to ensure its longevity.

Keep adding useful equipment to your shop (once you learn which articles really help save time and muscle). Instead of buying tools, always consider making them yourself. It may be less difficult than you think, and while you will derive pleasure in their making, you will simultaneously improve your skills and gain confidence for still more challenging projects.

Files will often remove steel as efficiently as grindstones — generally as quickly, and with greater control. Don't neglect learning how to file accurately. Make yourself carry out, at least once, the exercises recommended in dealing with hinged tools like tinsnips, shears, pliers, etc. Try to master the filing technique, for it will benefit all you plan to accomplish in toolmaking.

For the serious student, who wants to perfect his filing, write to file manufacturers for their instructive booklets.

When I was a student in Holland, we spent all the shop hours during our whole first year learning only *flat-filing,* so as to achieve machine-like precision. Only after we had acquired that hand skill were we permitted to learn blacksmithing, lathe-turning, drilling, milling, and so forth.

Never walk away from a live fire (forge, torch, or brazier) without a very real awareness of the possible danger. The same holds true for a sharp tool left clamped in a vise.

The sound toolmaker is always in absolute control. Distractions lead to accidents — to yourself, to others, and to your tools. Proper care and concentration will go a long way toward toolmaking that is neither hesitant nor mistake-ridden.

GLOSSARY

ADAPTER—A driven instrument made to fit one type (size) of tool at one end and another type (size) at the other end.

AGITATOR—*See* Paint mixer

ALIGNMENT—The state of being in line with another element, rather than askew in relation to it.

ALLOY—A compound or fusion of two or more metals.

ANGLE IRON—Steel bars that have a cross section of an angle (usually 90°). The iron, in this term, is a holdover from the days before iron was made into steel. Now all angle iron is actually steel, either mild or high-carbon steel.

ANNEAL—To soften steel through slow cooling after enough heat has made the steel lose its hardness.

APRON—In mechanics, an extended platform.

ARBOR—A wheel, axle, or shaft rotating in one or more bearings held by a frame that is bolted down.

AUGUR—A wood drill, as a rule over $1^1/_2$ inches in diameter. (Also may refer to augers to drill holes in earth.)

BALL RACE—The parts of a ball bearing between which the balls "race."

BASTARD FILE—A file with teeth coarser than a smooth file and less coarse than a coarse file.

BEVEL—In cutting tools, the facet that has been ground at the cutting edge (inside and outside bevels).

BICK—A part that fits into an anvil's hardy hole and acts as a horn or beck.

BLANK—The rough shape of a tool before filing, grinding, etc., has prepared the tool for tempering and assembly with the handle.

BOSS—A locally raised part of steel.

BRITTLE QUENCH—*See* Quench

BROACH—A boring tool, a reamer.

BUFFER—A cotton wheel used to polish surfaces.

BUFFING WHEEL—A motor-driven cotton wheel that rotates at high speed. A buffing compound rubbed into the cotton buffs (polishes) the steel held against the wheel.

BUNSEN BURNER—A gas burner with a single blue flame used in laboratories to heat liquids and objects.

BURIN—The cutting tool of an engraver.

BURR—A small rotary file, often used to take off a burr or curl left on the edge of steel by previous cutting. A burr may also be the "feather-edge" left on a tool's cutting edge in the final step of sharpening the tool.

BUSH HAMMER—A tool with a hammer face having 9 or more raised points which, on impact, crush or pulverize the surface of stone. From the French *boucher:* to crush, to eat, to bite. The *bush tool* also has 9 or more raised points which, when hammered upon, crush or pulverize the surface of stone.

BUTTERFLY-CENTER— A lathe-center insert placed in headstock that has four sharp wings and a center pin that press into the wood that is to be turned on a wood lathe.

CAP SCREW (OR TAP-BOLT)—A bolt (without its nut) screwed into a threaded hole of one part, to hold another part clamped onto the first.

CAPE CHISEL—A narrow chisel that cuts deep grooves, specifically key slots, in steel.

CARBIDE-TIP—An extremely hard tip soldered onto the end of a regular high-carbon steel bar used to turn wood or steel on a lathe.

CARRIAGE BOLT—A bolt that ties together wooden members in structures. It has a square section under the head to keep the bolt from turning.

CASE HARDENING—The process of applying a skin-deep hardness to the outer surface of mild steel in a forge fire.

CENTER-PUNCH—Tool used to make a "center" mark for locations to be drilled, or to mark off pattern outlines on steel.

CHASING TOOLS—Tools used to make marks (raised or depressed) in metal surfaces to create texture.

CHECKING (OF WOOD)—The splitting of wood during drying.

CHISEL—A metal tool with a blade having a sharp-edged end; used for cutting wood, stone, metal, or other material.

CHISEL, CARPENTER'S WOOD—A flat chisel for cutting wood.

CHISEL, COLD—A chisel that may be used on cold annealed steel to cut it.

CHISEL, HOT—A chisel used to cut yellow hot steel. The steel is cut with the hot chisel on the soft anvil table or a mild-steel plate placed over the hard anvil face. The chisel is either a hand-held long cold chisel or a sturdy chisel head fastened to a long wooden stem.

CHUCK—A clamp screwed on a rotating shaft to fasten drills, small grinders, etc.

CLAW—A multiple-toothed stonecarving tool used to refine the rough texture left by the one-point tool.

COEFFICIENT OF CONDUCTIVITY—A number that indicates the degree of speed at which heat is conducted from one spot to the next in a type of steel.

COIL SPRINGS—Springs made of long, high-carbon steel rods that are wound hot around a bar and afterward tempered to the hardness for which such springs are designed.

COKE—The substance fresh coal becomes after heat has driven out all elements that give off smoke and yellow flame. Coke resembles charcoal in that it gives off a blue flame and lights easily.

COLLAR—A steel ring, often mounted on a shaft with a set screw.

CONDUIT PIPE—Galvanized steel pipe through which electricians install electric wires.

COUNTERSINK—A cone-shaped, large *drill bit* used to bevel the edge of a sharp-edged cylindrical hole left by a smaller drill; a shallow cylindrical depression around a hole, larger than the hole in diameter.

CUTOFF WHEEL—A thin abrasive wheel that cuts steel too hard to cut with a hacksaw.

DIE—A two-part mold (male and female) used for making and reproducing a form one or more times. The material is held between the dies that are then forced together to produce the form or shape that the dies have at their contacting planes. Dies are used strictly to mass-produce articles or to make an article that is too complicated to make easily by hand. (Also, a matrix.)

DOG—A tool clamped on the workpiece to engage the lathe headstock so that the workpiece can turn.

DOWEL (STEEL)—A locking-pin that holds parts and keeps them from shifting their positions.

DRAWING TEMPER COLOR—Reheating brittle-quenched steel that has been polished to see the oxidation color spectrum (temper colors) clearly. Once this color spectrum appears and the wanted color, which corresponds to its *hardness,* has been "drawn," the tool is quenched.

DRAWING-OUT STEEL—*Stretching* steel, making it longer or wider or both. The opposite is to upset steel, making it thicker and/or shorter.

DRESSER—A tool that cuts or wears down the surface of grindstones.

DRESSING—Making an inaccurate grinding wheel accurate with a dresser by wearing the wheel surface down to exact shape.

DRIFT—A tapered steel pin that is driven into a hole in stone to split it. Another use is to pull together two slightly unmatched holes in two plates to align them perfectly.

DRILL BIT—Could be called a *drill,* but generally this term refers to a local *bit* at the end of a plain drill rod. Such bits may be of varied designs to meet various drilling problems.

DRILL PRESS—A machine for drilling holes in metal or other material.

EMBOSS—To raise steel locally with bosses. The *boss* is a form of die which, pressed or hammered into the steel plate from one side, raises the steel surface on the other side of the sheet.

EYEBOLT—A bolt that has a hole in a round, flattened end instead of the hexagon, or round, or square-bolt head.

FACE—Generally refers to a flat surface on the sides or top of a tool or machine part: an anvil face, side-face, the face of a disc, "to face" a surface, when grinding, milling, and cutting steel surfaces.

FERRULE—A metal ring, cap, or tube-section placed on the end of a handle to keep it from splitting.

FIREBRICK—A brick that withstands high temperatures as in brick-lined kilns and fireplaces.

FIRECLAY—A clay that will not crack when fired.

FLANGE—A projecting rim; *also,* a plate to close a pipe opening.

FLASH FIRE—A fire that starts suddenly when an inflammable liquid reaches a heat corresponding to its "flash" point, setting the liquid aflame.

FLATTER—A tool shaped like a hammer head but with an accurate, square, flat face at one end and a crowned end at the other that can be struck with a heavy hammer. The flatter's face, placed on a heated inaccurate flat section of a workpiece lying on the anvil, can flatten it out accurately.

FORGE—A furnace in which steel is heated.

FREEZING—The bonding together of two clamped-together steel parts that have corroded or have been forcefully locked together. To break this bond is a frequent chore when taking rusted machinery apart.

FULLER—A blacksmith's tool that fits in the hardy hole of the anvil (bottom fuller), or is fastened on a long wooden stem (top fuller) in order to groove steel, draw it out, or "set" rounded corners. Comes in various sizes.

GAUGE—A specific size in reference to steel sheet or bar thickness, nail size, etc.

GRINDING POINTS—Miniature high-speed rotary grindstones.

HACKSAW—A hand saw with narrow blade set in a metal frame, used to cut metal.

HARDY—An anvil insert that acts as a cutter of hot steel. Also called hardies are hardy-type tools that fit in the hardy hole, but have other special names, as a rule, i.e., *fullers.*

HARDY HOLE—The square hole in the anvil that the hardy fits into.

HEADING PLATE—A thick, flat piece of steel with a slightly tapered hole in the middle which receives a rod that has been upset at the end. The hot end can then be hammered into a head.

HEADSTOCK—The rotating driver end of a lathe.

HEAT—The period that the hot steel, removed from the fire, maintains its forging heat.

HEATING—The period of heating the steel.

HEAT-TREATING—The process of *tempering* steel for a specific hardness; can also refer to treating steel to bring about a specific softness.

HIGH-CARBON STEEL—A temperable steel, primarily used to harden such steels for specific hardness in the process called "tempering." In industry, steel of over 0.2% carbon.

HOLD-DOWN OR HOLD-FAST—A contrivance for holding the heated workpiece to be forged when the smith needs both hands free to manipulate his tools. One end of the hold-down is rammed in the anvil's hardy or pritchel hole so that its other end will hold down (or hold fast) the workpiece.

HOLLOW-GRIND—To grind the bevel of a cutting tool concave.

HONING—Grinding a steel surface with a *honing stone.* This stone leaves an almost-polished surface.

HONING STONE—The finest hand-grinding stone; used for final polishing.

HP (HORSEPOWER)—A unit of power, used in stating the power required to drive machinery.

JIG—A device that acts as a guide to accurately machine-file, fold, bend, or form a workpiece. This is used if lack of skill handicaps the worker in making the workpiece. Such jigs guide him and also save time in mass production of tools.

KEEPER—The part of a door latch through which the latch bolt slides.

LAPPING—An abrasive action in which a grinding compound is used between two surfaces that, when held pressed together in movement, grind themselves into one another.

LATHE—A machine for shaping articles that causes them to revolve while acting upon a cutting tool.

LEAF SPRING—A spring with an oblong cross-section and a sufficient length to act as a spring. Automobiles as a rule use such springs singly or in graduated layers to suspend the car body over the axles.

LOW-CARBON STEEL—A steel that contains less than 0.2% carbon, and is not temperable.

MALL—A larger hammer with wooden head, sometimes steel-weighted, used to drive stakes in the ground.

MALLEABLE—Capable of being shaped or worked by hammering, etc.

MALLET—A wooden hammer-head on a short handle used to hammer on wood-carving gouges. Sometimes the mallet head is made of rawhide, plastic, or hard rubber.

MANDREL—A bar inserted in the workpiece to shape, hold, or grind it, as in a lathe.

MATRIX—A female die in which a malleable substance may be formed by pressing to fill it. A cavity in which anything is formed or cast.

MILD STEEL—A low-carbon steel. It is not temperable.

MILLING CUTTER—*See* Seating cutter

MORSE TAPER—A type of taper, named after its inventor, Mr. Morse, that holds fast to its seating in a clutch-like action without freezing to its surface and can be knocked loose easily when required.

NAIL SET—A tool resembling a center-punch but with a hardened-cup-shaped end instead of a ground point. This cup-shaped end, placed on the nail head center, keeps that tool from slipping sideways while "setting" the nail.

OFFSET—The step down (or up) in a bar or plate from its original alignment into another alignment, as a rule, parallel with it.

ONE-POINT TOOL—The basic stonecarving tool that "chips" stone in the first roughing-out action of stonecarving.

OXIDATION COLOR SPECTRUM—The color spectrum that results from the oxidation of cold steel as it gradually gets hot. The polished metal sheen shows that colors as clearly as the color spectrum in rainbows.

PACKING—Compacting a high-carbon steel to improve its quality by striking with heavy hammerblows when the steel is at cherry red heat.

PAINT MIXER (AGITATOR)—A rod with a crooked end which, when rotated in the paint, mixes it.

PATINA—The colored oxidation on metal surfaces. It results during the process of tempering the metal. (On bronze and many other metals, a patina comes about after long exposure to the oxygen in the air and to chemicals.)

PEARL GRAY—A typical file color. When high-carbon steel emerges from a quench, pearl gray indicates "file hardness."

PEEN END—A hammer with a wedge-shaped, round-edged end or a half-sphere ball end used to strengthen steel by indentation. A *cross peen* hammer has the rounded edge of the peen at 90° to the hammer stem.

PILLOW—In mechanics, a supporting block that makes up the lower half of a split bearing.

PINION—A small gear that drives, or is driven by, a larger gear.

PRITCHEL HOLE—The round hole next to the square hardy hole in the anvil.

PUSH-ROD—The rod in and engine which "pushes" a valve to open the cylinder for the intake or expulsion of its gases.

QUENCH—To cool hot steel in a liquid. *Brittle quenching* is the act of cooling high-carbon steel at its critical heat at its fastest so that it will emerge brittle-hard.

QUENCHING BATH—The liquid into which the hot steel is dipped or immersed to cool it.

RASP—A coarse file used mostly to grate or tear softer materials such as wood, horn, plaster of Paris, and soft stones that are not abrasive.

RELIEF—A projection of a design or figure upward from a plane surface.

REPOUSSÉ—A raised surface on a flat plane achieved by pushing out from below a sheet the portions to be raised.

RHEOSTAT—An instrument that may be adjusted to let more, or less, electric current pass; for instance, to regulate the speed of an electric motor or to dim or brighten a light bulb.

ROUT—To cut or scoop out material with a router tool.

SADDLE—A rounded piece of steel on which to form another piece in its shape.

SCRIBE—A sharp-pointed steel marking pin used to scratch a line onto a work-piece.

SEATING CUTTER—Tool used to cut a *seat* in a part onto which another part fits exactly. The cutting also may be called *milling,* and the cutter then would be a *milling cutter*.

SET HAMMER—A hammer head fastened on a long wooden stem resembling a flatter. Placed on a partly formed section of a workpiece, it sets it into its final position when a regular hammer delivers a blow on it.

SET SCREW—A screw that clamps or *sets* one part onto another part.

SHANK—The part of a tool between tang and blade.

SHOULDER—In craft usage, an abrupt wider or thicker dimension in rod or shaft against which another part rests.

SLAG—A melted mix of noncombustible matter in coal. It lumps together.

SLEDGE—A heavy (6-pound or more) hammer on a long stem, used by the blacksmith's helper using both his hands, for heavy hammering of heavy steel.

SLEDGING—Striking with a sledge hammer.

SLEEVE—In bearings, the bushing; a precisely honed bronze tube that fits the shaft it bears.

SOCKET—An open part into which another part fits.

SPECTRUM—A division of colors occurring on the shiny part of steel when it is heated for tempering; similar to the rainbow colors seen through a prism.

SPRING STEEL—A high-carbon steel tempered so that it will act as a spring.

STEEL PLATE—Refers to flat sheets of steel thicker than $3/_{16}$ of an inch. It is optional at what thickness or thinness metal may be called sheet metal. Example: boilers are made of plate steel; stovepipes are made of sheet metal.

STEEL STOCK—The supply of steel from which an item is selected to forge or machine or grind the workpiece to be made.

STEP PULLEY—A ratio-increasing multiple pulley.

STROPPING—The final step in sharpening a cutting edge on a leather strop.

STUB SHAFT—A short, stubby shaft.

SWAGE—The "saddle" that is grooved to form steel and fits into the anvil's hardy hole (bottom swage), and a similar tool, fastened to a long wooden stem, placed *over* steel to form it (top swage).

TAILSTOCK—The center pin in the stationary end of the lathe that holds the rotating metal, wood, or other material between the two lathe centers.

TANG—The part of the tool blank that is locked into the tool handle.

TEMPERABLE STEEL—A steel of a higher than 0.2% carbon quality, which can be *tempered*.

TEMPERING—In forging metal, the process to arrive at a specific hardness of high-carbon steel.

TEMPLATE—A pattern, often made from cardboard or sheet metal, to serve as a model, the outline of which is scribed on the steel to be cut.

TINSNIPS—Sturdy, short-bladed shears that cut sheet metal.

TOOL REST (OR TOOL POST)—As a rule, those parts on machines onto which a tool is held down firmly to be ground down. Also may refer to the clamp on a machine to hold a cutting tool.

TRIP-HAMMER—A mechanical hammer that is activated by a foot control.

TRIPOLI—An abrasive-impregnated wax compound that, when rubbed into a rotating cotton buffing wheel, acts as the finest steel polisher.

UPSETTING—The process of making a piece of steel shorter and thicker.

VEINING TOOL—A V-shaped gouge that cuts V grooves referred to as "veins."

VISE—A two-jawed screw clamp bolted to the workbench to hold things steady while they are being worked.

VISEGRIP PLIERS—Self-locking pliers.

WELD—To fuse metals together under heat.

WROUGHT IRON—Iron that has been worked in a "puddling" process to purify it. It contains no carbon and is least subject to rusting. It is rarely used today, and hence not found in scrap piles. It can be welded easily and will not burn during melting as does steel.

ABOUT THE AUTHOR

Alexander G. Weygers—scientist, inventor, artist, author, and raconteur—was a true modern Renaissance man. He was born in Java and educated in Holland as an engineer, a profession he practiced for some years in Indonesia and the United States before devoting himself exclusively to the fine arts. He was apprenticed to sculptor Lorado Taft in Chicago, and then studied wood engraving in Paris, marble carving in Florence, and human figure drawing in The Hague. With his early school training in blacksmithing, he was able to begin making his own stone- and wood-carving and engraving tools, and soon he was making custom-designed tools for artists and craftsmen all over the United States. During the latter part of his life, he lived and worked in California's Carmel Valley with his wife.

ALSO FROM TEN SPEED PRESS

THE COMPLETE WOODWORKER
edited by Bernard E. Jones

Over 1,000 illustrations and diagrams help the craftsman or the beginner understand the tools and techniques of traditional woodworking. A revised edition of a great classic in the field.
6 x 9 inches, 416 pages, ISBN 0-89815-022-1

THE PRACTICAL WOODWORKER
edited by Bernard E. Jones

A companion book to THE COMPLETE WOODWORKER, with detailed instructions on advanced woodworking techniques and many pages of plans and designs for everything from tables to greenhouses.
6 x 9 inches, 600 pages, ISBN 0-89815-106-6

THE HANDYMAN'S BOOK
Tools, Materials, and Processes Employed in Woodworking
by Paul N. Hasluck

Originally published in 1903, this extraordinary book is an encyclopedic treatment of the woodworking techniques and tools of the day. With over 2500 illustrations.
6 x 9 inches, 800 pages, ISBN 0-89815-203-8

HOMEWORK
Ten Steps to Foolproof Planning Before Building
by Peter Jeswald

This all-in-one workbook, planner, and sourcebook presents the planning process before construction in 10 easy steps that are sure to make building, remodeling, or renovating a home fun, affordable, and successful.
8 1/2 x 11 inches, 240 pages, ISBN 0-89815-744-7

BEFORE YOU BUILD
A Preconstruction Guide
by Robert Roskind

A comprehensive guidebook and housebuilding checklist based on the experience of the recognized leaders in the field of "hands-on" building education in the country. A selection of the Self Sufficiency Book Club and the Popular Science Book Club.
8 1/2 x 11 inches, 240 pages, ISBN 0-89815-036-1

BUILDING YOUR OWN HOUSE
by Robert Roskind

Profusely illustrated with diagrams, drawings, and hundreds of photos, this selection of the Self Sufficiency Book Club and the Popular Science Book Club presents a wealth of solid information in a concise form.
8 1/2 x 11 inches, 448 pages, ISBN 0-89815-110-4

BUILDING YOUR OWN HOUSE II
Interiors
by Robert Roskind

This hands-on guide takes up where the first volume left off, guiding the builder through finishing a house once the framing is done. A selection of the Popular Science Book Club.
8 1/2 x 11 inches, 288 pages, ISBN 0-89815-358-1

COHOUSING (REVISED)
A Contemporary Approach to Housing Ourselves
by Kathryn McCamant and Charles Durrett

Expanded to include the exciting new communities being lived in and developed in the United States, COHOUSING is an invaluable guide and resource for those looking for affordable, neighborly, and safe housing solutions. Illustrated.
8 1/2 x 9 1/2 inches, 288 pages, ISBN 0-89815-539-8

REHAB RIGHT
How to Realize the Full Value of Your Old House
by Helaine Kaplan Prentice, Blair Prentice, and
    the City of Oakland Planning Department

A highly illustrated, sensible guidebook for rehabbers who want to preserve the valuable architectural features of their old house. Covers codes, permits, and financing, and presents a full array of repair solutions that fit the architectural style of just about any house built since the late 19th century.
11 x 8 1/2 inches, 144 pages, ISBN 0-89815-172-4

To order, or to receive our free catalog of over 500 books, posters, and tapes, write to

TEN SPEED PRESS
P.O. Box 7123
Berkeley, CA 94707
Or call (800) 841-2665